THE UNIVERSITY OF WESTERN ONTARIO
SERIES IN PHILOSOPHY OF SCIENCE

A SERIES OF BOOKS

ON PHILOSOPHY OF SCIENCE, METHODOLOGY,

AND EPISTEMOLOGY

PUBLISHED IN CONNECTION WITH

THE UNIVERSITY OF WESTERN ONTARIO

PHILOSOPHY OF SCIENCE PROGRAMME

VOLUME 7 ·

PHYSICAL THEORY AS LOGICO-OPERATIONAL STRUCTURE

Edited by

C. A. HOOKER

University of Western Ontario, Ontario, Canada

D. REIDEL PUBLISHING COMPANY

DORDRECHT : HOLLAND / BOSTON : U.S.A.

LONDON : ENGLAND

Library of Congress Cataloging in Publication Data

Main entry under title:

CIP

Physical theory as logico-operational structure.

(The University of Western Ontario series in philosophy of science ; v. 7)
Includes bibliographies and index.
1. Quantum theory. 2. Physics—Philosophy. I. Hooker, Clifford Alan. II. Title. III. Series: University of Western Ontario. The University of Western Ontario series in philosophy of science ; v. 7.
QC174.12.P45   530.1′.2   78-12481
ISBN 90-277-0711-1

Published by D. Reidel Publishing Company,
P.O. Box 17, Dordrecht, Holland

Sold and distributed in the U.S.A., Canada, and Mexico
by D. Reidel Publishing Company, Inc.
Lincoln Building, 160 Old Derby Street, Hingham,
Mass. 02043, U.S.A.

TABLE OF CONTENTS

PREFACE

In two earlier volumes, entitled *The Logico-Algebraic Approach to Quantum Mechanics* (hereafter LAA I, II), I have presented collections of research papers which trace out the historical development and contemporary flowering of a particular approach to physical theory. One might characterise this approach as the extraction of an abstract logico-algebraic skeleton from each physical theory and the reconstruction of the physical theory as construction of mathematical and interpretive 'flesh' (e.g., measures, operators, mappings etc.) on this skeleton. The idea is to show how the specific features of a theory that are easily seen in application (e.g., 'interference' among observables in quantum mechanics) arise out of the character of its core abstract structure. In this fashion both the deeper nature of a theory (e.g., in what precise sense quantum mechanics is strongly statistical) and the deeper differences between theories (e.g. classical mechanics, though also a 'mechanics', is not strongly statistical) are penetratingly illuminated. What I would describe as the 'mainstream' logico-algebraic tradition is captured in these two collections of papers (LAA I, II).

The abstract, structural approach to the characterisation of physical theory has been the basis of a striking transformation, in this century, in the understanding of theories in mathematical physics. There has emerged clearly the idea that physical theories are most significantly characterised by their abstract structural components. This view is in opposition to the traditional view that one should look to the applications of a theory in order to understand it. (There is no suggestion that either view captures the whole truth, only that the traditional view has been more than counterbalanced by the emergence of deep structurally-based insight.)

Quantum theory itself has gone through an evolution in this century which illustrates the shifting perspective. From a collection of intuitive physical maneouvres under Bohr, Born and others, it moved through a formative mathematical stage in which the structural framework was bifurcated (between Schrödinger and Heisenberg) to an elegant culmination in von Neumann's Hilbert space formulation. This latter version has in fact proved to be a major source of structural insight into mathematical

vii

Hooker (ed.), Physical Theory as Logico-Operational Structure, vii–xviii,
All Rights Reserved.
Copyright © 1978 by D. Reidel Publishing Company, Dordrecht, Holland.

physics. Even so it is still flanked by ill-understood formalisms for the relativistic particle and field-theoretic versions and it has been criticised for being insufficiently general, either for mathematical or physical purposes. We are faced today with a steady elaboration of abstract mathematical frameworks for physical theory; and theoretical physics is strongly oriented towards manipulation of these frameworks. (Perhaps, in fact, too strongly oriented since many of the manipulations seem to loose all sight of the physical intuitions which are at their basis and which are ultimately likely to lead to fruitful mathematical innovations. An account of the development of quantum theory is found in Jammer, M., *The Conceptual Development of Quantum Theory*, McGraw-Hill, 1966 and the *Philosophy of Quantum Mechanics*, Wiley, 1974. Incidentally, a similar story holds for the evolution of relativistic theories in this century, and of course for attempts to consistently combine the two.)

While it is undoubtedly the historical combination of emerging mathematical and conceptual insight through the elaboration of formal systems which has made possible the development of the analytic traditions presented in these volumes, there has emerged out of this general historical development an approach which is importantly different from the logico-algebraic approach as I have characterised it. While the aim of this approach is still to reveal physical theory as a logico-algebraic core plus constructions, the core itself is to be constructed precisely on the basis of an analysis of the theory's specific applications. The general idea is that the 'logic' of the physical domain in question is in some way contained in, can be 'read off' from, the experimental results in the domain. The theory can then be entirely reconstructed from this core plus experimental results in the domain. To some writers this idea of 'reading off' the logic is quite explicit, most explicit perhaps in the examples of Giles and in the early writing of Finkelstein (see LAA II). To other writers the particular logico-algebraic structure of a theory is postulated as the simplest theoretical structure, perhaps of a sufficiently generalisable sort, sufficient to account for the operationally-demonstrated relations among experimental results (cf. Mielnik). What all of these authors evidently share in common is a conviction that the structure of a theory should in some strong sense be dictated by an operational analysis of the experimental domain of the theory.

In this respect it is worthwhile (and long overdue) to distinguish the logico-operational approach from the logico-algebraic approach, on several counts: (1) Historically, the focus of the mainstream logico-algebraic analysis has been on formal reconstruction and clarification of various physical formalisms, accepting these more or less as given. By contrast, the strongly operational analysis has invariably led to formalisms for which a given theory, e.g. quantum mechanics, is but a special case (e.g. Randall/Foulis). (2) While the core of the historical mainstream might be described as clarification and even justification (e.g. of a theory's consistency and completeness), often enough the more generalised analyses of the operational approach have led naturally into criticisms of theory and/or to 'generalisations' of theories (e.g. Finkelstein, Mielnik, Kupczynski). (3) Though by no means universal, philosophical thinking in the mainstream logico-algebraic tradition is dominated by non-operationalist thought, much of it explicitly realist (Bub, Putnam—see LAA II) and hence anti-operationalist.

Historically, however, the operational approach has always been perceived as closely linked with the mainstream logico-algebraic tradition, and even been submerged in it. It is only very recently that a separate tradition has emerged. In large part this must be due to the fact that the formalisms involved are only just now reaching mature development. Indeed, in this sense the present volume represents a series of 'firsts': The first mature presentation of the dialogic approach, i.e. with probability theory, completeness results etc., included and a sufficiently rich logical structure to handle mathematical physics (but see also Mittelstaedt's *Philosophical Problems of Physics*, Dordrecht: Reidel, 1976, chapter VI). The first comprehensive presentation of Randall and Foulis' operational manifolds approach, with a new enrichment of the manifold structure permitting a large extension of their analysis. The first comprehensive statement of the conceptual foundations of Finkelstein's new theoretical constructions. It is my intention, and hope, that this volume will serve to clearly separate out this logico-operational tradition and establish it on its very considerable merits.

Operationalism, as a philosophical theory of science, has had a mixed reception, a chequered career. Spawned, in its modern garb, by nineteenth century positivist thought, it emerged as a recognisable philosophy of science under Bridgeman in the 1920's. Bridgeman was a physicist and his philosophical theory was closely linked to the actual practices of working scientists, especially physicists. It emphasised the importance of actual experimental procedures for obtaining an empirical grip on theoretical ideas. The essence of Bridgeman's idea was that all meaningful theoretical concepts should be 'operationally defined', which was taken to mean that the meanings of theoretical terms were determined by their defining operations.

For a variety of reasons, operationalism proved popular among scientists themselves and is still a strong component of the informal culture of working scientists today. Among the undoubted reasons for its popularity are: (1) It offered a view of theory from the day-to-day perspective of the working scientist; (2) it promised to remove from science all speculative terms and theories, thereby appealing to the perennial suspicion of 'going beyond the facts', of not proceeding strictly inductively; (3) by restricting existing theoretical formalisms to operationally justified principles it freed up the possibilities for future theorising (by showing how few constraints the facts impose).

Philosophically, operationalism quickly became linked to positivism, that most austere form of empiricism. (For a review see my papers in *Synthese* **26** (1974), 409 and **32** (1975), 177.) At the time Bridgeman introduced his doctrine, positivism was rising to philosophical ascendancy in the philosophy of science because of the logical and theoretical prowess of its proponents, its elegant coherence and its powerful anti-metaphysical (read anti-dogmatic) drive. Operationalism enjoyed a certain philosophical currency as a part of this development.

Now the philosophic picture has changed quite dramatically. Empiricism, and with it operationalism, have been in steep decline for over two decades. Roughly, these doctrines have been charged with offering seriously inadequate conceptions of science, particularly with respect to the degree to which theory can be, or is, determined by the 'facts' and the extent to which the actual practice and progress of science is confined by strict induction from the facts. Within the sciences, too, there has been a marked decline in attachment to operationalist/positivist principles, within physics and closely allied fields (e.g. quantum chemistry) as the extraordinary theoretical ideas of quantum and relativistic theories become increasingly apparent, in biochemistry and neurophysiology e.g. as the complexities of capturing an adequate theoretical picture of living processes becomes more apparent. On the other hand, one receives the impression (without having conducted a poll) that amongst strongly experimentally oriented scientists in any field, and among social 'scientists' more generally, there is still a strong presumption in favour of operationalist/positivist principles.

One might offer essentially strategic explanations for these historical phenomena. Operationalism protects the interests of the experimentally oriented and it justifies the caution and vagueness of much social science theorising in a field where few penetrating theoretical ideas have emerged as yet. More than this, the obvious tension between the second and third reasons I offered above for its popularity, allows scientists to move between conservative caution and daring speculation as the occasion suggests. (Einstein, e.g., employed this bi-polarity of operationalist thought supremely well in his analyses of dynamical concepts.)

Certainly a central idea behind the studies presented in this volume has been to obtain significant 'generalisations' of existing theory by first employing an operational clearing of the way. (See also the papers by Jauch in LAA I and Gudder in LAA II and their references.) What is historically new here is the operational exploitation of logico-algebraic structure. Hitherto operational analyses had applied basically to descriptive terms. Logical terms were taken, following the positivist tradition, as *a priori*, *qua* conventional. With the operational analysis extended to include logical structure, and the logic thereby rendered in some sense empirical (*a posteriori*), all of the power of the mainstream logico-algebraic analysis of theory became accessible also to the operationalist analyses. And the fact is that this approach has led to new insights into physical theory and to significant 'generalisations' of it, as the studies presented here demonstrate.

All of which raises a much more interesting, if much more difficult, question. As a pragmatic strategy for theory development, operational bipolarity is a philosophically harmless heuristic for science. But the pragmatic success of the approach constitutes one good reason for believing that there is more to it than this – that it is, after all, an important truth about theorising. (To many, e.g., Einstein's operational analyses demonstrate the relativity of distant simultaneity, and not merely open the way to postulate it.) These considerations may be bolstered by a purely philosophical argument to the effect that the structure of science as conceived from an operationalist viewpoint is essentially indistinguishable from that provided by a hypothetical realism (i.e. one admitting that contemporary theories are only guesses at the truth). I do not myself find these arguments, or the operationalist philosophy, ultimately convincing (see again *Synthese* 1974, 1975 as above), nonetheless I grant that the success of operational analyses points to a deeper significance for its main ideas than merely heuristic attractiveness. In my view, an adequate theory of concept formation in biological organisms of our constitution and circumstances can accommodate the important semantical insights of operationalism compatibly with a thoroughgoing realism over-all.

Moreover, it ought to be added that not all of the authors represented here would be willing to call themselves operationalists in anything like the Bridgeman/positivist tradition (e.g., cf. the opening disclaimer by Randall/Foulis). It might well be claimed, in fact, that the papers present-ed here represent a distinctive re-flowering of the root idea of operation-alism, but in a wider, deeper fashion, freed from its narrow positivist beginnings. But deeply operationalist I believe each of these papers to be and, my opinions aside, they stand in their own right as an eloquent case for taking the approach seriously.

They also make an eloquent case for re-opening the question of the nature and status of logic. Hitherto in virtually all philosophical traditions, including empiricism, logic has been treated as *a priori*, as fixed, given. (The reasons for this have, however, differed widely, from the purely conventional status accorded it by positivism to the ultimate form of reality of platonic idealism.) Now, in common with many writers of the logico-algebraic mainstream, the logico-operationalist insists on the empirical character of logic. Logic is conceived as having a role in theory roughly akin to that of space-time geometry. Of course, this does not require a realist interpretation of logic (logic as objective structure), it is quite compatible e.g., with logic being interpreted as a system of empiri-cally induced conventions.

In this connection it is to be noted that the authors herein, and those of LAA I, II, by no means agree on the relationship which they see be-tween their work and the nature and status of logic. Some of them do not emphasise logical structure at all, concentrating instead on algebraic structure—logic enters only as a formal analogue of the algebraic struc-ture (see e.g., Kupczynski, Mielnik, cf. Gudder/Greechie in LAA I and Ruttiman and Gudder in LAA II). In other writers the focus is explicitly upon the abstract structure as logical, its being so is crucial to the analysis (Giles, Mittelstaedt, Stachow, cf. Bub, Putnam in LAA II). Thus the question of the empirical status of logic is raised, rather than settled, by the papers in these volumes. Indeed, the answer is inextricably bound up with the question of the proper understanding of the role of logic in physical theory. It is this latter issue which these papers, and the con-trasting papers of the mainstream tradition, so acutely raise.

One of the answers provided by writers in the mainstream tradition is "Not at all, except as mere formal analogy, logic is a purely linguistic affair". Although this (traditional philosophic) position can be construed traditionally in terms of linguistic rules and conventions, the dialogic approach, e.g., offers a radically different version of it: logic as a set of

reasoning maneuvres based in the actual physical possibilities of the
reasoning situation. Logic as a representation of the strategic choices of
a rational human being, where strategies are dictated, at least in part, by
circumstance. This is a newly interesting (if not wholly new) approach to
logic because of its powerful reconstruction of physical theory, and it is
one which may find a natural place in a thoroughly evolutionary, nat-
uralistic approach to language and to reasoning in general. An alternative and fundamental distinction made by others in the field is that
between a representation of linguistic reasoning transitions and a representation of the structure of physical possibilities (Bub, – see LAA II – and
possibly Randall/Foulis here). Developed in such a fashion that the two
structures may be incompatible with one another (e.g., respectively
Boolean and quantum), the distinction runs directly counter to the dialogic approach and yet may be consistent with the traditional position on
the nature and status of logic. On the other hand the two structures may
be identified (e.g., Putnam, Finkelstein in LAA II, Finkelstein here)
resulting in a quite different logical realist position.

As in the cases of LAA I, II, I do not intend to offer a detailed analytical
commentary on the papers. Again, this would only be to emphasise one
point of view in a rapidly growing field which deserves to be left open and
free from closed systems of ideas and it would in any case require writing
a major treatise of my own which would defeat the purpose of this book.
The papers must stand on their own merits – which they are easily capable
of doing. However I shall offer some brief orientating comments.

There is no doubt about the operationally-inspired origins of Finkelstein's work. A glance at his earlier papers, reprinted in LAA II, will
confirm this. In those papers he writes sometimes as if the 'working logic'
of a theory were little more than a shorthand description of the properties
of experimental machines, later on (e.g., in his paper in volume III of
Harper, W. and Hooker C.A., *Probability, Statistics and Statistical
Theories of Science*, Dordrecht: Reidel 1975) he writes as if uncovering
the logic were uncovering the objective possibility structure of nature.
In any event in my view Finkelstein has taken the logico-algebraic/
operational approach further than anyone else by pressing on into relativistic quantum theory. All other writers have hitherto been willing (or
forced) to stop at non-relativistic theory, where dynamics can be separated
from non-temporal structure and the former represented as a transformation on the latter (cf. Gudder in LAA II). This separation is no longer
possible in a relativistic theory and so relativistic theories pose a funda-

mental challenge to the logico-algebraic tradition to re-think the relation between dynamics and logico-algebraic structure.

Finkelstein's first encounter with this problem came indirectly and within the logico-algebraic tradition when he investigated quaternionic quantum mechanics (in conjunction with Speiser and Jauch—see LAA II). This led to a preliminary analysis of 'second quantisation', the quantisation of the field, in particular the relativistic field, as the introduction of quantification to quantum logic. Rather than pursuing this latter idea further in this form—it remains a still neglected topic—he advanced instead a more radical proposal: the 'state' was to be abandoned in favour of 'process' as the fundamental building block of theory and dynamics was to precede a-temporal structure in the order of formal reconstruction —both changes which, he insisted, were demanded by a relativistic theory. The traditional logico-algebraic reconstruction of a non-relativistic theory may be constrasted with Finkelstein's priorities for relativistic theories by listing the components in order of priority from top to bottom:

Non-relativistic	Relativistic
logic	dynamics–applied physics
algebra	geometry
geometry	logic
constructions	algebra
dynamics	constructions
applied physics	applied physics

Here "constructions" is used to refer to probability measures, operators and so on constructed on the algebra or geometry. In the non-relativistic case dynamics is realised as a set of transformations on the logic, in the relativistic case logic is constructed as a representation of the structure of dynamical processes. One could already see the importance of the question of the relation of logical to dynamical structure long before Finkelstein's work—e.g., through reflection on the roles of space and time in classical phase space, cf. my paper in Hooker, C.A. (ed.) *Contemporary Research in the Foundations and Philosophy of Quantum Mechanics*, Dordrecht: Reidel, 1973, and even in the earlier work of Bohr on 'rationalising' quantum theory and electrodynamics, cf. my paper in Colodny, R. (ed.) *Paradigms and Paradoxes*, Pittsburgh: University of Pittsburgh Press, 1972—but Finkelstein has served to raise it in a new and acute form. So far as I am aware, Finkelstein's work is alone in the logico-algebraic/operational tradition in pressing this issue, though both the dialogic and operational manifold approaches are now developed to the point where they seem capable of developing their own treatments of the issue. Note, though, that their treatments, proceeding through a representation of the spacetime structure in the relations among operational descriptions and results, reflected in logical structure, will be of a quite different sort, at least superficially, than Finkelstein's. (But it will not be until the constructions have been carried out–after all, no satisfactory relativistic quantum theory exists, anywhere, yet–and the constructions compared, that anything penetrating will be able to be said.)

One of the features illustrated *en passant* in Finkelstein's work is the achievement of generalisation through simplification, the simplification being achieved through operationally-sanctioned elimination. This idea is not new even to the mainstream logico-algebraic tradition; it is e.g., what led Gudder, Greechie and others to concentrate on partially ordered sets rather than lattices as the appropriate structure for non-relativistic quantum theory and which led in turn to a number of significant generalisations arising out of the weaker structure imposed (see their papers in LAA I and II, and also the papers by Gudder and by Finch in Harper/Hooker, vol. III op. cit.). But the process is clearly displayed in Kupczynski's paper in the context of a wide range of issues fundamental to the operational reconstruction of quantum theory. Moreover, his paper is a clear illustration of the manner in which operational analyses generate new programmes of research, in this respect he finds common ground with Finkelstein.

For Mielnik too, operational analysis offers the opportunity to generalise quantum mechanical structure. Unlike Finkelstein and Kupczynski, however, the point of departure is not a simplified structure, but a careful operational reconstruction of the full traditional mathematical formalism. In two earlier papers, discussed briefly in Mielnik's opening section, he has developed the formalism on an operational base of production, filtering and detection operations (cf. Finkelstein's similar base). It would have been pleasing to also have included those papers here but size limitations prevent it. The present work is of more relevance to include because it traces the development of Mielnik's position from the stage of elementary operational analysis to its culmination in the sophisticated convex structures approach and the possibilities for extending quantum theory beyond the existing formalism are explored. The reader can read this article in conjunction with that by Gudder in LAA II and the opening section of that of Randall/Foulis here in order to place the convex structures approach in a wider perspective of alternative constructions. *Conversely the beginning student might do well to commence with these latter two papers before pursuing any one approach in detail.*

Of the dialogic approach, here presented in its full flowering for the first time, I shall make only these brief further remarks, the corpus itself presenting a remarkably coherent statement of the position which it would be redundant to repeat here. To Mittelstaedt goes the historical credit for developing what was largely a purely logical formalism into an analytical tool adequate for modern mathematical physics, Mittelstaedt's approach being brought to its present maturity and completeness by his student Stachow. Despite some important differences of formulation, the work of Giles in many ways complements that of Mittelstaedt/Stachow. Giles himself, however, approached the dialogic from a different operational background, one more oriented to the algebraic analysis of physical theory – see his paper in LAA II.

What is more intriguing and of more importance is to compare and contrast the dialogic approach with the operational manuals approach developed by Randall and Foulis. As I remarked earlier, this latter approach is presented here in comprehensive form for the first time also. Hitherto, e.g., this approach had not been enriched by the notions of morphisms and compounds of manuals, both of which prove necessary to permit the re-capturing of many features of physical theory and its interpretation (e.g., operations corresponding to conditional probability statements and questions of the embedding of one structure in another – both crucial for the interpretation of theory). Here these additions are made, though, as with so much in the dialogic approach also, their exploitation has hardly begun. In any event with these additions the operational manuals approach acquires an analogue of the conceptual apparatus present in Giles' treatment of instructions for operations and their intimate connection with the subsequent logical structure of argument. These parallels in treatment are the more striking because it is currently so unusual among theories of physical theory to take the specification of operations so seriously. Despite these parallels, and the impressive ability of both theories to develop sufficiently rich mathematical structures to provide precise translations (in their own terms) of most or all of the major interpretive concepts and issues of modern physical theory, the two theories are developed in quite different manners and often in widely divergent terms. Yet the intuition suggests itself that these two approaches are developing much the same structures, though employing different mathematical means. (Very roughly: what the manifolds approach presents directly in terms of constructs on the manifolds of operations the dialogic approach first reflects as logical structure, subsequently erecting its constructs on that structure.) One of the important challenges, now that these approaches have reached some maturity, will be to find penetrating translation theorems linking them.

Beside the works listed in the extensive bibliographies to these papers, and the references of the papers of LAA I, II, there are important papers in Harper/Hooker, vol. III, op. cit., bearing on the logico-operational approach. In particular, there is an earlier paper by Randall and Foulis which provides a more extensive discussion of the logical structures arising in their approach and relating it to a theory of inductive support for statistical assertions. In addition, the papers by Finch and by van Fraassen/Hooker have different, but significant, connections with the logico-operational approach.

London, Canada
February 1977.

ACKNOWLEDGEMENTS

I gratefully acknowledge the assistance of my secretary Mrs. P. Switzer in completing the editorial task and I am particularly grateful to Mr. David Holdsworth for his careful and patient compilation of the indexes.

'Is the Hilbert Space Language too Rich?' by M. Kupczyński in *International Journal of Theoretical Physics*, **10** (1974), 297–316. Reprinted by permission of the author and *International Journal of Theoretical Physics*.

'Generalized Quantum Mechanics' by B. Mielnik in *Communications in Mathematical Physics*, **37** (1974) 221–256. Reprinted by permission of the author and *Communications in Mathematical Physics*.

'Completeness of Quantum Logic' by E-W. Stachow in *Journal of Philosophical Logic* **5** (1976) 237–280. Reprinted by permission of the author and D. Reidel Publishing Co.

'Quantum Logical Calculi and Lattice Structures' by E-W. Stachow in *Journal of Philosophical Logic* **7** (1978) 245–284. Reprinted by permission of the author and D. Reidel Publishing Co.

DAVID FINKELSTEIN

PROCESS PHILOSOPHY AND QUANTUM DYNAMICS*

Today we are apt to take the concepts of logic as fixed. For Boole, underlying every class or proposition was a mental act he called an *election*: the operation of selecting from an arbitrary class the members belonging to the given class, or of which the proposition is true. His laws all concern the effects of these acts of election performed in succession or alternation. He posited that elections are 1. distributive, 2. commutative, and 3. idempotent. He goes on to say that some will challenge the a priori truth of these laws perhaps, and some might even suspend them, but the resulting system of logic will then be a very different logic from that which we know. First I want to show these laws have indeed been challenged and suspended. So I brought along hardware for physical, not mental, elections for photons; three polarizers, x, y, z at angles of $\theta = 0$, $\pi/2$, $\pi/4$. Look:

In the order xyz, they are black: they select no photon. Now zxy, still black. Then yzx; still black. But in the order xzy, 25% light transmission: y and z don't commute.

Moreover the resultant of two elections yz is not an election, violates Boole's 3rd law, idempotency. But it takes 4 polarizers to show that $yzyz \neq yz$.

The order of these actual elections is significant in the microscopic quantum world, and this is the typical case.

We can't discover quantum logic with strips of polarizer. We only get half of it, the other half being the existence of quanta themselves. But any source and any sufficiently sensitive detector of light will do for that – the dark-adapted eye, photographic film, or photomultipliers, say. The combination leads to giving up the idea that acts of election commute and from that it's not far to the noncommutative observables of Heisenberg.

The quantum logical formulation of that is in terms of a lattice of acts of preparation and detection in which the cap and cup, the *and* and *or* operations, are nondistributive. There is a simple correspondence between nondistributive lattice and noncommutative algebra.

*Edited transcript of a talk given at the Department of Philosophy, University of Western Ontario, London, Ontario, November 29, 1974

1

Hooker (ed.), Physical Theory as Logico-Operational Structure, 1–18,
All Rights Reserved,
Copyright © 1978 by D. Reidel Publishing Company, Dordrecht, Holland.

Quantum logic is much simpler than classical. Classical systems A, B (sets) can be added, multiplied, and exponentiated:

$$A + B, \; AB, \; A^B.$$

Quantum sets can only be added and multiplied. Much of set theory goes away. But sum and product are enough for a predicate calculus with quantifiers.

That's about all I need to say about quantum logic because I'm concerned with the next step in physical theory, one that's been facing us for over half a century now. We have relativity, we have quantum theory, and we have no concept of the universe. We have instead a sublunary sphere at around a centimeter, dividing gravitational world from quantum. It's probable that the very concept of time and evolution of quantum mechanics is wrong. Instead of arguing for this on philosophical grounds, I would like to give a formal substitute–a quantum logical theory of a different space-time structure. Perhaps I should spend a few minutes giving what I think is the language of the future physics, and some examples of theories in it. I'll translate four classic texts into this language: Newtonian mechanics, Schroedinger quantum mechanics, a relativistic theory of a trivial world not fit for life, and then a fourth of a theory which is not yet a candidate for the world we live in, but at least it's not trivial, it's a world with interaction and mass spectra, built just with the quantum logical sum and product.

PLEXOR ALGEBRA

First a bit of mathematics. What I'm doing now is formalizing the calculus of Feynman diagrams. I think the world is a Feynman diagram. Feynman diagrams constitute a category blending quantum ideas, linear structure, with topological ideas, algebraic topological structure. The world we live in has strong elements of both in it. To emphasize that Feynman diagrams are a new category I'll give them a new name, *plexors*. A plexor involves first of all a graph, a set of ordered pairs. I'll call an ordered pair a dyad in order to get into a uniform terminology, and I'll write an ordered pair so there's never any ambiguity about which comes first and which comes second, by putting the two members of the pair with an arrow between them: $\alpha \rightarrow \beta$, or $\beta \leftarrow \alpha$. And the order of the arrow is the order that counts. As a set of pairs, a graph is simply a relation. It's also a

1-complex, so we can speak about its being connected, its algebraic topology and homology; and we can also apply all the operations of set theory to it.

The other element required to define a plexor is going to be a linear space L and this will be used in just the way that linear spaces are in quantum theory: the vectors of L correspond to certain processes of control, namely acts of production, of preparation, sources of the microsystem in question; and the vectors in the dual space L^T correspond to acts of detection, registration, recording, sinks for the microsystem under consideration. The linear space is over the complex field.

If δ is a dyad, $\delta = (\alpha \to \beta)$, let $L(\delta)$ be the space of maps $L(\alpha) \to L(\beta)$, where $L(\alpha)$ means a replica of L labeled by α. Then *a plexor of $L(g)$ is* simply a vector in the tensor product $\prod_{\delta \in g} L(\delta)$. The things we tensor-multiply are algebras of maps. Since $L(\delta) \approx L \otimes L^T$, in imagination we may populate the points of the graph with replicas of the linear space.

A tensor product has more structure than just a vector space: The index set, here a graph; and associated projections and imbeddings.

This is a discrete analog of a fiber bundle, in which the linear space is the fiber and the graph is the base. And it's going to be used in the theory much as we use a fiber bundle in physics. However the graph represents not a causal order–causality is a higher level concept–but an incidence relation; it gives only the propinquities of space-time. The graph is a flow chart, expresses the process structure of the world, and the linear space describes the elementary quantum processes linked by the graph.

The world is not just one graph. We consider a family of graphs, which we can think of as quantum possibilities for the structure of the world. This family of graphs will have an additional internal structure. For example, in some of the cases, there'll be graphs of this kind: $\cdot \mapsto \cdots \mapsto$, the graph of a linear ordering. In that case, we can take two and put them together and get another; so the set of graphs has an internal law of composition, it's more than just a set of graphs, it's a graph algebra. The set of graphs is typically provided with several relations or operations within it, and becomes a graph algebra, using the word algebra in the sense of universal algebra, not linear. I define the *plexors over a graph algebra* simply as a direct sum of those over all the graphs in the algebra. In all cases I'll only look at finite graphs.

The dimension of L is infinite in quantum mechanics and quantum field theory. In this talk I reduce L to dimension 2. One step more and no L, just a direct sum over graphs.

On the other hand, I started from a still more general concept of plexor, with linear spaces associated with the points of the graph as well as the lines. So we are specializing the language more and more as we go. The work is to do this without loss of expressiveness. That we can go as far as we have in this direction shows again that the usual language is enormously redundant. This brings us to the next element of a physical theory, how the formalism is used to map experience; from syntax to semantics; Leibniz would say from *calculus ratiocinator* to *characteristica universalis*.

There is an elementary time τ associated with the elementary dyad of a graph. Poincaré called the elementary time a *chronon*. Finer divisions of time arise only as averages.

SEMANTIC PROCESS

That was about formalism. Now meaning. In each of the quantum theories that I'll describe, interpretation is as follows. In principle, we go into a laboratory, look at what's being done, the process, and associate with that process an element of the plexor algebra; or conversely carry out a process associated with a plexor. I call this the *semantic process*. It associates with each physical process an element of the plexor algebra in a way that is faithful to the incidence relations among processes. The basic principle, *fidelity to process*, is to represent correctly the relations that the processes have to each other; then representations of matter, space-time, particles, and so on, must follow.

We may look back at elementary courses in quantum mechanics and see how they teach us, for example, to associate with a preparation of positions a particular (singular) vector of a Hilbert space $L^2(x)$; with a preparation of a momentum, another vector in the same Hilbert space. With another process, dynamical evolution in time, goes a vector in a bigger space $L(x) \otimes L(x)^T$, a vector called a unitary operator. And so forth.

Notice that the semantics need not make sense of all the vectors in the linear space. Quantum mechanics uses all, but classical mechanics uses only the vectors of a basis. Much of the structure of the theory is determined by what vectors are used; in L and in the larger spaces too, like $L \otimes L^\tau$.

In the four examples of theories, I try to make explicit the semantic process as well as the particular graph algebra and linear space. We can't formalize the semantic process. At best we can give a dictionary, go back to another language in which we've already learned the semantic process.

But ultimately it's an unformalizable thing, it's the interaction between the human being and the world. Our human ability to abstract, to associate different events with the same symbol in a way we agree on, is being called into play at this point of any physical theory. Our machines do it too.

More terminology: In the semantic process, it's customary to single out a particular set of processes called preparation. These always form a lattice; general processes form a larger lattice. If these lattices are Boolean, we speak of as classical system; if irreducible, a quantum system. Ordinary quantum mechanics, Schroedinger-type quantum mechanics, is not as far as we can go in the direction of a quantum theory. The larger lattice is still reducible. There are quantities characterizing the general process, even when it's most fully described, that commute with everything, belong to the center; classical variables. Time, in ordinary Schroedinger theory, is such a classical variable. It commutes with all process observables in ordinary quantum theory. We never superpose processes of different duration when summing Feynman diagrams. They always go between the same two times. Whereas we superimpose diagrams of different x, say, for a single particle.

You can think of t as an operator by going to a bigger ψ space, and Piron has formalized the theory years ago. Almost everyone learning Schroedinger theory is struck by the special nature of time in it, and tries to dissolve the usual algebra in a bigger one which includes t and d/dt. But I think we need a concept of quantum history.

QUANTUM HISTORY

There is no problem with a quantum mechanical trajectory. The usual way of describing processes done one after the other is to multiply the operators representing them. Then that product belongs to no larger a Hilbert space than the operators from which we started. Here I *don't* form the operator product to describe two processes done after each other. Two processes in succession are a more complex process than their ordinary operator products, there is more going on. We lose information by forming the operator product. Instead I form a tensor product that contains complete information about the first process and also complete information about the second process. There's a quantum logical correspondence between the tensor product for quantum mechanics and the ordinary direct product in classical physics. In classical physics a trajectory is the direct product of what goes on at one time, what goes on at the next time, and so on. We don't look just at the map from the beginning point to the end point of the curve and wonder where the trajectory went. But quantum mechanics has been first washing out history and then wondering where it went; merely by using inner products instead of outer. I'll use the outer product, as in classical physics.

This means the linear space of the process grows in dimension with the passage of time. In fact time could be introduced at this point as a logarithm of linear space dimension, since with each chronon its dimension goes up by a factor $n = |L|$, the dimension of the linear space L. This was remarked by Weizsäcker.

We can always go over to the usual formalism by taking a gigantic trace which converts outer products to inner products at all points. And since the process is not closed at the ends, the trace is just the ordinary unitary evolution operator for the entire process ignoring the intermediate stages. But if we want to discuss the intermediate stages, they are all present as commutative factors in the giant tensor product.

In a certain sense it's meaningless to speak of the energy of the system in cases where we discuss its trajectory. There is a complementarity between energy and time as operators in this theory, a complementarity lost in the usual theory, where energy and time have been treated in a way different from that between p and x. By putting everything in one algebra at the start, as has been recognized by many people, we can see energy and time as complementary in the same way as momentum and space. That seems a prerequisite for a relativistic theory.

TRACE

Finally the return to experience. The purpose of theory is to tell whether a process is possible from its symbolic description, to evaluate the possibility of possibilities. This is done as follows.

Plexors which describe a complete physical process, with all its inputs and outputs, are called *closed*. Such a plexor has a unique *trace*. Take the trace. If it's zero, that process doesn't go, is *forbidden*. Otherwise it's *allowed*. The law of nature is a selection rule: Processes of trace zero can't happen.

How do we compute this trace? One way of thinking of a plexor is as a great tensor, with an index for every member of every dyad. (If a point p

occurs in several dyads it's associated with several indices.) A tensor, but not just a tensor. There are lines joining indices upstairs with indices downstairs, pairing its indices. The tensor has an algebraic topology on it indices. Instead of the usual line or even two lines of indices, the tensor has a topological 1-complex of indices. Then there's a unique concept of trace in which you equate each index to the one it's paired with, and sum in the Einstein convention, over equated indices. We get a unique number which depends both on the topology and the linear structure. That's the amplitude for that process. If it's zero the process doesn't go, is forbidden, unlawful. The law of nature has the universal form

$$A = \text{Tr}P.$$

The amplitude of a process is its trace.

MONADS

It's handy to think of the graph $p \rightarrow q$ as a combination of two semigraphs, $p \rightarrow$ and $\rightarrow q$. So I introduce an ideal element, a null point a special point aside from all the others in the graph, and think of $p\rightarrow$ as a special or improper dyad, with one point of the original collection and the null point. I call $p\rightarrow$ and $\rightarrow q$ *monads*. A dyad containing the null point is a monad. Not only will I not write the null point, I won't count it. And $p \rightarrow q$ is obtained by stringing $p\rightarrow$ and $\rightarrow q$ in series and cancelling out the null point in between. A graph is *closed*, if all the dyads are proper dyads, with no null point. A graph in the original sense is closed. A graph which terminates in null points is not. Only closed graphs have numerical traces. We can apply the trace to non-closed graphs, but we'll be left with indices at the end.

As a dyad represents a quantum jump, a monad represents a process of creation or annihilation; it doesn't say which, that depends on a microscopic time direction. The word monad is used by Aristotle, Bruno, and Leibniz for elementary energetic entity, as opposed to atom for elementary material entity.

NEWTONIAN MECHANICS

Now let's express Newtonian mechanics in this language.

(The success of the old theory isn't an accident, the new one doesn't see the world in a totally different way. It's often a matter of enlargment of domain. I wouldn't trust this language if there were working theories with sizable domains of validity that did't fit into it. So I insisted in this program that each element of the final theory have a correspondent in the previous theories. This is just Bohr's correspondence principle generalized slightly and used as a tool in making up this theory.)

For classical mechanics the graph algebra is the linear graphs. The linear space in the case of classical mechanics is L^2 on the phase space, $L^2(p, x)$, as in Koopman's mechanics (1936).

Now the semantics. If we prepare the system to lie within a certain set of the phase space (notice this mix of formal and informal terms. *To prepare* is an informal physical art. *Phase space* is a mathematical concept. This is typical of any semantic correspondence. It can't be formalized. It links mathematics to the laboratory. And yet I'll continue to talk as if the billiard balls in the laboratory or the planets in the sky were points in phase space because of the habits I have from early training.) If we prepare the system to lie at a certain point (a, b) in phase space, I'll associate that process with the singular vector $\delta(p-a, x-b)$, a plexor with only one index, a monadic.

In Newtonian mechanics, control processes, whether emission or immission, are assigned to delta functions in phase space. $L^2(p, x)$ is a linear space whose basis vectors represent maximal controls. Superpositions are not given meaning.

For nonmaximal control, we go to statistical operators, or maps of that linear space. In that algebra of maps is the commutative subalgebra of functions, generated by the delta functions. The projections in this subalgebra are associated with the most general control of position or momentum. That's supposed to give the rules for representing all control processes.

There's only one other process in the game, and that's the dynamical evolution, the flow in phase space. This is the exponential $\exp[H]_P\tau$, whose exponent is the Poisson bracket with the Hamiltonian multiplied by a time, the duration of the monad. Now each arrow in a graph is associated with a factor, an element in the algebra, this exponential.

Then it's trivial to verify that forbidden processes are those for which the trace of the plexor associated in this way vanishes.

Classical Measurement Problem

Nevertheless a certain problem is raised. Control and evolution are described in two very different ways. Control provides a commutative algebra, which isn't enough to describe evolution. Evolution has to change the p's and x's. So we had to add a new structure, the differential structure, to describe evolution. Now this is going to leave a paradox. If we look at a process of measurement from outside, we see an evolutionary process of the system and the observer, the metasystem. The process of control is internal to the metasystem even though it's external to the system. Yet we're saying that internal processes belong to a commutative algebra and external ones to the noncommutative. Moreover controls have no inverses, while evolution is unitary and has inverse. How can the one process be both unitary and nonunitary? This is the problem of measurement. It is a legitimate problem of classical mechanics, and is inherited by quantum mechanics. But it is a paradox only if ψ is regarded as an element of reality. I prefer the following alternative.

I think of the vector describing a process not as something that exists out there in the Platonic sense, rather as an extension of ordinary English. In many ways the logic of quantum physics is much closer to ordinary language than that of classical physics. We can think of it as having a subject-predicate structure, in fact I think I need to for this analysis. I'll use Dirac's stroke | as the symbol for the subject. It represents the system carrying out the measurement, regarded from its own point of view, the equivalent to the word I in the English language. It's the simplest possible description of the observing system: it exists. I observe, therefore I exist. And the lattice of this system is the simplest possible one, one line, the lattice of the Aristotelian monad:

$$1$$
$$|$$
$$0$$

Dirac then describes a specific control process by a symbol that corresponds to the predicate of the sentence and stands next to the stroke, as in $|p\rangle$; within the predicate is the object of the sentence, an angle bracket representing the microsystem, or the exit port for the microsystem, if you like.

We can just as well think of $|p\rangle$ as a black box in a flow chart. The graph is the flow chart and $|p\rangle$ and $\rangle p|$ tell us what happens at points in the flow chart

SCHROEDINGER QUANTUM MECHANICS

The graph family is the linear graphs as before.

The linear space is $L^2(x)$.

The semantic process assigns to control of position $x = a$ the (singular) $|x = a\rangle = \delta(x - a)$; to control of momentum $p = b$ the vector $|p = b\rangle = \exp(ibx/\hbar)$; to dynamical evolution, some operator $\rangle T\rangle = \exp(iH\tau/\hbar)$.

For example, the plexor describing emission ϕ, transmission for time t, and immission ϕ^* is the tensor product

$$p = \phi^* \otimes T \otimes \ldots \otimes T \otimes \phi$$

with t/τ factors of T. Its trace is the usual probability amplitude $A = \phi^* U \phi$, with $U = \exp(iHt/\hbar)$.

A RELATIVISTIC THEORY

Graph family as before, the same line structure because I'm not going to do interactions yet.

For linear space I take the simple linear space that can support relativity. The essential group of relativity here is the Lorentz group, not the Poincare group. This is important. The world seems to have a hierarchical structure. Each level has its own characteristic group structure. Typically as we go down into the small, into fine details, we lose invariance; as we go up, we gain invariance, and ignore detail. If we have a collection of things and combine them by truly quantum logical operations, just adding and multiplying, we may gain but never lose invariance. Classically, if we have a group acting on points, and we put them together, we might make an orbit of the group. Then even if the points are not invariant, we've made something which is invariant, created an invariance by abstraction. Blake's aphorism: To abstract is to be an idiot. And another one: To abstract is to ignore. In this case we're ignoring in the technical sense of ignoring coordinates and creating symmetry.

I suppose then that the translation group operates at a higher level in the hierarchy than the Lorentz group, simply because it makes sense to speak of the Lorentz group as acting locally in ordinary field theory, acting on the vectors and the fields at a space-time point. And that point corresponds to one of the points of our graph. We can't discuss the translation properties of a physical system in such a local way. So I'm going to suppose that the Lorentz group goes way down at the bottom of the world, or since $2 = 1 + 1$, maybe one rung from the bottom, whereas the translation group first arises after we've put the points together and made the graph and smoothed it out. Then we can move the graph along, shifting one point to its nearest neighbor at the higher level in this logical construction.

The simplest space that admits the Lorentz group is the two-dimensional spin space $2C$. I think that the recognition that the Dirac particle moves in the direction of (loosely speaking) its spin, more accurately $dx^\mu/d\tau = \gamma^\mu$, is a clue. Think of the spin of an elementary particle as the growing tip of its world graph, where the next instance of creation is going to take place. So I take for linear space L just the two-dimensional complex linear space $2C$. I believe this is the origin of the Lorentz group: It's the automorphism group of the logic of a binary decision. Notice however that this logic only uses a linear structure. I haven't mentioned Hilbert space structure anywhere. That omission was in preparation for this step. If I'd introduced negation into the logic, then I would have to look at the invariance group of that, and I'd have the unitary group, the rotation group. I'd have lost relativity. So I had to build up a negationless logic.

The idea of a negationless logic is an ancient one from India as well as a modern one from lattice theory. I don't know whether one is a formalization of the other. But I have found it helpful to trace ideas back along the other branch of world thought, of cosmology, the process one.

Now the semantics. I have to build space and time and energy and momentum and so on out of this machinery. I use the fact that the infinite dimensional Hilbert space we usually associate with a particle moving in space could be the result simply of a great many factors two, not an infinite number but a large number of factors, so large we lose count. Then translation processes are not a continuum but just have a huge number of discrete steps. By keeping track of the intermediate processes, by taking the outer product instead of the inner, I don't have to start with infinite dimensional Hilbert spaces. I can start with two-dimensional spaces, and have the dimension grow until we reach processes that take so long that as far as we can tell they form an infinite dimensional space.

This is the reason this kind of a formalism can avoid from the start all the infinity problems of field theory. It's a finite theory in the small. Notice the two previous examples of Newton and Schroedinger had to use infinite dimensional linear spaces. That's where the continuum's built in, and where all the divergences come from, the fact that two points can get arbitrarily close.

Now semantics. I lapse into the normal procedure for quantum mechanics, which is not to go to the yes-or-no, the projections, but just give operators for particular quantities. Then we use spectral theory to get out the spectrum and the projections. We can think of the operator for a quantity as a generating function for a spectrum. It's a way of giving a whole family of mutually exclusive, together exhaustive, projections, expressing the logic of that degree of freedom, of that observable.

One rule I use is that space and time differences associated with processes are additive. If we do processes in succession, we add up their space-time increments. We have to read this backwards now: space and time arise as additive measures on the processes that are being carried out. They therefore play much the same kind of role in this game as the statistical operator, the density matrix in ordinary quantum mechanics, which also composes in an additive way etc. One of the things that gives me the feeling this might be on the right track, is that the statistical operators or hermitian forms in this space are in one to one correspondence with the future vectors of Minkowski space-time. I think then this is the origin of Minkowski space-time. It encourages me then to say that the usual things we call classical space and time arise from deeper non-commuting operators for each dyad.

We can guess what the underlying operator is just by its transformation properties. There is only one way to make a vector in this space, even if I don't see its logical meaning yet. If we just use the usual quantum mechanics there aren't any. The usual vector we make in $2C$ is the Pauli spin operators σ^μ and they're not operators in a relativistic theory. They have a mixed index structure, go from one space to another, from a space to the complex conjugate space. They do not form an algebra. But if we insist on a vector of operators on $2C$, we must give up linearity. And then it's unique. Antilinear operators on this linear space are products of matrices x_B^A with the complex conjugation operator K, $x_B^A K$, and transform as 4-vectors. The vector is $\eta^\mu = \varepsilon \sigma^\mu K$.

I assume that the space-time coordinate of one end of a process relative to the other is

$$x^\mu = \tau \sum \eta^\mu$$

where η^μ is an antilinear vector operator associated with each dyad.

Then I need momentum. I'll show the provisional one that I've used so far. I abandon a particular operator for p. All I represent is the momentum control process. It's a dubious assumption that there really is an operator of momentum which goes all the way down, because momentum represents infinitesimal translation and that can't be meaningful at the

DESCRIPTION	CODE	QTY.	LIST PRICE	LOC.	BOX(ES)	
HOOKER/PHY LOGICO OPER STR	HOPHI	18%/1	48.50	28R	01	*
PACKING AND HANDLING	9998	1	1.00			

PACKING LIST

569624	0071916	44.84	1	30MAR79	PHML	A
YOUR ORDER	INVOICE NO.		UNITS	DATE		

H-10

fundamental level. But there must be something down there which looks enough like momentum to explain the importance of momentum at the higher level, something which *corresponds* to it in the Bohrian sense. It doesn't have to be an observable. It's enough if we get a family of projections.

The procedure I'm adopting here is that familiar from the theory of coherent state of linear oscillators where we deal with nonorthogonal overcomplete families of operators.

I call $\exp i k_\mu x^\mu = |k\rangle$ the plexor describing a momentum control process. Then it's obvious I'm building in the usual Weyl type commutation relation for momentum and coordinates in the limit of the macroscopic continuum.

[Q: Do all these different components of x commute? The η^μ don't commute. A: They don't commute, that's right. And therefore x^μ won't satisfy with itself the usual commutation relations. But notice that x^μ is a sum over all the dyads in a sequence. Therefore the nonzero terms in a commutator are very small in number, just the self terms. So in the continuum limit, we have the usual commutativity.]

Now I compute the energy momentum spectrum, the mass spectrum of this theory. And I guess what I should have done is paraphrase, do the computation in each of our theories so I could show you I was doing exactly the same thing here, not inventing new procedures. No new element is introduced.

[Q: Isn't this a little bit like what Penrose is doing? A: It's even more like what he used to do: spin networks.]

One computes the amplitude for propagation with energy-momentum k, and then the mass spectrum is the poles of this amplitude. We just compute the S-matrix directly, find its poles, and these are the physical particles, the mass spectrum. This is worked out elsewhere. The result is a family of spin-$\frac{1}{2}$ particles of masses.

$$m_n = 2\pi\hbar n/\tau, \quad n = 0, \pm 1, \pm 2, \dots .$$

This is a mass spectrum with a neutrino at $n = 0$, a particle with mass 0. And then surrounding it are massive particles with a linear spacing in masses, reflecting the triviality of our crystalline time axis.

In this graph algebra there are no interactions. But we can imagine a graph algebra of the following kind. In all cases the graphs are built up of a small number of basic things called vertices. Suppose the basic vertex is a tetrad, an X, and we only allow tetrads to link like XX...X. This is a spin 1 propagation process. It can be regarded as the genetic code of the photon. But we don't just get the photon, again we get a particle of mass 0 (spin 1 now) in the middle, and an excited state of mass $2\pi\hbar/\tau$ which of course we would hope is the intermediate boson, the carrier of the weak interaction. And unfortunately we get longitudinal photons too.

I've made model calculations of the weak interactions and so on, to see if they fit into this language. One can always carry over the structure of the usual theory: one kind of graph for propagation, and this is usually associated in continuum theory with a Feynman propagator; another kind of graph for interaction; and the relative importance of these two is assessed by a coupling constant. The arbitrary graph is a mix of propagators and interactions. This is the usual duality between propagating and interacting, between space-time and what goes on in it. This is what Einstein wanted to eliminate in forming a unified field theory. The thing that strikes one now is that there's no need for this. We can build a whole world out of interaction.

A UNIFIED QUANTUM THEORY

First the graph algebra:

I'll call a quantum theory in which the graph algebra consists of a single vertex and composites thereof, a unified quantum theory. In it all the processes in the world are built up of one process. We're not looking for particles. They are long thin nets, not elementary at all. We're taking the processes which go on in the world and breaking them down into their elementary parts.

The need for two kinds of constituents in the world was already felt by Aristotle. The elementary constituents of inert things or matter he called atoms and the elementary constituents of vital things, energetic things, he called monads. For him, the human soul was a monad, as it is in the usual description of a measurement process in quantum theory.

The first to insist that if a monad was really indivisible it had to be local, very, very small, was probably Bruno. I use his word monad for this ultimately local entity, which was later adopted by Leibniz. Then n monads make an n-ad.

So the graph algebra is a tetrad X and all graphs made out of it, heads to tails, like XX, XXX, ...

The meaning of the arrows is the source of some puzzlement. Feynman says forget about the direction of time, forget about time's arrow, it doesn't exist down there in the microworld, it's something the observer carries into the system. And so these arrows can't represent the causal order, and we don't demand that the graph be a partial ordering, like the causality relation of relativity is required to be. It took me years before I was willing to entertain that. I went along thinking in terms of general relativity. General relativity is a world founded on causal order. It leads us to look for a causal order down in the microcosm. But I could never find any reasonable way to build up partial orderings vertex by vertex and get any correspondence with Feynman, where the arrows don't represent the causal order at all but the flow of charge or the flow of some other conserved quantity, with the typical particle-antiparticle relation conserving TCP. So I'm going to suppose henceforth that Feynman was right, Einstein was wrong, there is no causal structure in the microcosm, these arrows have nothing to do with the time ordering, they represent currents. And that means we make the most general graph conserving arrows.

For linear space L I keep the same linear space, the spinor space, and for interpretation, the same formula for x, but I no longer have a unique path going between two points.

Now this non-uniqueness is on the one hand a puzzle, but on the other hand it's something to be grateful for because we have somehow to see general relativity looming on the horizon. As we go up in the hierarchy, in the ladder of types, we lose information, lose structure, ignore coordinates, and the group gets bigger and bigger. Somewhere way up there we'll reach the Einstein coordinate group, or some approximation to it.

There are two ways of losing a symmetry: we break it by going down the ladder, dissolve it by going up. For example when we ascend to general relativity Poincaré invariance is dissolved into the BMS group, a bigger group of which the Poincaré is not even a normal subgroup, so speaking of Lorentz invariance loses invariant meaning. It's not that we break the conservation. We have lost it among an infinite number of candidates, a continuum from which it can't be distinguished. I assume that general relativity will come way up the ladder just because of the huge invariance group of the theory. It's the only way to account for such a big group in a hierarchical structure.

Yet even down here we want to see something like the noncommutativity of displacements in space and time, something funny has to happen when we come to loops, and here is a funny thing happening: If we have a loop of two arcs like (), which do we use to define the coordinate going from end to end? The coordinate will depend on which of the two arcs we use, and the sum of the relative coordinates around a closed path isn't zero identically. Our coordinates are non-holonomic.

However in the models I've made, the discrepancy comes out to be zero on the average. So the hope is that we'll get Minkowski space-time in a simple average, and when we look at departures from this average, which may take larger ensembles, we may find general relativity.

To proceed I average over all paths between two points, in forming the coordinate, just to have a definite coordinate available for trial computations. Averaging over all paths is motivated by the resemblance between the coordinate and the statistical operator. The statistical operator too is averaged over parallel processes.

Now the computation of the mass spectrum is hard; I will not attempt it until I have studied the internal consistency of these ideas more. There are still *ad hoc* assumptions that might conflict.

[*Q*: What about statistics? *A*: Feynman would say that you ignore statistics in summing over the diagrams and at the very end you antisymmetrize. In quantum logical terms this is to say that the structures we look at are sets, the processes you're looking at are sets of monads, not sequences, permutations or what have you. There are many methods of aggregation for going from an individual to a collective in classical logic and most of them have correspondents in quantum logic. The Fermi-Dirac ensemble is a quantum correspondent of the classical concept of set. The clue is that you speak of something being in a set or not, you don't speak of being in a set twice. This is *ad hoc*.

Q: What about the Bose-Einstein statistics?

A: In this game bosons appear as assemblages of an even number of fermions.

Q: Somewhere at the beginning of your talk you spoke about a classical measurement problem, a measurement problem in classical mechanics. Is that changed?

A: One aspect is inherited with no change in this evolution: the fact that a control is a singular projection and dynamical evolution is a non-singular operator. In firing bullets at a target we follow the flow in phase space, and chop out that part of it which passes through the target to describe the ordinary process of observation, of selection. No unitary evolution will do that exactly. In quantum mechanics people speak of a collapse of the wave function to describe this irreversible process.

Q: How do you account for internal symmetries?

A: As in any hierarchically structured theory. To put things together is to increase the symmetry. I have no calculations to go with this at all. But try to make simple propagating structures out of tetrads. You can't make a linear one because that's not in the graph algebra. The first one you can make that's simple is probably the double helix XXX... X which represents a spin 1 propagation. The simplest propagation of spin $\frac{1}{2}$ you can make is something like three intertwined strands. This would have the symmetry of something made of three particles but there's no way of getting the particles to separate, because they only propagate by interacting. Maybe they are quarks.

Q: Does this mean that the arbitrary division between internal and space-time symmetries disappears in that theory?

A: In principle you can translate any kind of quantum theory into this language. If you want to play the game with separate internal symmetry degrees of freedom, a separate isospace, isospin space, you can. You take the original linear space and multiply it by as many other factors as you like. But that would spoil the game. The hope is that a straightforward calculation of the whole structure will automatically produce these internal degrees of freedom. They won't be of a space-time nature, because we're way down inside space and time, but intermediate between the space-time level and the monadic.]

The main things we've learned so far are that nets of binary quantum decisions are a rich enough language for formulating relativistic quantum dynamical theories with interactions; and that while a cutoff length in the large just leads to a small mass-shift, the cut-off length in the small leads to a new spectrum of particles wherever the continuum has only one particle.

ACKNOWLEDGMENT

I have benefitted from discussions of this work with G. McCollum.

Yeshiva University

POSTCRIPT (JUNE, 1975)

Subsequent simplifications include the elimination of spinors and the imaginary unit, leaving real scalar plexors; at the expense of providing the graph with a *flow*, a 1–1 mapping of each dyad $p \to q$ entering a vertex q to one $q \to r$ leaving the vertex q, as though the graph were populated by atoms undergoing the process. Now the space-time structure is derived from the tetradic nature of the vertex, not from independently postulated spinors on the lines.

It seems worthwhile computing the masses and coupling constants of this theory.

EDITOR'S NOTE.

For the reader's convenience, Finkelstein's theory is developed in the following papers: 'Space-Time Code', and 'Space-Time Code N' for N = II, III, IV, V, respectively in *Physics Review* **184** (1969), 126; *D5* (1972), 320; *D5* (1972), 2922; *D9* (1974), 2219; (with G. Frye and W. Susskind) *D9* (1974), 2231. See also 'Unified Quantum Theory' (with G. McCollum) in Castell (ed.) *Quantum Theory and the Structures of Time and Space*, Munich: Hans Carl Verlag, 1975.

FORMAL LANGUAGES AND THE FOUNDATIONS OF PHYSICS

0. INTRODUCTION

When I first began to take a serious interest in theoretical physics I was attracted by quantum field theory, a mysterious subject which – at least in the form of quantum electrodynamics – nevertheless seemed to be extraordinarily successful in predicting experimental results. It was an obvious challenge to try to develop a formulation of this theory which would be both mathematically satisfactory and, from a physical point of view, *self-contained* in the sense of not depending for its interpretation on prior physical theories. However, I soon found that this task was far from easy; so much so that it seemed advisable first to aim at a fully satisfactory account of ordinary quantum mechanics, a subject of which quantum electrodynamics might be described as a very sophisticated descendent.

Of course, a number of *mathematically* adequate accounts of quantum mechanics already existed, and others have appeared since: I mention in particular those of Von Neumann [34] and Mackey [24]. However, in common with nearly all current presentations of physical theories, these formulations are deficient in one important respect: while adequately describing the mathematical structure of the theory, they fail to give a comparably precise account of its physical interpretation. At first I tried to minimize this defect by devising formulations – of quantum mechanics and other theories – in which only *very simple* experimental concepts were referred to [8, 9, 10]. However, as a result of these efforts, I became increasingly convinced that such ameliorative measures were inadequate, that the deficiency must be overcome once and for all, and that this could only be done by finding some way for employing a formal language to describe physical experiments. There thus arose a program of research which is still in progress and which is reviewed here up to 1975.

Much of the work reported here has appeared in previous papers. The procedure for defining the interpretation of a formal language by ostensive definition, explained in Section 2 and exemplified in Sections 6 and 7, is a development of the 'documentary interpretation' which was first presented in [12] and, in abbreviated form, in [11]. These two papers also

19

Hooker (ed.), Physical Theory as Logico-Operational Structure, 19–87.
All Rights Reserved.
Copyright © 1978 by D. Reidel Publishing Company, Dordrecht, Holland.

contain the basic logical development–the new concept of a proposition (Section 9) and the dialogue interpretation of logic (Section 10). Concise descriptions of this new approach to logic have appeared in [15] and [16]. Finally, the connection with probability, described in Section 12, was first given in [13] and has been reviewed along with earlier work in [14].

This work has been supported throughout by grants from the National Research Council of Canada.

1. COMPLETE THEORIES

The principal defect of most formulations of physical theories is the failure to give an adequate account of how the theory is supposed to be applied in practice. Admittedly, this situation conforms to the usual meaning of the word 'theory', by which experimental details are explicitly excluded. However, insofar as a theory is intended to have some empirical meaning, such a formulation must be regarded as being incomplete. It is not possible to assess the validity of a theory that is formulated solely in this way; we are unable to say whether it works in practice, simply because we have not been informed how it is *intended* to work. Information of the sort required is usually provided not as part of the theory but through a course of practical training, normally extending over many years and ranging far beyond the specific concerns of the theory in question. Nevertheless, this information is required before we can say that we understand the theory as a *physical* theory: i.e. before we can make any practical use of it. Accordingly, let us refer to a theory in the conventional narrow sense as a *deductive theory*, and introduce the concept of a *complete theory* as one that includes all the experimental information necessary to apply the theory in practice in the intended way. We shall refer to this information as the *rules of interpretation* of the theory and to the rest as its *deductive part*.

To give a formulation of a complete theory both the deductive part and the rules of interpretation must be precisely described. In the case of the deductive theory this is usually done–if it is done at all–by means of an *axiomatization*: certain *proper axioms* are postulated and the assertions of the theory are then deduced as *theorems* by means of the axioms and rules of inference of classical logic. This is a straightforward process and examples abound in the literature [5, 24, 25, 34]. Little attention has been paid, however, to the problem of the precise specification of the rules of interpretation. It is with this question that we shall be largely concerned here. The problem reduces to that of making precise reference to practical situations. For this purpose some language must be employed, which we shall call the *primary language* of the formulation. This language may, but need not, coincide with that used in the deductive part of the theory. That part, being essentially just a mathematical theory, is usually expressed in the language of classical mathematics; in any case, we shall call the language used there the *secondary* or *theoretical* language of the formulation. We thus have the following situation. The deductive part of the theory, written in the secondary language, determines a set of statements (theorems) which are *asserted* by the theory. Next, by certain *rules of translation*, some at least of the theorems are converted into sentences in the primary language. (Of course, if the primary language should coincide with–or simply *contain*–the secondary language this step does not occur.) Lastly, the practical meaning of these sentences is determined via the semantics of the primary language.

The common–indeed the universal–practice is to use for the primary language a natural language, e.g. English. This procedure has obvious practical advantages, but it places a severe limit *in principle* on the precision with which the intended interpretation of the theory can be described. Indeed, every natural language is deficient in two respects: first, its syntax is ill-defined and at times ambiguous; secondly, the meanings of the terms used are unclear. The first deficiency can easily be overcome. There is no reason why the syntax of a language should not be defined with the same precision as is found in mathematics, where the only unavoidable ambiguities arise from the misinterpretation of individual symbols (or of 'rules of construction', see below). One then has an artificial or *formal* language. To overcome the second deficiency the semantics of the primary language must be defined in a precise way. What this entails will be discussed in the next section.

2. THE DEFINITION OF PHYSICAL CONCEPTS

Suppose that we have a formulation of a complete physical theory, as described above, with a formal primary language. We must now consider how the rules of interpretation can be given: i.e. how precise physical meanings can be assigned to the terms in the primary language. Of course, it would defeat the purpose of introducing a formal primary language to use a natural language to explain these meanings: if the meanings are to be explained verbally at all this must be done in terms of the primary language itself. The situation is similar to that faced by a teacher of experimental physics, in which the meanings of terms must be explained using only the language, say English, to which the terms belong. We get an exact analogue if we assume that the pupil arrives without any knowledge of English, so that the work has to be carried out *ab initio*. Let us then first discuss this situation; we can afterwards consider what changes may be necessary in the case of a formal primary language.

It is immediately evident that the method used in teaching the meaning of a term depends very much on the 'degree of sophistication' of the concept involved. In this respect physical concepts vary widely. At the one extreme there are terms such as *magnetic field, entropy, polarized beam of neutrons*, which are significant only to a physicist; then there are others, like *heat, speed, weight,* which have a meaning for the average adult; and still others, like *water, ball,* that can be at least partially understood by a child; finally, at the other extreme there are very simple concepts like *bang, dark, hot,* which can be appreciated even by an infant. In the teaching of physics concepts are introduced in increasing order of sophistication, each new term being normally explained verbally with the aid of previously understood terms. However, this does not apply to *direct concepts* such as the last three examples – that is, concepts which relate to direct experience; such concepts are more easily *demonstrated* than explained. In fact it is possible to recognize in this way a rough sort of partial ordering among physical concepts. We imagine ourselves giving an explanation of the term denoting a given concept A. The concepts which arise in the explanation are then regarded as *more direct* than A. It is most important, in applying this definition, to beware of merely giving an explanation of A as a theoretical concept; it is the empirical concept with which we are concerned. For example, it would be inappropriate to 'explain' a metal as being 'a material in which electrons are free to move from atom to atom'; this description is not aimed at clarifying the experimental significance of the term 'metal'. It may be helpful to regard such explanations as being addressed to a young but intelligent child, their object being to enable the child to understand sentences employing the terms, in the sense of being able to recognize in practice the phenomena to which these sentences refer.[1]

With this notion of the ordering of physical concepts by directness let us return to the question of the definition of physical terms. As we have seen, there are two ways in which the physical interpretation of a term may be given: either verbally, by means of an explanation in the primary language itself, employing terms that are more direct than that being defined; or *ostensively*[2]– i.e. without using language but simply by *pointing out* the object referred to, *demonstrating* the procedure called for, and so on. Of course, in practice any term–even one referring to an individual object–is applicable to experience on many occasions. For an ostensive definition it is necessary for the teacher to point out a number of these instances before the pupil will 'catch on' and demonstrate by his response that the meaning of the term has been grasped.

Verbal definition is the more familiar type. When available it is generally more convenient than ostensive definition and it is commonly used for all sophisticated terms. However, since any verbal explanation assumes prior understanding of other terms it is clear that not every term can be defined in this way (see also [28], p. 29): ostensive definition must be used, at least in the early stages of setting up a language. Ostensive definition is the means normally adopted for the definition of direct terms, but it may be used for other terms also: for instance, the meaning of 'cup' is generally conveyed to a child ostensively. However, it is clear that the less direct a term is the more difficult it will be to give its meaning in this way. In selecting terms to be defined ostensively one should therefore choose those which are as direct as possible. Note also that, since each term requires individual treatment, only a finite number of terms (in a given language) can be defined ostensively. In contrast, verbal definition is not limited in this way; linguistic tricks, such as recursion, can be used to explain an infinite number of terms.

With this analysis of the possibilities of definition for terms in a natural language let us now consider the case of a formal primary language. Any formal language consists of a collection of meaningful expressions[3] which are formed from certain *primitive expressions* by the repeated use of stated *rules of construction*. Usually, each primitive expression consists of a single symbol and each rule of construction is also denoted by a symbol. Any *composite* (i.e. nonprimitive) expression is formed by some rule of construction from simpler expressions. In assigning an interpretation to the language the natural thing is to arrange that the meaning of each composite expression should be determined by the meanings assigned to its components and to the rule of construction involved. If the components are themselves composite expressions we may assume their meanings to have been established in a similar way, but the meaning of the rule of construction must be established ostensively: by giving simple instances of its use, together with–in each case–ostensive definitions both of the composite expression formed and of its components.[4] In this way, with the aid of an ostensive definition for each rule of construction, the meaning of every expression is established in terms of those of the primitive expressions. If these too are defined ostensively the interpretation of the language is complete.[5]

Notice that for the possibility of establishing an interpretation in the above way two features of the primary language are necessary. First, there must be only a finite number of *primitive* concepts (by which I mean primitive expressions and rules of construction); otherwise the task of giving their ostensive definitions would never end. Secondly, to facilitate this task, the physical interpretation of the primitive concepts should be as direct as possible.

Examples of formal primary languages interpreted in this way will be given in Sections 6 and 7, but first let us look at another boon which a physical theory may derive from such an interpretation.

3. THE JUSTIFICATION OF PHYSICAL THEORIES

The semantics of the primary language of a complete physical theory can serve a purpose other than the obvious one of specifying precisely the empirical meaning of the theory: it may provide a *justification* of the theory as a whole–the theory is justified insofar as it works. It is often claimed that no further justification of a theory can be expected. In fact, this is true in the case of most current formulations, for it is usually the case that individual axioms–and indeed many other sentences of the theoretical language–have no empirical interpretation. However, if a theory can be formulated in such a way that all, or at least some, of the axioms have an interpretation (via the semantics of the primary language) then these axioms may acquire an *individual* justification in that they are seen to be 'obviously true' in virtue of the experimental meanings of the terms involved. The presence of such self-evident axioms in a formulation of a theory results in two advantages. The first is that insofar as it is justified in this way the theory may be regarded as providing not merely a description but an *explanation* of the phenomena it refers to. Indeed, even if the axioms are not entirely justified by the interpretation, the very fact that they *have* an experimental meaning helps to give the physicist a stronger feeling of understanding these phenomena than he would otherwise obtain. The second advantage is more tangible: when – as it will in due course – the theory needs to be modified or extended, the degree and nature of the justification of the individual axioms provides welcome guidance in the difficult question of deciding which axioms to alter and how.

It is sometimes claimed[6] that, on the contrary, the desire for self-evident axioms is misguided in that it leads to an undesirable rigidity in the theory: roughly speaking, if all axioms were entirely self-evident then no change at all could be contemplated. This is true, of course; in fact, such a theory would be correct, so no change would be necessary. However, as anyone who has worked on the development of new formulations is aware, one is never in this happy state: no axioms are entirely self-evident, and the difficulty in the selection of an appropriate modification of a theory is, in practice, always due to the range of choice being too great, never too small. Nor is it ever likely to be otherwise. The axioms of a theory can never be perfectly self-evident, if only because of the residual vagueness in the rules of interpretation. (Self-evidence should therefore be regarded as a quality which may be possessed to varying degrees, but never totally, by a sentence.)

In order to permit even the possibility of obtaining self-evident axioms attention must be paid to the way the deductive theory is formulated (in the secondary language). The choice of axioms must be matched to the construction of the primary language in such a way that the terms employed in the axioms – axiomatic terms, for short – are associated by rules of translation with terms in the primary language and hence by rules of interpretation with physical concepts. Moreover, these concepts must be so direct – so closely related to direct experience – that any empirical relations that exist between them are quite plain and unequivocal. We shall refer to a formulation in which this is done as a *satisfactory formulation* of the theory.

There is a simple way to ensure that these conditions are satisfied. Suppose that the deductive theory is formulated directly in the primary language, so that the secondary language either coincides with or is a sublanguage of the primary language.[7] Then every sentence of the theory will have an empirical meaning. Thus, since the primitive concepts of the primary language have already been chosen to be 'direct', it only remains to select the axioms so that they are expressed (as far as possible) directly in terms of these primitive concepts.

Certainly, physical theories are not normally formulated like this, but we may try to reformulate particular theories in this way. The first step in this process is that of selecting, for use as primitive concepts, the most direct physical concepts that are referred to by the theory. In the next section I shall illustrate this process with some concrete examples.

4. THE SELECTION OF DIRECT CONCEPTS

The reformulation of a physical theory in accordance with the proposals of the last two sections involves a number of considerations that are not related to the use of a formal primary language. First among these is the selection of appropriate primitive concepts.

As an important example of this process consider the case of a *physical quantity* or *observable*. Examples of observables occur in almost all physical theories–for instance, temperature, mass, energy, etc.–and such quantities are used as primitive concepts in most formulations of physical theories. Ought such a notion to be retained as a primitive concept of the primary language if the principles of Section 2 are adopted? To answer this question we must decide whether there is any 'more direct' concept that could be used instead. Suppose then that we wish to explain, to a child say, the empirical meaning of a term denoting some observable. About the best we can do is to demonstrate at least one way in which the quantity in question can be measured. In most cases such a demonstration will end by our obtaining, perhaps from the position of a pointer on a scale, the successive digits in the decimal representation of the quantity required. Now, from the child's point of view, this process of reading a scale is certainly nontrivial, and for a complete theory some account of it must be given. Let us imagine, then, that we wish to explain how such a pointer reading should be taken. Since we cannot assume an ability to read we decide to simplify the task by regraduating the scale in binary rather than decimal notation (fig. 1); only two characters need then be learned. But while engaged on this task we realize that the whole process can be vastly simplified as follows. We provide the child with the sequence of masks shown in fig. 1, together with the following instructions: Apply the first mask. If the pointer is visible write 1 (after a binary point); if not write 0. Apply the second mask and add a second digit in the same way, and so on. In this way the taking of the pointer reading is replaced by a sequence of measurements of a simpler type. Indeed, the scale-with-a-mask-attached represents an observable which can take only two values

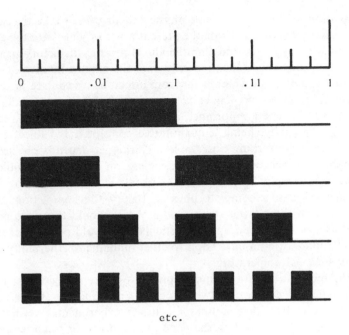

Fig. 1

(pointer invisible or pointer visible). We shall call such an observable a *test*, and designate the two possible values *yes* and *no*. (0 and 1 have undesirable numerical connotations, and the other possibility, *true* and *false*, would be unwise since we shall later be deeply concerned with the meaning of truth.)

The discussion shows that the concept of a test arises when one tries to explain the empirical meaning of an observable. Consequently, by the argument in Section 2, *test* is the more direct concept, and it should be adopted as a primitive concept in preference to *observable*. Similarly, 'experiment' should be replaced by 'elementary experiment' where the words are used in the following sense: An *experiment* is any procedure which starts from scratch and culminates in the measurement of an observable, and an *elementary experiment* is one in which this observable is a test. Thus an elementary experiment is a complete experimental procedure that results in one of two outcomes, *yes* or *no*.

As a second example of the choice of new primitive concepts for a reformulation, this time of a whole physical theory, let us take the case of classical thermodynamics. From the present point of view both the classical formulations (Kelvin and Carathéodory) are unsatisfactory in many respects. For simplicity I mention only one: the axioms involve the concept of *work*, a concept which is far from direct, as we can see in several ways. First, it can not be defined *ab initio* in any simple way; indeed, in spite of its fundamental importance, it is barely covered in the high school curriculum. Secondly, being a quantitative concept it is subject to the above criticism concerning observables. Thirdly, it depends on classical mechanics, and hence (through kinematics) on geometry; should we attempt, in a straightforward way, to embed a classical formulation of thermodynamics in a complete physical theory the completion would have to contain these latter theories too. Work should therefore not be adopted as a primitive concept; in particular, it should not appear in the axioms. Similar criticisms also apply to other fundamental thermodynamic concepts such as temperature.

For the recognition of appropriate primitive concepts a good principle is to consider those assertions which on the one hand are typical of the theory and on the other admit of a direct experimental verification. Through the second property the concepts occurring in such assertions will be direct: through the first they will be fundamental for the subject. In the case of thermodynamics such an assertion is typically of the form (consider, for instance, the first and second laws): such and such a process (i.e. change of state) is, or is not, possible. Two candidates for primitive concepts appear here: first, the notion of a *state* (of a system); secondly, a relation between states which holds when the first state a can 'transform' into the second b – let us then write $a \to b$. A little further thought along these lines yields another primitive concept. Frequently a system is a 'union' of two other systems, and its state c is then determined by the states, a and b, of the parts; we write $c = a + b$. This operation $+$ is our third primitive concept. It turns out [8] that these three concepts alone suffice for the formulation of a large part of classical thermodynamics. [Notice that none of them is quantitative and none depends on mechanics.]

The choice of suitable axioms involves more thought. One tries to select properties of the primitive concepts that are simply stated – and of course true. Some such properties are self-evident in the sense explained in Section 3; three examples are: $a + b = b + a$, $a + (b + c) = (a + b) + c$, if $a \to b$ and $b \to c$ then $a \to c$ (where a, b, c are arbitrary states). Others are not – for example, $a \to b$ and $a \to c$ implies $b \to c$ or $c \to b$ – but, owing to the relative directness of the concepts involved, they all have a direct experimental meaning.

In [8] these and a few other properties are adopted as axioms for thermodynamics. These axioms, of course, embody only the most general principles of classical thermodynamics. They do not explain–as statistical mechanics does–the natural occurrence of particular kinds of system (e.g. the perfect gas). Nevertheless, one can deduce from them (a) the existence of 'sufficiently many' *components of content*–quantities which are additive and always conserved, familiar examples being *entropy*, mass, components of momentum, etc–and of one quantity, entropy, which is additive, is conserved in any reversible process, and increases in any irreversible process; and (b) the existence of *equilibrium states*, each of which is characterized by a set of *components of potential* (corresponding to quantities like temperature, pressure, chemical potential, etc.) each of which determines the natural direction of flow of the corresponding component of content. By showing explicitly how such concepts can be defined in terms of the three chosen primitive concepts this formulation brings out their fundamental role in the subject and in particular their independence of the principles of mechanics.

Being a neat self-contained theory classical thermodynamics is an ideal subject for the sort of reformulation advocated here, but it cannot be denied that mechanics, and particularly quantum mechanics, is a much more detailed, powerful, and comprehensive theory. In the next section we make an initial survey of this territory, with the object of identifying and explicating a suitable set of primitive concepts. We shall find that this exercise introduces a new philosophical perspective, differing substantially from the orthodox view.

5. THE CONCEPTS OF QUANTUM MECHANICS

The most important application of the process of reformulating a physical theory arises in the case of quantum mechanics, for although this theory provides the foundation for the bulk of modern physics it is beset by serious problems of interpretation which obscure the meaning of the formalism and retard its further development. The program laid down in Sections 2 and 3 will serve as a framework for the fundamental concepts of the theory and allow us to recognize and examine some of the philosophical presuppositions that are implicit in the conventional approaches to the subject. In addition, a special motivation applies in this case: although quantum mechanics may be regarded as a generalization of classical mechanics there are features of the latter theory, for instance the topology, and especially the manifold structure, of phase space, which have no analogue in quantum mechanics, and because the standard formulations of classical mechanics and quantum mechanics bear little resemblance (the phase space of classical mechanics is in no way a special case of the Hilbert space of quantum mechanics) it is very hard to see how quantum analogues of these features should be introduced. However, if the two subjects can be reformulated in accordance with the principles advocated here the experimental meanings associated with the concepts and axioms of each theory should allow us to recognize the appropriate analogues.

As in the case of thermodynamics the first step in the reformulation of quantum mechanics is to select a set of direct physical concepts to serve as primitive concepts. In elementary formulations of quantum mechanics the primitive concepts include (pure) *state* and *observable*, represented respectively by a wave function (or more abstractly a vector or a ray in Hilbert space) and a Hermitian operator. Is a *state* a direct concept? The answer depends, of course, on the physical interpretation assigned to the term. Using the customary *objective interpretation*, in which 'ground state of a helium atom' (to take a simple example) refers to an actually existing (although invisibly small) object, the term is not in the least direct: it would be quite impracticable to explain adequately and in unsophisticated language what particular state was intended. However, if 'state' is understood in terms of the *operational interpretation*, as referring to a particular *method of preparation* (of a given type of system), then 'ground state of a helium atom' may be 'explained' by *teaching by demonstration* some procedure for preparing such atoms, for instance in the form of a beam passing into a vacuum from a porthole in a cylinder containing helium at some low temperature and pressure. Insofar as such a procedure can in time be taught to an intelligent but uneducated apprentice, and in principle without the use of words, we have an effective ostensive definition; this notion of state is thus sufficiently direct, and may be adopted as a primitive concept.[8]

Now a physicist might criticize the above example as follows: 'Not all of the prepared atoms will be in the ground state. A certain fraction (depending on the temperature of the cylinder) will be in excited states.' His language shows that he is using the objective interpretation of 'state'; with the operational interpretation each atom is *by definition* in the same state, for all have been prepared in the same way. Nevertheless, he is right: to be precise, our method of preparation corresponds to a *mixed state*[9] which is a probability superposition of the ground state and the excited states–although the probability associated with excited states is, for any reasonable temperature of the cylinder, extremely small. In fact, the example brings out a second point: the fact that, strictly speaking, only mixed states can be prepared; a pure state can be obtained experimentally only as a limiting case of mixed states. In other words, an ostensive definition is actually possible only for a mixed state; pure states must be defined verbally in terms of mixed states. [Of course, only *certain* mixed states can be prepared. Consideration of examples suggests that these include (a) states of thermal equilibrium (Maxwell-Boltzmann distribution) associated with nonzero absolute temperature T, and (b) such states as can be obtained from the former by allowing the system to evolve in a controlled environment, again at a nonzero temperature.]

It should be noted that the above arguments apply equally to classical mechanics. If, for instance, we try to 'release a pendulum from rest in a given position' we must first hold it in that position. Yet while we attempt to do so both the position and velocity of the pendulum will 'be subject to thermal fluctuations' (i.e. neither will have a precise value) corresponding to the nonzero temperature of the apparatus used. Thus a pure state can be prepared only by using apparatus at absolute zero temperature–and this is in principle unattainable.

The above discussion shows that, both in classical and in quantum mechanics, *pure state* is a less direct concept than *mixed state*; the latter should therefore be preferred as a primitive concept. Notice that this is the direct opposite of what is done in conventional formulations. There, pure states are introduced first, and the concept of a mixed state is brought in later (if at all) as a relatively sophisticated notion which is useful for certain statistical purposes. In addition, the objective interpretation of 'state' is tacitly assumed, and with it a certain ontology which may be described as the *pure state philosophy*: it is assumed (or taken as obvious?) that any system is at any instant 'really' in some definite pure state; the notion of a mixed state is merely a way of representing the observer's incomplete information regarding the system. The reasons for adopting this attitude are obscure; no justification is normally offered. However, one may conjecture that the *origin* of the pure state philosophy lies in a certain technical advantage possessed by pure states in *classical* mechanics. There, it is not only the case that pure states of the parts of a compound system uniquely determine a pure state of the whole (which is true also in quantum mechanics) but also that conversely *every* pure state of the whole uniquely determines (and arises from) pure states for the parts. Because of this 'decomposition property' the pure state philosophy can be adopted in classical mechanics without contradiction – and indeed with advantage, since the decomposition property is not valid for mixed states.

An important consequence of the decomposition property is the justification it affords for a second basic philosophical principle, the 'decomposition principle', according to which any compound system is to be explained in terms of the behavior of its parts; indeed this follows (in the presence of the decomposition property) as an almost inevitable consequence of the pure state philosophy. There is no need to emphasize the fundamental role played by the decomposition principle throughout the whole of modern science.

One may conjecture that the fact that these two principles, the pure state philosophy and the decomposition principle, have become so deeply entrenched in physical thought is due to the fundamental role played by classical mechanics in all pre-quantum physics. It is unfortunate that these principles have been unthinkingly carried over to quantum physics, in spite of the fact that the decomposition property, on which their justification depends, no longer holds.

This concludes our discussion of the concept of a state. The treatment of observables can be briefer since the arguments are in many ways similar. As in the case of states the notion of an observable can be admitted as a direct concept only if we adopt the operational interpretation: an observable is to be regarded not as a property possessed by the system but rather as an *experimental procedure* that, applied to a system 'in' (i.e. prepared by) a given state, culminates in the determination of a numerical value. Moreover, as we saw at the beginning of Section 4, the general concept of an observable is not as direct as that of a two-valued observable, or test; the latter is therefore to be preferred as a primitive concept. Tests play a major role in the so-called 'quantum logic' approach to quantum mechanics [19, 24, 33]. They are variously described as *questions, yes-no experiments*, or *propositions*. The last term is somewhat misleading. Unlike a proposition, a test can never be described as true or false *per se*, but only when taken together with a particular state. Thus a test is better regarded as a (unary) *predicate*, i.e. a function that assigns to each state a proposition. As we shall see later, this interpretation is also to be preferred on syntactic grounds.

The tests used by the quantum logicians are pure tests. It is possible to introduce *mixed tests*, which are related to pure tests in just the same way as mixed states are related to pure states.[10] Indeed, this is not only possible but desirable, for (as in the case of states) the concept of a mixed test is more direct than that of a pure test: tests which can actually be realized in practice are always mixed tests.

Since the concept of a mixed test is not well known it seems wise to digress briefly to explain it and to justify these claims.

In the usual mathematical formalism of quantum mechanics a pure state of a system is represented by a unit vector ϕ in a Hilbert space H and an observable by a hermitian operator A, the significance of the correspondence being given by the two rules:

(1)   Any measurement of the observable (represented by the operator) A will yield an eigenvalue (or, better, a point in the spectrum) of A.

(2)   The *expectation value* of (the result of a measurement of) the observable A in the state ϕ is the inner product $(\phi, A\phi)$

Any mixed state is represented by a density matrix: i.e. a positive hermitian operator ρ of trace 1. For the mixed state ρ the expectation value of A is $\text{Tr}(\rho A)$, where Tr denotes the trace. This expression agrees with (2) if we assign, as the density matrix corresponding to a pure state ϕ, the projection on the 1-dimensional space spanned by ϕ.

In the case of a (pure) test it is customary to call the measured values 1 and 0 (corresponding to outcomes *yes* and *no*) so that, by (1), a pure test is represented by a projection operator T in H. It then follows from (2) that for any mixed state ρ,

(3)   probability of outcome 'yes' $= \text{Tr}(\rho T)$.

Now consider a measurement of the energy E of an electron, for instance by a magnetic spectrometer. At the far end of the instrument a detector sits behind a slit whose two edges determine the lower and upper limits, a and b, of the energy band to which the instrument responds. Thus, ideally, the detector responds to electrons with energy in the interval $[a, b]$: i.e. the spectrometer plus detector constitutes a test represented by the projection operator $P = f(E)$, where f is the idempotent function given by $f(\lambda) = 1$ for $a < \lambda < b, f(\lambda) = 0$ otherwise. [If $dP(\lambda)$ is the spectral resolution corresponding to E, so that $E = \int \lambda dP(\lambda)$, then $P = \int f(\lambda) dP(\lambda)$.]

The function f might be described as the *response function* of the arrangement: $f(\lambda)$ gives the probability of response to an electron of energy λ. But this interpretation makes it clear that the above account is not really applicable to a practical laboratory arrangement: not only does $f(\lambda)$ never strictly attain the ideal values 1 and 0, but, more important, it does not change *discontinuously* at the critical values a and b; all sorts of defects prevent such ideal behaviour–imperfections in the detector, inexact collimation of the incident beam, inaccuracies in the magnetic focussing, and so on. Thus the test is not strictly speaking represented by the projection operator P. It might be claimed that it is nevertheless represented by *some* projection operator P' 'nearly equal'[11] to P. However, there is an unavoidable circumstance that precludes any such claim: namely, the existence of 'thermal fluctuations' in the spectrometer, which will affect the magnetic field, the location of the slit edges, and so on. Conventional quantum mechanics is not really adapted to describe this situation, but we can attempt to do so by supposing that the projection P' varies randomly with time: i.e. P' becomes a (projection-valued) random variable. The probability of outcome 'yes' when the above test is applied to the mixed state ρ is now the expectation value $\langle \mathrm{Tr}(\rho P') \rangle$ of $\mathrm{Tr}(\rho P')$, which, since Tr is a linear function, may be written $\mathrm{Tr}(\rho \langle P' \rangle)$. Thus the probability of response for our test is still given by (3), provided we set $T = \langle P' \rangle$, the expectation value of P'. Now, T is not a projection. For $\langle P' \rangle$ may be regarded as a weighted mean of projections, and as such is a hermitian operator satisfying $0 \leq \langle P' \rangle \leq 1$ [0 and 1 are the zero and identity operators on H and the ordering is the usual one: $A \geq 0$ means $(\psi, A\psi) \geq 0$ for all ψ.]; such an operator can be a projection only in the trivial case where the random variable P' is constant (see [7], p. 13), and this can happen only at temperature absolute zero.

A test whose response to an arbitrary state ρ is given by (3) with an operator T which is not a projection (but satisfies $0 \leq T \leq 1$) we call a *mixed test*. Although it is represented by a hermitian operator, a mixed test is (unlike a pure test) not a special case of an observable. In particular, (1) does not apply: mixed or not, a test has only two outcomes, *yes* and *no*.

The above discussion indicates that tests occurring in practice are always mixed; pure tests are idealizations that can only be approximated in practice. Accordingly *mixed test* is a more direct concept than *pure test*.

This fact, plus the observation that certain mixed tests (like some mixed states) can be defined ostensively, indicates that a satisfactory formulation of quantum mechanics should employ *mixed test* and *mixed state* (with the operational interpretation) as axiomatic concepts.[12] This was the starting point for the work reported in [10]. The actual choice of axioms involved various considerations. In the following I give only a sketch. First, there are a number of 'self-evident' axioms: i.e. assertions whose validity follows from the meanings of the terms 'test' and 'state'. Roughly speaking, these have the effect of setting up a dual pair of partially ordered vector spaces, the states lying in one and the tests in the other. In each space the ordering is determined by a set of *positive vectors*, which forms a cone. In addition, the ordering in the state space may be defined in terms of that in the test space. The crucial question is: What kind of partially ordered linear space may occur as a test space?

Superficial considerations of the nature of tests and states (such as were used in setting up the dual pair of spaces) seem to provide no significant restriction on the possible kinds of test space. Yet, further restrictions are necessary; without any we obtain so broad a generalization of classical and quantum mechanics as to be practically useless. At this stage there appear to be two possibilities for further progress:

1. Abandoning epistemological considerations, we may seek as strong an axiom as possible, subject to the condition that it be mathematically attractive and compatible with both classical mechanics and elementary quantum mechanics, and try to motivate it by the consideration of particular examples. This is what was done in [10]. This approach has certain virtues. As was hoped, it does throw light on the relationship between classical and quantum mechanics; in particular, it suggests how one can erect in quantum mechanics a mathematical structure analogous to the topology (or even the manifold structure) of phase space in classical mechanics. However, the formulation cannot be regarded as conclusive, since the final axiom is rather sophisticated and in no way self-evident; thus the theory provides no 'justification' of quantum mechanics. [I should perhaps point out that the same sort of thing occurs in other current formulations. Usually, however, the axiomatic concepts are not selected for directness, so that even the *possibility* of justification is absent.]

2. Alternatively, if we really wish to obtain a justification of quantum mechanics, we must attempt a much deeper analysis of the concepts of state and test in the hope of revealing hidden properties which were not evident in our first superficial treatment. After all, what *is* a 'method of

preparation' of a system? What, indeed, is a 'system'? The analysis of such questions presents formidable difficulties which are due in part to the fact that the language in which they are expressed is not precise enough for the task in question. Thus we meet again the need for a formal language capable of describing the experimental interpretation of the theory: i.e. for a 'primary language' in the sense of Section 1. We now turn to a consideration of the structure of such a language.

6. A PRIMARY LANGUAGE FOR ARITHMETIC

As we saw in Section 2, to describe the semantics involved in a physical theory while avoiding the vagueness inherent in the syntax of a natural language such as English it is necessary to resort to a formal language. In the case of a mathematical theory there is an established procedure for this, which can be illustrated by an example. Let's consider a very simple theory T concerned with arithmetic operations on numbers. However, to give the theory an experimental flavour (as befits a prototype for physics) let us assume that it deals not with the abstract 'natural numbers' but with *numerals*, where for definiteness a *numeral* is defined to be a concrete expression consisting of a row of vertical strokes–chalk marks on a blackboard, for instance.

Let us suppose that the theory is so simple that it employs only the concepts of *sum, successor,* and *even*. For instance, some assertions in the theory might be (for any numerals x and y): *if x and y are even then so is* $x + y$; $x + x$ *is always even*; *if x is even its successor is not even*; (but not, for example, $x + y = y + x$, since ' $=$ ' is not a concept of the theory). If we introduce a formal language L_1 with the three primitive symbols, S for *successor*, A for *sum*, and E for *even*, together with parentheses, symbols x, y, z, ... for variables, and the usual logical symbols: $\wedge$ (and), $\vee$ (or), $\rightarrow$ (implies), $\neg$ (not), $\forall$ (for all), $\exists$ (for some), then the above three assertions can be written:

(i) $\forall x \forall y((E(x) \wedge E(y)) \rightarrow E(A(x, y)))$,
(ii) $\forall x E(A(x, x))$,
(iii) $\forall x(E(x) \rightarrow \neg E(S(x)))$.

We shall assume that the theory T is such that every theorem can be formally expressed in this language. (It will not be necessary to specify the theory more precisely.) Owing to their concrete interpretation, numerals may be regarded as physical objects and S, A, E as denoting certain experimental operations that may be carried out on these objects. In this way T becomes a (very simple) *physical* theory and L_1 serves as a secondary language for T.

Before discussing the question of a primary language for T let us note two relevant features of the language L_1. First, it has the virtue that the primitive concepts (associated with the symbols S, A, E) are direct: i.e. they have a simple experimental significance (which will be spelled out below). On the other hand, it has a feature which makes it inadequate as a primary language: it contains no expressions that denote *particular* numerals. This is a common characteristic of a theory. In Euclidean geometry, for instance, arbitrary points may be denoted by letters A, B, C, ... but there are no symbols for particular points (for example, *the centre of the earth* in physical space or *the origin* in a mathematical space). Indeed, such symbols are unnecessary, for particular points are never mentioned in theorems.

Now consider the problem of describing precisely the rules of interpretation for T. A solution to this problem is offered in the branch of logic known as *model theory*. According to it, the sentence (ii), for example, means (very roughly) that $E(A(n, n))$ holds for every numeral n. This explanation is satisfactory for model theory, in which it is assumed that the set of all numerals is 'given' (together with the appropriate interpretations of E, S, and A as applied to this set), but it is of no help in our case since one of our main concerns is the question of *how* this set is to be given.[13] Indeed, it would seem that we must first explain what a numeral is, for only then can we try to show how the symbols S, A, E should be interpreted in terms of operations on numerals.

Now, we may hope to 'explain' a *particular* numeral, for instance the single stroke | (corresponding to the number 1), ostensively: by simply exhibiting a variety of instances of this numeral – that is, individual occurrences which may differ in position, in size, in the nature (pen, pencil, chalk, etc.) of the stroke, and so on. But if we try to give a similar ostensive definition of the generic term, 'numeral', itself we face two difficulties. First, the concept is more complicated than that of a particular numeral, since the instances vary in an additional respect (number of strokes). Secondly, and more important, it is not sufficient merely to give instances of numerals; one must also show how to tell when two expressions are instances of *the same* numeral. [Formally, the reason that this is necessary arises from the nature of the *theory*. When any of the assertions (i) –

(iii) above is applied, it is necessary that the variable x (for instance) be replaced, in both of its occurrences, by (instances of) the 'same' numeral.]

For these reasons it would not be easy to give an adequate account of the rules of interpretation for L_1 directly. However, these difficulties can be avoided. Suppose we enlarge the language L_1 by introducing a new primitive symbol I to denote the numeral |. In the new language L so formed there is a term for each of the numerals |, ||, |||, ... (for instance, I, $S(I)$, $S(S(I))$, ...). Now suppose that, following the procedure outlined in Section 2, ostensive definitions are given of the meanings of the four primitive symbols I, S, A, E. For present purposes it will be convenient to represent these ostensive definitions by four 'primitive documents', one labelled by each of the primitive symbols, inscribed as follows: (Provided we are confident that the meaning of each document *could in principle* be conveyed by demonstration there is no harm in using this concrete representation.)

I: "Draw a stroke".

I labels a document that represents a particular numeral.

$S(---)$: "Add a stroke to the expression ---."

Here '---' represents a blank space on the document labelled S. When any document representing a numeral, and labelled by an expression t say, is pasted into this space the whole, now labelled $S(t)$, becomes a document representing the successor of the numeral represented by t. Similarly, the documents labelled A and E have two and one blank spaces respectively and the following inscriptions:

$A(---, ...)$: "Place the numeral (labelled by the expression) ... immediately to the right of the numeral --- so as to form a single numeral."

$E(\ \ )$: "Taking the strokes in the numeral --- from the left delete the first, leave the next, delete the next, leave the next, and so on. If the last stroke is deleted the outcome is 'no'; otherwise it is 'yes'."

Given any suitable expression it is easy to see how a straightforward pasting operation constructs from (copies of) the primitive documents a uniquely determined document, labelled by the given expression and describing either the construction of a numeral or (if the symbol E is involved) an experimental procedure leading to one of two outcomes, *yes* or *no* – i.e. an *elementary experiment*, as defined in Section 4. 'Suitable' expressions are thus of two types. An expression of the first type, labelling a particular numeral, is called a *constant term*. [Of course, the labelling is not unique–for instance $A(I, I)$ and $S(I)$ both label the numeral ‖. But there is no harm in this. It simply draws our attention to the *empirical* nature of the fact that the procedures represented by these terms result in indistinguishable objects–with different interpretations of the primitive symbols this might not have been the case.] An expression of the second type, which with our practical interpretation labels an elementary experiment, is called an *atomic sentence*.[14] For example, the expression $E(A(I, S(I)))$ labels a document which says, essentially, "Draw two strokes. Beside one draw another stroke. Lay these expressions together. Then test the result for evenness according to the instructions for E." [We get '‖‖', of course, and the last instruction then gives the outcome 'no'.]

To make clear precisely which expressions are 'suitable' we must specify the syntax of the language L, or rather of its atomic part $L^{(a)}$ –the fragment of L in which logical symbols and variables do not appear. This may be done formally as follows:

The primitive symbols of $L^{(a)}$ are I, S, A, E together with parentheses and commas. The meaningful expressions of $L^{(a)}$ are those *constant terms* and *atomic sentences* which can be formed by the (repeated) use of these two rules:

(a) I is a constant term. If x and y are constant terms then so are $S(x)$ and $A(x, y)$.

(b)   If x is a constant term then $E(x)$ is an atomic sentence.

By means of this syntax and the ostensive definitions of the primitive symbols the question 'What is a numeral?' may be answered: *a numeral is a figure that is produced by following the procedure given by a constant term.* But we can do better than this. The interpretation of the theory T can now be given without reference to the concept of a numeral. Indeed, thanks to the fact that the language L is an *expansion* of L_1 big enough to provide a name for every object, the difficulty mentioned above–that arises in the application of orthodox model theory–can be avoided. We can now say, for instance, that the sentence (ii) (a sentence in L_1) means that the sentence $E(A(t, t))$ (a sentence in L) holds for every constant term t (in L). A similar treatment applies to sentences in L_1 involving the existential quantifier. In this way the interpretation of an arbitrary sentence in L_1 is reduced to that of atomic sentences in L, and–as we have seen–every such sentence itself supplies a precise description of the corresponding elementary experiment, in that each step in the procedure has been defined ostensively and the way the steps are combined is given by the syntactic structure of the sentence.

In this way the language L serves as a primary language for the theory T.

Note the following features of the language L. These may be taken as general properties of any primary language that is interpreted in the above way. A formal language with these properties will be called a *standard interpreted language*.

(1) The atomic part of the language employs a finite number of primitive symbols, the meaning of each symbol being defined ostensively. (Note that this is possible only insofar as the primitive symbols denote *direct physical concepts*.)

(2) From these symbols the syntax gives rules for the construction of expressions of a certain type, called atomic sentences. (In general there will also be expressions of other types, called constant terms.)

(3) Through the definitions of the primitive symbols there corresponds to every atomic sentence a definite elementary experiment.

If the language is to serve as a primary language for some theory then every elementary experiment to which the theory refers must be labelled in this way by some atomic sentence, and distinct experiments (i.e. experiments which are not treated alike by the theory) must never be labelled by the same sentence. [In addition, the constant terms normally label physical objects. However, 'object' is a vague notion which need not be introduced formally at all.]

Notice that, by (2), the primary language can serve as a means of recording the results of experiments. Indeed, logically we might expect that the primary language (e.g. L) would initially be designed and used for this purpose; only after a theory (e.g. T) had been conjectured might a particular sublanguage (e.g. L_1) be recognized as being sufficient to express the generalizations embodied in the theory.

Notice that insofar as the use of the primary language is concerned, all that matters is the correct interpretation of the atomic sentences. What is necessary is that, whenever a purported trial of the elementary experiment corresponding to a given atomic sentence is carried out, all observers are agreed as to (a) whether it *was* a trial (i.e. whether the procedure satisfied the conditions stated on the document labelled by the sentence), and (b) what the outcome was. Insofar as the users are confident that they will always agree on these matters the interpretation of the language has been adequately described.

7. A PRIMARY LANGUAGE FOR LIGHT POLARIZATION

Although it was drawn from mathematics, the example discussed in the last section was presented as a theory about 'experimental arithmetic', concerned with practical operations on certain concrete expressions called numerals. As such, it can be regarded as a very simple example of a physical theory. We saw how a formalized primary language could be set up in this case and how its interpretation could be given ostensively, without relying on the use of any other language. However, it remains to be seen what new features may appear when one attempts to formulate a more realistic physical theory in this way.

To examine this situation it is best to discuss a concrete example. Owing to the fundamental role of quantum mechanics in modern physics and the wellknown difficulties associated with its interpretation it is natural to choose for this purpose a quantum-mechanical system, and the simplest possibility is a spin system described by a two-dimensional Hilbert space. Such a system has various practical realizations, the most accessible from an experimental point of view being that which describes the state of polarization of a light photon.

Let us first consider the theoretical structure of this system. We saw in Section 5 that the basic concepts are *mixed state* and *mixed test*. In this case (2-dimensional Hilbert space) the mathematical representatives of these concepts can be given a simple geometrical presentation as follows.

Each mixed state is represented by a 2×2 positive Hermitian matrix ρ of trace 1, and therefore can be written in the form

$$\rho = \tfrac{1}{2}[\rho_x \sigma_x + \rho_y \sigma_y + \rho_z \sigma_z + \mathbf{1}]$$

where

$$\sigma_x = \left(\begin{smallmatrix} 0 & 1 \\ 1 & 0 \end{smallmatrix}\right), \ \sigma_y = \left(\begin{smallmatrix} 0 & -i \\ i & 0 \end{smallmatrix}\right), \ \sigma_z = \left(\begin{smallmatrix} 1 & 0 \\ 0 & -1 \end{smallmatrix}\right), \ \mathbf{1} = \left(\begin{smallmatrix} 1 & 0 \\ 0 & 1 \end{smallmatrix}\right),$$

are the Pauli spin matrices and the identity matrix, and ρ_x, ρ_y, ρ_z are real with $\rho_x^2 + \rho_y^2 + \rho_z^2 \leq 1$. Here $\boldsymbol{\rho} = (\rho_x, \rho_y, \rho_z)$ may be regarded as a point in an auxiliary (3-dimensional Euclidean) space $\mathbf{P}$. In this way the mixed states are set into one to one correspondence with the points of a sphere in $\mathbf{P}$ called the *Poincaré sphere* [30]; moreover, this correspondence is affine: i.e. it preserves weighted means.[15] Each point on the surface of the Poincaré sphere represents a state of perfect polarization. For instance, assuming a horizontally moving photon, we may arrange that the points $(0, 0, 1)$, $(0, 0, -1)$ represent right-handed and left-handed circular polarization respectively, and that (for any θ) the point $(\cos 2\theta, \sin 2\theta, 0)$ represents plane polarization in a plane making an angle θ with the vertical. Other points on the surface of the sphere represent intermediate states of elliptic polarization. The centre of the sphere represents an unpolarized photon,[16] and other internal points correspond to various degrees of partial polarization.

Similarly, each mixed test (see Section 5) is represented by a hermitian operator T such that $0 \le T \le 1$. Writing T in the form

$$T = \tfrac{1}{2}[T_x\sigma_x + T_y\sigma_y + T_z\sigma_z + T_t\mathbf{1}]$$

these conditions become $|T| < T_t < 2 - |T|$, where T is the vector (T_x, T_y, T_z) in $\mathbf{P}$. The mixed tests are thus represented by the points of a region $\mathbf{T}$, which is the intersection of a pair of cones with common but oppositely directed axes, in the 4-dimensional space $\mathbf{P} \times \mathbf{R}$ which we shall call *polarization space*. [$\mathbf{R}$ denotes the 1-dimensional space of real numbers.] The pure tests are the extreme points of this set. They consist of $(0, 0, 0, 2)$ (representing $\mathbf{1}$), $(0, 0, 0, 0)$ (representing $\mathbf{0}$), and those points $(T, 1)$ for which $|T| = 1$. The latter are in a natural one-to-one correspondence with the pure states, and this correspondence can conveniently be used to embed the Poincaré sphere (via the mapping $\rho \to (\rho, 1)$) into polarization space.

A mixed state ρ and a mixed test T together determine an elementary experiment. To this experiment the theory assigns (see Section 5):

(1)   probability of outcome 'yes' $= \mathrm{Tr}(\rho T) = \tfrac{1}{2}(\rho \cdot T + T_t)$

This will suffice as an outline of the theoretical structure of a quantum-mechanical system represented by a 2-dimensional Hilbert space.

Leaving aside the question of axiomatization for the present, let us consider now the problem of giving–via a primary language–a set of rules of interpretation for this embryonic theory. If the theory is to be taken literally this will involve giving a physical interpretation for a light photon in an arbitrary state of polarization, and this interpretation must (see Section 5) describe a *method of preparation* for such a state. A difficulty immediately arises: there is no practicable procedure which will on demand produce with certainty a single light photon and never more than one. Nevertheless, let us pretend for simplicity that this can be done;[17] more precisely, that we have in the lab a device which will, on pressing a button, emit in a given (horizontal) direction, a single vertically polarized photon.[18] By pointing out–to an apprentice, say–how to recognize and use this device we furnish an ostensive definition of a certain state S, say. Colloquially, S would be described as "the state of vertical polarization", but formally it is simply the method of preparation that has thus been ostensively defined. It will be convenient in the following to use S also to refer informally to the device involved.

It is not possible, of course, even in principle, to provide similar ostensive definitions for *all* states, since there are infinitely many of them. In any case, it would in practice be foolish to attempt to supply a separate device for the preparation of every state. One observes instead that a nonabsorbing *filter* placed before the device S yields an arrangement which acts as a method of preparation for photons in another state. For instance, a (diagonally oriented) quarter-wave plate Q will yield a device that produces a circularly polarized photon, while a saccharimeter cell C filled with sugar solution will, by rotating the plane of polarization, furnish a device which yields a photon polarized in an arbitrary plane, depending on the strength of the solution. Such filters may be used in series. In fact, it is not hard to show that, starting with S and using only filters of these two types, every *pure* state of polarization can be obtained. [In effect, the cell C induces an arbitrary rotation of the Poincaré sphere about the z-axis, while the quarter-wave plate Q (used as above) effects a rotation by $\pi/2$ about the y-axis. But the smallest group of rotations containing these elements is the group of all rotations of the Poincaré sphere–and this group of course acts transitively on the surface of the sphere.] No ordinary filter, however, will produce in this way a state of *partial* polarization. In principle, such a state may be produced from S by using a more sophisticated sort of filter,[19] but to illustrate how the syntax and semantics of a primary language may be defined it will be enough to discuss the pure states. In concrete terms the problem reduces to that of (a) giving a formal name (term) for each such state, and (b) so instructing an apprentice that he is able, given any such term, to prepare the corresponding state. We have already assumed that the lab possesses the device S; by demonstrating it to our apprentice we provide an ostensive definition of the corresponding state. In the same way let us assume it is equipped with a supply of quarter-wave plates and of (clean dry) saccharimeter cells. By showing our apprentice how to recognize a quarter-wave plate we furnish an ostensive definition of Q. We can also show him how to recognize a saccharimeter cell, but more explanation is required for its use since this involves the preparation of a sugar solution of an arbitrary given strength. This can be done as follows.

We first show him how to measure a unit quantity of sugar, say 1 gm. Let 1 denote this, ostensively defined, procedure. Next, let us introduce a function $H(\ )$, the idea being that if x is a term denoting the method of preparation of any quantity of sugar then $H(x)$ shall denote the procedure in which the quantity x is prepared *and then divided in half, one half being discarded.* The symbol $H(\ )$ is defined ostensively by demonstrating the procedure $H(a)$ for various particular quantities a until its meaning is grasped (with sufficient accuracy), not only for the particular quantities involved in the demonstration but for all other (reasonable) quantities. [In the terminology of Section 2 $H(\ )$ is not a 'meaningful expression' but denotes a *rule of construction* by which from any expression a (for a quantity of sugar) the expression $H(a)$ is formed.] Similarly, we may introduce and define ostensively a binary function $A(\ ,\ )$, where $A(x, y)$ denotes the procedure of preparing quantities x and y and then combining them into a single quantity.

We now have terms for arbitrary quantities of sugar–more precisely, for any quantity of the form $m2^{-n}$ gm, where m and n are integers ($m > 0$). For instance the preparation of 3/4 gm sugar is represented by (among others) the term $H(A(1, H(1)))$. Let q denote any such term. We next show our apprentice how to carry out the procedure, denoted $C(q)$, of preparing the quantity q, dissolving it in a fixed amount of water, and filling a cell with the resulting solution. By doing this for various values of q we provide an ostensive definition of the function (rule of construction) $C(\ )$, by which a filter is formed corresponding to any allowed quantity of sugar. The term Q, together with all terms of the form $C(q)$, form a second 'type' of term, terms for filters, To obtain an adequate number of terms of a third type, terms for *states*, it now remains only to introduce a binary function $F(\ ,\ )$ to denote the procedure of *joining together* a (device for preparing a) state s and a filter f to obtain a new (device for preparing a) state $F(s, f)$. Again, ostensive definition of F is accomplished by repeated demonstration of this procedure of 'joining together', for various particular values of s and f.

So much for states. A similar discussion applies to tests. As we saw in Section 5, a test is represented by a detector which, on the arrival of a photon, either responds (outcome *yes*) or does not respond (outcome *no*). We must assume the availability of at least one such detector, and for our present purpose it will be convenient to assume that this is a device D that always responds to a vertically polarized photon and never to a horizontally polarized one.[20] By placing before this detector filters of the types already introduced we can produce a device that responds always to a photon of any desired pure state of polarization (represented by a point on the surface of the Poincaré sphere) and never to one in the 'orthogonal' state (represented by the diametrically opposite point). To represent this process we introduce a function $G(,)$: for any filter f and any test (i.e. method of detection) a, $G(f, a)$ will denote the test which is realized by placing the filter f in front of (the device which realizes) a. The interpretation of G can be defined ostensively in the same way as for F above. Finally, we introduce a similar process $E(,)$ of adjoining a state and a test: $E(x, a)$ will denote the procedure of constructing the state (= method of preparation) x and the test (= method of detection) a, placing the devices so obtained together (in correct alignment), and pressing the button on the device x causing it to emit a photon. An outcome, *yes* or *no*, will then be obtained from the detector.

This completes the description of the semantics of a primary language L' for light polarization experiments. The syntax of L', which was established in the process, may be defined formally as follows:

The meaningful expressions of the language L' are formed from the primitive symbols 1, H, A, Q, C, S, F, D, G, E, (plus parentheses and commas) and are of five types: as well as atomic sentences there are constant terms for quantities of sugar, filters, states, and tests; let's call them for short *sugar-terms, filter-terms,* etc. The syntax is given by the following rules (cf. fig. 2):

(a)   1 is a sugar-term. If p and q are sugar-terms then so are $H(p)$ and $A(p, q)$.

(b) Q is a filter-term. If q is a sugar-term then $C(q)$ is a filter-term.

(c) S is a state-term. If s is a state-term and f is a filter-term then $F(s, f)$ is a state-term.

(d) D is a test-term. If f is a filter-term and d is a test-term then $G(f, d)$ is a test-term.

(e)   If s is a state-term and d is a test-term then $E(s, d)$ is an atomic sentence.

It is convenient to adopt the following modified notation, which does not alter the formal structure of the language. Whenever p, q, s, f, d are terms of the appropriate types,

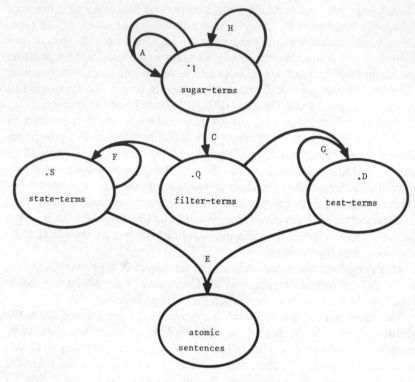

Fig. 2

$p/2$ will denote $H(p)$,
$(p + q)$ will denote $A(p, q)$,
C_p will denote $C(p)$,
(sf) will denote $F(s, f)$,
(fd) will denote $G(f, d)$,
(sd) will denote $E(s, d)$;

in addition outer parentheses may be omitted. This makes for easier reading. For instance, the atomic sentence

$$E\left(F\left(F\left(S, C\left(1\right)\right), Q\right), G\left(C\left(A\left(1, H(1)\right)\right), D\right)\right)$$

now takes the form $((SC_1) Q) (C_{1+1/2}D)$.

Like the language L, which was introduced in the last section as a primary language for certain aspects of arithmetic, L' is a 'standard interpreted language': its primitive symbols are all defined ostensively in such a way that each atomic sentence denotes an elementary experiment. For instance, the sentence just given denotes the following procedure: "Fill a cell with a solution made with 1 gm of sugar. Place it in front of the standard source S and in front of this combination place a quarter-wave plate. Now measure a second gram of sugar, divide it in half, and place half along with a third measured gram. Dissolve this combination in the standard quantity of water and fill another cell, placing this cell before the standard detector D. Finally, place together the two assemblages so formed and press the button." An outcome, *yes* or *no*, will then be obtained. However, these experiments differ essentially from those denoted by the atomic sentences in L: they show *dispersion*, in that repetition of an experiment does not in general result in the same outcome being obtained. Consider, for instance, the experiment named by the atomic sentence $(SQ)D$. Here a circularly polarized photon produced by the combination SQ enters the detector, and by (1) above the probability of response is $1/2$. What happens on an individual occasion is not predicted by the theory; nor does it seem to be determined in any physical way.

This new feature has dire consequences for the further development of the formalization programme. In the example of arithmetic (Section 6) no dispersion occurs, and we can associate with each atomic sentence a truth value, *true* or *false*, according to whether the outcome of the corresponding experiment is *yes* or *no*. Moreover, a typical theory is concerned with relations between these truth values, and such relations can be expressed formally by means of the primary language together with classical logic, as in the examples given at the beginning of Section 6. However, in the case of the language L' for polarized light experiments, although each atomic sentence still denotes an elementary experiment there is no natural way of assigning truth values to the atomic sentences – it is unreasonable to call a sentence true simply because an outcome *yes* has just been obtained, since the next trial may well yield the outcome *no*. Moreover, it would scarcely help to do so, since a quantum-mechanical theory hardly ever makes definite assertions about the outcomes of elementary experiments. Instead its theorems have the effect of assigning certain *probabilities* to these outcomes, or of asserting *relations* between such probabilities. For example, the following might occur as theorems of the theory of light polarization given at the beginning of this section (supplemented, in the case of 7.1, by a few constant terms). [In these examples elementary experiments are named in the language L', and $p(E)$ denotes the probability that the experiment E yields the outcome 'yes'.]

7.1.   EXAMPLE. $p((SQ)D) = 1/2.$

7.2.   EXAMPLE.   For any state x, filter f, and test a,
$$p((xf)a) = p(x(fa)).$$

7.3.   EXAMPLE.   For any test a there is a test b such that, for
every state x,
$$p(xa) + p(xb) = 1.$$

Note that although the language L' is *used* here these assertions are not
sentences in L': apart from the unformalized sections in 7.2 and 7.3 they
involve real numbers, the equality relation $=$, and the function symbol
$p(\quad)$. Two problems then arise. First, there is the question of assigning a
precise interpretation to such probability assertions. Then–when this is
solved–there remains the problem of how to express such assertions in
some formal language; it would be nice if this could be done using the
primary language itself, as was possible in the dispersion-free case (cf. the
assertions (i)–(iii) in Section 6). In Section 9 we shall see how these problems can be dealt with, but first let us examine in more detail the phenomenon of dispersion and certain related concepts.

8. PERFECT ELEMENTARY EXPERIMENTS

We have seen how a formal primary language can be set up and how its
interpretation can be established by giving ostensive definitions for the
primitive symbols. The notion of a language presupposes the existence of
a body of *participants* or *speakers* who are confident of each other's ability
to use the language without disagreement. In particular, this means they
are confident that, *given any atomic sentence and any procedure that purports to be a 'trial' of the corresponding elementary experiment, (a) each
speaker will be able to come to a decision as to whether it really is a trial
(i.e. accords with the conditions associated with the given sentence) and if
so what the outcome is, and (b) all speakers will agree in their conclusions.*[21]
This is our basic assumption concerning the interpretation of atomic
sentences.

We shall need the notion of a *refinement*, which is reached as follows.
The conditions associated with an atomic sentence E must permit some
latitude regarding the execution of the corresponding elementary experiment; otherwise these conditions would (almost) never be satisfied and
any assertion or prediction concerning the experiment E would be empty.[22]
(Of course, the degree of latitude permitted is itself determined by the

sentence E.) A *refinement* of E is obtained when the conditions are made more stringent, thus reducing this latitude:

8.1. DEFINITION. An elementary experiment E' is a *refinement* of an elementary experiment E if every trial of E' (a) is at the same time a trial of E and (b) yields the same outcome under either interpretation. Two elementary experiments are *disjoint* if they have no common refinement.

Illustrations of these concepts will arise below, in connection with the discussion of dispersive and dispersion-free experiments. We start with the latter notion. An elementary experiment E is *dispersion-free* if trials of E always give the same result.[23] The prime examples occur with the atomic sentences of a primary language for a mathematical theory (e.g. the language L of Section 6). Consider, for instance, an atomic sentence which expresses formally the statement "7001 is a prime number". The corresponding elementary experiment will consist in some well-defined procedure for testing this assertion. Clearly, every trial of this experiment will give the same outcome.[24] The same is true of experiments in classical mechanics. Consider, for instance, the following:

E_1: Displace the bob of a simple pendulum, of length 1 metre, 10 cm to the right. Release it and let it swing freely. Look after a given time t. If the bob is then to the right of the equilibrium position the outcome is 'yes'.

In principle this experiment will always give the same outcome.[25] Even in practice this will usually be so if t is not too large. The same applies, again 'in principle', to most experiments in classical physics. For instance:

E_2: Heat a solid iron sphere of diameter 10 cm to 100°C and allow it to cool for 10 minutes in a room at 20°C. If the final temperature exceeds 50°C the outcome is 'yes'.

E_3: A horizontal wire is held over a pivoted compass needle. A current is caused to flow northwards through the wire. If the north pole of the needle is deflected eastwards the outcome is 'yes'.

An elementary experiment is *dispersive*[26] if the outcome is not the same on every trial. Examples of dispersive experiments arise in several ways. First, it may be that the latitude admitted in the specification of the procedure for executing the experiment is so great that the outcome may be affected by the *way* in which this procedure is followed. For instance:

E_4: Think of a number (i.e. a positive integer). Square it. If the result exceeds 100 the outcome is 'yes'.

E_5: Drop an egg. If it breaks the outcome is 'yes'.

These satisfy the definition of an elementary experiment but they are not the sort of experiment one wants to refer to in a scientific theory. To exclude such examples we must first pin down what is unsatisfactory about them. Consider the following refinement of E_4:

E'_4: Think of a number over 10. Square it. If the result exceeds 100 the outcome is 'yes'.

Let E''_4 be the refinement obtained by replacing 'over 10' by 'under 10'. (Notice that E'_4 and E''_4 are *disjoint* refinements.) Clearly, these refinements lead to different outcomes. Similarly, if we introduce the refinements of E_5:

E'_5: Drop an egg from a height of 2 cm on to a cushion, ...;

E''_5: Drop an egg from a height of 2 metres on to a concrete floor, ...;

then, although these refinements may still show dispersion, they will surely differ in the probability[26] of outcome 'yes'. Such refinements will be called *significant*[26]. It is the presence of significant refinements that causes us to regard the descriptions of the elementary experiments E_4 and E_5 as incomplete, and to reject these experiments as unsuitable for scientific discussion. An elementary experiment that has no significant refinement will be called *perfect*[26].

It is clear that every dispersion-free elementary experiment is perfect. At first sight it may seem that the converse applies also. For whenever a physicist is confronted with an elementary experiment which is not dispersion-free his immediate reaction is to assume it is not perfect: he seeks an 'explanation' of the variation in outcome–i.e. he tries to discover a significant refinement.

As an example consider the elementary experiment E_1 in the case when the time t is large. If in practice we did not always obtain the same outcome we might suspect that it was affected by some condition, perhaps the room temperature (expansion of the pendulum wire affecting the period), not specified in the rules of interpretation for the atomic sentence that names the experiment. If further experience substantiated this belief we would have identified a significant refinement E'_1 of E : namely, that given by stipulating a definite temperature (or rather temperature interval) for the room. If E_1' again showed dispersion we would seek a further refinement, perhaps by imposing a condition on the altitude of the laboratory, and so on.

When this process of successive refinement is carried out on an arbitrary elementary experiment E one of two things may happen: (a) we may eventually reach – or at least "approach arbitrarily closely" – a refinement of E which is dispersion-free, or (b) the successive refinements may instead approach an experiment which is not dispersion-free but for which no further significant refinement can be found. In both cases the refinement E_0 so reached is perfect, but only in the first case is it dispersion-free.

Case (a) has been illustrated by the experiments $E_1 - E_3$. It is clear that a similar discussion could be given for almost any elementary experiment in classical physics. Indeed, that some perfect refinement is what is intended when a physical experiment is referred to is tacitly understood in normal communication between physicists; for example, this is what was involved in the reference above to the elementary experiments E_1, E_2, E_3, as being 'in principle' dispersion-free.

Examples of Case (b) originate in phenomena of two types, *quantum* and *thermal*. The experiments on light polarization named by the atomic sentences in the language L' of Section 7 provide examples of the quantum type; a simple instance, mentioned at the end of Section 7, is the experiment E_6 named by the sentence $(SQ)D$. A similar example is provided by the *double Stern-Gerlach experiment* (see, for instance, [27] p. 291):

> E_7: A neutral silver atom moves in the direction of the positive Y-axis. Two Stern-Gerlach experiments are performed, the first measuring the x-component, S_x, of its spin and the second measuring the z-component, S_z. A trial is admissible only when the result of the first measurement is positive, and the outcome of the trial is 'yes' if and only if the result of the second measurement is positive.

Allowing, perhaps, some minor refinements both E_6 and E_7 are perfect elementary experiments. [At least, orthodox quantum mechanics asserts them to be so, and nearly all physicists would concur, although a thoroughgoing believer in 'hidden variables' would disagree.] However, they are by no means dispersion-free: indeed (according to theory) the probability of outcome 'yes' is in each case 1/2.

E_6 and E_7 are typical of the elementary experiments discussed in theoretical quantum mechanics; such experiments are perfect, but in general dispersive. Similarly, perfect refinements of E_1, E_2, E_3 are typical elementary experiments of classical physics and support the claim that such experiments are (perfect and) dispersion-free. These examples also show, however, that elementary experiments *in practice* are rarely perfect. Indeed, the strategy of experimental physics consists to a large extent in refining imperfect elementary experiments until they approach perfection.

I pass now to the second type of dispersive experiments falling under Case (b): those which (if necessary after refinement) are perfect but still dispersive and whose dispersive nature is not due in any way to quantum phenomena. I give first an example from statistical physics.

> E_8: A dust particle, large enough to be seen with a microscope but small enough to exhibit Brownian motion, is enclosed in a hollow cylinder. The cylinder is held for a long time T in a thermostatically controlled environment at temperature θ (one might say 'in thermal contact with a heat reservoir'), all other interaction with the cylinder and its contents being forbidden. The final position of the particle is then observed in the microscope. If it lies in the right hand half of the cylinder the outcome is 'yes', otherwise 'no'.[27]

This experiment clearly exhibits dispersion; the probability of outcome 'yes' may be expected to be 1/2. I claim that it is (almost) perfect. In support of the claim I shall refute several proposals which purport to yield significant refinements. For simplicity I assume that no quantum effects need be considered.

One obvious possibility is to assign a definite initial position, say at the right hand end of the cylinder, to the dust particle. However, since the time T is long, this will have very little effect on the probability. As a more crafty proposal a 'refinement' might be suggested involving observing the position of the particle just prior to the end of the period T and adding, as a condition of admissibility of the trial, the requirement that the particle should then lie in the right hand half of the tube. Naturally, this modified experiment will have a much smaller dispersion. However, it is *not* a refinement of the original experiment since the condition of no interaction imposed in E_8 has been violated: even in classical physics no information can be obtained without interaction.

I consider lastly a proposal that originates in an attitude that is so common that one might call it the 'official view' of the classical physicist. The argument may be expressed as follows:

"Let K denote the system consisting of heat reservoir, cylinder, and contents. During the experiment K is isolated. Consequently, the variation in outcome (from one trial to another) must reflect a variation in the initial state X." [Our critic is regarding K as a mechanical system, and he appeals to the principle of causality (valid in classical mechanics), *every effect must have a cause*.] "*In the specification of the experiment nothing is said concerning this initial state*. Let us then introduce a refinement E_8' of E_8 by requiring an exact observation of the initial state X; i.e. an exact determination of the initial position and velocity of every particle in the system K. (Although this is totally impracticable it is possible, at least classically, in principle.) The statistical description of the Brownian motion can now be replaced (in principle) by an exact calculation based on classical mechanics, and the final position to be expected for the particle found. Finally we require, as a condition of admissibility of the trial, that the initial state of the whole system be such as to lead to a right-hand position of the particle. The refinement E_8' so constructed is dispersion-free and therefore certainly significant, in contradiction with the claim."

The flaw in this argument is revealed by the sentence in italics. Our critic is evidently imbued with the "pure state philosophy": for him any system is at any time in a definite pure state. In the case of the heat reservoir no such state is mentioned; hence the italicized remark.

However, here–as generally in thermodynamics–the term 'heat reservoir' is intended to denote a particular kind of system *in a state of thermodynamic equilibrium at a definite temperature* (here θ). Such a state is not a pure state but–if it must be represented in the formalism of classical mechanics–a mixed state given by a Boltzmann distribution. However, we must remember that, since E_8 is put forward as an elementary experiment, the atomic sentence defining it must determine a *practical procedure* that results in outcome 'yes' or 'no'. No theoretical description of an equilibrium state is appropriate; we need an account of how an equilibrium state is to be prepared in practice. Such an account is not difficult to give. It may conveniently be assumed to end with the stipulation of a long 'waiting time' during which the system 'settles down'–a time so long that the details of any previous history of the system are 'forgotten' (the system attains a *stationary state* in the sense of [8] pp. 17–18). With this clarification of E_8 it is clear that the specification itself precludes any observations of the kind required in the description of E_8', so that E_8' is not a refinement of E_8.

Of course, this interpretation of E_8 is incompatible with the pure state philosophy. In fact the example further illustrates the argument, put forward in Section 5, that the preparation of an equilibrium state corresponds much more closely to a practical procedure than does the preparation of a pure state.

Consider now the following elementary experiment which, unlike E_8, is taken from the everyday world:

> E_9: An opaque urn containing a coin is shaken for a long time. It is then put down and the coin is observed through the neck of the urn. If 'heads' is seen the outcome is 'yes'.

With certain further stipulations this elementary experiment is perfect, although it is certainly not dispersion-free. Although the stipulations are natural and would normally be taken as implicit in the description of E_9, it is desirable for the sake of clarity to make them explicit. First, we must preclude the possibility of any further refinement of the following sort: "peep in at the coin after the shaking but before the outcome is formally observed and add as a condition of admissibility of the trial that the side of the coin then seen be 'heads'." To prevent this obvious form of 'cheating' it is not sufficient merely to say 'No looking'; the position of the coin might be determined in other ways–by electric or magnetic fields, X-rays, and so on. It is necessary to stipulate that *no interaction* with the coin be permitted after the shaking has commenced (except, of course, the interaction between coin and urn.)

Next it is possible to imagine a (clearly significant) refinement consisting in shaking the urn in a precisely determined way that causes the coin to move in a calculable manner. To prevent this we need to ensure that the shaking be 'at random'. However, there is no use in simply prescribing that this be the case, since the notion of randomness is singularly elusive and in any case a theoretical definition would not help: we need to stipulate some practical procedure (i.e. a procedure that can eventually be *demonstrated* in the course of ostensive definition) which fulfills the intuitive requirements associated with randomness. We might, for instance, arrange that the shaking be carried out mechanically by a machine controlled by (a) the output of a Geiger counter responding to α-particles from a radioactive source, or (b) the output of an amplifier responding to thermal noise in a resistor.

There is a particularly simple solution, related to (b), in which the shaking machine is not needed. If we use a 'coin' of microscopic dimensions we may rely on Brownian motion to do the shaking for us. We start with the urn (and 'coin') cold, raise its temperature until a significant amount of

Brownian motion is occurring, and then cool it again. The two stable positions of the 'coin' form a degenerate ground state and end with equal occupation probabilities.

The clarification of E_9 that we obtain in this way is almost identical to E_8. It is clear that a similar discussion could be given of other cases associated with instructions like 'Spin a coin' 'Throw a die', 'Shuffle and cut a pack of cards', and so on. This suggests that there are only two factors which give rise to perfect but dispersive elementary experiments, quantum phenomena and statistical (or thermal) phenomena. [I do not intend to imply by this that these factors differ in some essential way.] Naturally, both factors may be present at once.

9. TROUBLE WITH TRUTH

We now return to the problem which arose at the end of Section 7, of how to formalize a theory involving dispersive elementary experiments. Let us first recall how classical logic deals with the dispersion-free case.

Classical logic is based on the concept of truth. It is assumed that every atomic proposition[28] either has or has not the property of *being true*. Moreover, this is supposed to be an objective property, in no way a matter of opinion, although it is admitted that erroneous views regarding the truth of propositions may on occasion be held and also that there may be propositions for which there is no way – or at least no known way – for determining the truth value.

Technically, the importance of truth for classical logic arises from the role which it plays in determining the meaning of compound propositions. The definition:

9.1. A *proposition* is a statement that is either true or false.

leads to the basic principle:

9.2. The *meaning* of a proposition is determined by laying down the conditions under which it is true.

This principle is then applied to determine by recursion the meaning of every compound proposition, through the definitions:

9.3. $A \wedge B$ is true iff A is true and B is true,
$A \vee B$ is true iff A is true or B is true,
$A \rightarrow B$ is true iff A is false or B is true,
$\neg A$ is true iff A is not true,

$\forall x A(x)$ is true iff $A(t)$ is true for every constant term t,
$\exists x A(x)$ is true iff $A(t)$ is true for some constant term t,

where A and B are arbitrary propositions and $A(x)$ is an expression involving a variable x which becomes a proposition $A(t)$ whenever x is replaced by any constant term t.[29]

In the case of a dispersion-free physical theory like that of Section 6 there is no difficulty in implementing this procedure. Indeed, we are even in the favourable situation in which every atomic proposition P in the primary language is *truth-definite*: i.e. there is an experimental procedure which will determine its truth – namely, the elementary experiment which is described on the document labelled by P. (Of course, compound propositions involving quantifiers are not usually truth-definite, but in the course of a mathematical training one is brainwashed into not worrying about this!)

However, if we have a dispersive physical theory then, even if we have constructed a primary language as in the polarization example of Section 7, we are in deep trouble. As before, there corresponds to each atomic sentence an elementary experiment. But, as we have seen, we can no longer use this correspondence to assign a truth value to the sentence. Consider for instance the atomic sentence $(SQ)D$ of Section 7. We can hardly say that this sentence is true simply on the grounds that the last trial yielded 'yes', since the next may very well yield 'no'.[30] Without truth values we are in no position to apply classical logic.

One might try to avoid this difficulty by defining an atomic sentence to be true if the outcome of the corresponding experiment is *always* 'yes', but this procedure is of little use. First, in the case of a typical dispersive theory hardly any sentences would be true, which means that the result of employing classical logic would be trivial, and it would certainly not allow us to express the probability assertions involved in typical theorems. Secondly, no sentence would be truth-definite, since there is no procedure for testing *every trial*.

It thus seems that one must either reject classical logic or abandon the notion of a 'standard interpreted language', that functioned effectively in the dispersion-free case. If we follow the latter course the natural thing – since a typical theorem refers to probabilities assigned to atomic sentences – is to introduce, as the 'atomic sentences' of a new language, expressions of the form $p(E) = \lambda$, [read: E *has probability* λ *(of outcome 'yes')*], where E is an atomic sentence in the original sense and $\lambda \in [0, 1]$; for instance

$p((SQ)D) = \frac{1}{2}$ (Example 7.1) is now an 'atomic sentence'. One can then form 'sentences' from 'atomic sentences' and logical symbols in the usual way. This certainly allows one to express as formal 'sentences' typical theorems of a dispersive theory, but (a) it severely mutilates the syntax, (b) it introduces ad hoc real numbers (e.g. λ) as a new type of constant terms, and (c) it leaves unanswered the question of the interpretation of probability. [What is one to conclude from the assertion, $p((SQ)D) = \frac{1}{2}$? Not simply (one hopes) that the proportion of 'yes' outcomes in an infinite series of trials would tend to $\frac{1}{2}$, since this would be an untestable assertion. And yet what else?] Moreover, the 'atomic sentences' in the new sense are again not truth-definite: no amount of experimentation will allow one to conclude definitely that (for instance) the sentence $p((SQ)D) = \frac{1}{2}$ is true [or, indeed, that the more easily verified sentence, $p((SQ)D) > \frac{1}{2}$, is true].

We are driven, then, to investigate the possibility of abandoning classical logic itself, and in particular the procedure on which it depends of assigning truth values (true or false) to sentences. This means that we must give up the fundamental principle 9.2, by which the meaning of a proposition is determined by giving the conditions under which it is true, and we are faced with the crucial question: What can be used to replace it?

When one is faced with an abstract problem to which no immediate answer is apparent it often pays to be very simple and practical. What are we really trying to do? We are concerned with a language, and we seek a new concept of meaning for (indicative) sentences. Now a language is used for communication, so the function of a sentence is *to be asserted*, such an assertion having the effect of expressing a belief. This raises the question: How in practice is a belief expressed? Well, a typical assertion in the sort of theory with which we are concerned is an assertion of a probability, for instance *(SQ)D has probability* $\frac{1}{2}$, and in practice such a belief may be expressed in the form of a *bet*. What is a bet? It is a particular sort of *commitment*. (One might say *an offer to assume a commitment*, but such an offer is itself a commitment.) This line of thought suggests that we replace the classical definition of a proposition (Definition 9.1) by the following:

9.4. DEFINITION. A *proposition* is an expression whose assertion entails a definite commitment on the part of the speaker.

9.2 must then of course be replaced by:

9.5. PRINCIPLE OF TANGIBILITY. The *meaning* of any proposition is determined by laying down the commitment which is assumed by him who asserts it. (A meaning in this sense will be called a *tangible meaning*.)

Let us now implement this approach. We assume a standard interpreted language as defined in Section 6 and illustrated by the examples in Sections 6 and 7. To any atomic sentence A there corresponds an elementary experiment. Now, an assertion of A should surely express the belief that a trial of this experiment would yield the outcome 'yes'. So if we are to associate some commitment with this assertion it had better take the form of a penalty should such a trial yield 'no'. Conceivably the penalty could vary from case to case, but for simplicity let us assume it has the form of a constant 'fine' which for notational convenience we may assume to be $1.[32] This gives:

9.6. DEFINITION. He who asserts an atomic sentence undertakes to pay $1 should a trial of the corresponding elementary experiment yield the outcome 'no'.

It is to be understood that (a) the trial may be set up by the 'opponent',[33] (b) the opponent must announce before commencing a trial whether it is to be used in fulfillment of the obligation incurred under 9.5, and (c) when one such trial has been carried out (and the resulting debt, if any, paid) then the obligation has been fully discharged.

Two features of the new approach are already illustrated by Definition 9.6. First, an unwise assertion of an atomic proposition can result in a loss of no more than $1. This 'principle of limited liability' will later extend to all propositions. Secondly, to assert a proposition twice is not equivalent to asserting it once only. Of course these features have no analogues in classical logic, where the significance of *asserting* a proposition is not discussed.

It is clear from Definition 9.6 that an atomic proposition will rarely be willingly asserted.[34] (But this does not apply to a compound proposition, as we shall see later.) Indeed, a speaker S will willingly assert an atomic proposition A only if he feels sure that a trial of the corresponding elementary experiment (which we may also denote by A) will yield outcome 'yes'. In this case we shall say that A is *true for* S (or *relative to* S) and write $S \models A$. Similarly, if S is sure that A will yield outcome 'no' we say A is *false for* S. In general an atomic proposition will be neither true nor false for a speaker.

In this way the term 'true' reenters the theory. As in the case of classical logic, it is a term of the *metalanguage*, not of the language used by the speakers. (Though we could extend that language to include the term 'true' by making "*A* is true" mean roughly "I am willing to assert *A* as often as you like".) However, it now expresses a *relation* between a sentence and a speaker–it represents *subjective*, not obiective, truth.[35]

It may be claimed perhaps that a physical theory purports to represent absolute truth, and that our language should do so too since its purpose is to interpret such a theory. However, a physical theory need not be regarded as propounding absolute truth but rather as *recommending a certain code of belief*. Indeed, this has the advantage of according more closely with the common practice of physicists, who may well use in succession a number of seemingly incompatible theories. Moreover, the present approach to propositions is in no way *incompatible* with the notion of absolute truth–should anyone feel able to give a precise definition of this concept he is welcome to do so; the situation is rather that the concept is *unnecessary* for our purposes. We are concerned with the construction of a language, and a language is used to express beliefs, never to announce absolute truths. (Even a realist must admit that he uses language in this way!) Admittedly, absolute truth seems to be implicit in a statement like "I believe *A* is true, but I may be wrong." However, he who says this is apparently not prepared to assert "*A* is true." His remark indicates a *degree of belief* in *A*. In fact, in everyday life one frequently wants to express a degree of belief, and this is merely one of the devices employed in common language for this purpose. Classical formal logic, be it noted, provides no facilities for this. However, the logic we shall develop on the basis of Definition 9.4 is fully equal to the task. We now pass on to this development.

10. THE LOGICAL CONNECTIVES

We have now arrived at a new concept of a proposition and have assigned a meaning, in accordance with it, to each atomic sentence in any standard interpreted language. In order that the language can fulfill its purpose of acting as a primary language for the interpretation of a physical theory, however, it must contain a variety of logical symbols sufficient to ensure that the theorems can be expressed by means of suitable compound sentences. Since the new logic should surely reduce (in some sense) to classical logic when applied in the dispersion-free case, it is natural to assume that it contains connectives $\wedge$, $\vee$, $\neg$, $\rightarrow$, $\forall$, $\exists$ which reduce in that case to the usual ones (see Sec. 6). (It should not be inferred that each classical connective will have only one 'extension' to the general case.) Of course, other connectives may also be needed, but we shall find that these suffice for the present.

In the classical case the meanings of the logical symbols are given by means of a series of definitions (9.3) which together assign by recursion a meaning to every compound proposition. It is natural to proceed in the same way with the new logic, but with 'meaning' now interpreted in accordance with 9.5. Thus we must define the obligation associated with any compound proposition in terms of those associated with its principal components. For a workable logic we must have simple and intuitively acceptable definitions, and we ought at the same time to bear in mind that the language under construction should be able to express the probability assertions which are typical of dispersive physical theories like quantum mechanics and statistical mechanics. This provides the motivation for the following definitions (10.1–10.7). Throughout, A and B denote arbitrary propositions.

For $\neg$ we adopt:

10.1. DEFINITION. He who asserts $\neg A$ offers to pay \$1 to his opponent if he will assert A.[33]

That this definition is intuitively reasonable is clear, at least in the case when A is an atomic proposition. For the proponent will then recover his money from the opponent iff the trial of A yields 'no'. The justification of the definition of $\vee$ is even simpler:

10.2. DEFINITION. He who asserts $A \vee B$ undertakes to assert either A or B at his own choice.

It should be noted that with this definition a speaker will not willingly assert $A \vee \neg A$ for an atomic proposition A (i.e. $A \vee \neg A$ is not *true* for him) unless A is, in his opinion, dispersion-free–nor will he, even in that case, unless he knows (or is willing to find out) whether a trial of A will give 'yes' or 'no'.

We now come to the connective $\rightarrow$ (implies). We could adopt the classical procedure, introducing $A \rightarrow B$ simply as an abbreviation for $B \vee \neg A$. This leads to a structure equivalent to classical logic–it is essentially Hintikka's dialogue interpretation [18] for classical logic–but we have already seen that classical logic is inadequate for our purposes. Now, notice that

$A \rightarrow B$ could be construed, classically, as "B is at least as true as A". Regarding this as a qualitative comparison of 'trueness'—i.e. likelihood of producing the outcome 'yes'—we are led to suppose that he who believes $A \rightarrow B$ should be *willing to assert B if his opponent will assert A*. Moreover, we catch here a glimpse of how the language might be able to express relations between (subjective) probabilities. So let us adopt:

10.3. DEFINITION. He who asserts $A \rightarrow B$ offers to assert B if his opponent will assert A.

To be more precise, let us agree on the following. If either speaker asserts a proposition of the form $A \rightarrow B$ the other speaker may choose to *admit* the assertion, in which case it is simply annulled. Alternatively, he may *challenge* it by asserting A. The original speaker is then obliged to assert B, whereupon his original assertion of $A \rightarrow B$ is annulled, the associated obligation having been discharged. (An assertion may be challenged only once.)

The definition of $\rightarrow$ is a crucial feature of the logic. A similar definition was introduced by Lorenzen [22] in giving a dialogue interpretation of intuitionistic logic. The resulting logics are quite different, however, mainly because Lorenzen did not use Definition 9.6. (There are other differences as well.)

Let us add to the formal language the symbol F to stand for the particular prime proposition whose tangible meaning is given by:

10.4. DEFINITION. He who asserts F agrees to pay his opponent $1.

Then $\neg A$ may be identified with $A \rightarrow F$: these assertions incur the same commitment. The device of replacing $\neg$ by F is technically convenient. Let us therefore ignore 10.1 and introduce instead:

10.5. DEFINITION. $\neg A$ is an abbreviation for $A \rightarrow F$.

In selecting a definition for $\wedge$ (and) the obvious choice would be to let the assertion of $A \wedge B$ be equivalent to the assertion of both A and B. However, there are several reasons why this is not suitable here. Firstly, $A \wedge A$ would not be equivalent to A. Secondly, the analogous definition of $\forall x A(x)$ (see 10.7 below) is clearly not practicable. Finally, such a choice would mean abandoning the principle of limited liability. So instead we adopt an obvious analogue of 10.2:

10.6. DEFINITION. He who asserts $A \wedge B$ undertakes to assert either A or B at his opponent's choice.

Observe that, although the speaker is not obliged to assert *both* A and B, he must be prepared to assert either since he cannot tell which his opponent may choose.

As definitions of the quantifiers ∀ and ∃ we choose natural generalizations of 10.2 and 10.6:

10.7. DEFINITION. He who asserts $\exists x A(x)$ undertakes to assert $A(t)$ for some constant term t of his own choice. He who asserts $\forall x A(x)$ undertakes to assert $A(t)$ for some constant term t of his opponent's choice.

Here $A(x)$ denotes a formula in which x is the only free variable, and $A(t)$ is the proposition that results when x is replaced by t in $A(x)$.[29]

By recursion, the above definitions attach a tangible meaning to every sentence: he who asserts any sentence incurs, in so doing, a definite obligation. The general nature of this obligation is clear. Suppose that one speaker asserts a compound sentence, say $A \rightarrow B$. His opponent may if he wishes challenge it by asserting A whereupon the first speaker must assert B, his original assertion then being annulled. If A and B are themselves compound sentences the debate will continue in accordance with the definitions—which now act as *rules of debate*—with new sentences being asserted and previous assertions being annulled. At any time during the debate a *position* will obtain, in which each speaker is committed to a finite collection of assertions which we shall call his *tenet* (repetitions of assertions may occur in a tenet). As the debate continues the position changes, the asserted sentences becoming simpler, until a *final position* is reached in which only atomic sentences are asserted. The appropriate trials are then carried out, and each speaker pays the other in accordance with 9.6 and 10.4.

We assume, of course, that every assertion must in due course be dealt with in accordance with the rules of debate. In principle it is necessary to lay down *rules of order* governing the order in which the commitments incurred by the two participants shall be discharged. Moreover, *rules of trial* should also be given governing the question of how the trials should be carried out: who is to set them up, in what order, and so on. For example, consider the elementary experiment (E_5 of Section 8) "Drop an egg. If it breaks the outcome is 'yes'; otherwise 'no'." Clearly, the way in which these instructions are carried out does matter: the odds I might offer on the outcome would depend (for instance) on whether I or my opponent were to do the dropping. Of course, the difficulty here arises because the experiment E_5 has refinements which differ 'significantly' (in 'my' opinion): i.e. because it is not 'perfect'. However, we saw in Section 8 that only perfect elementary experiments are referred to in physical theories. If we assume that all the experiments involved are perfect then any speaker will be indifferent to the way in which trials are carried out so that rules of trial become unnecessary. It turns out that rules of order then become redundant too. Since this results in a great simplification in the logic we henceforth adopt as a standing assumption:[36]

10.8. ASSUMPTION. Every elementary experiment is perfect.

I conclude this section with a simple example that will illustrate the rules of debate and the differences from classical logic. For convenience the proponent and opponent will be referred to as 'I' and 'you' respectively, and each position will be denoted by an expression consisting of my tenet on the right and yours on the left of a vertical line.[37]

Consider the sentences $A \rightarrow B$ and $B \vee \neg A$, in which A and B are atomic sentences. In classical logic these sentences 'have the same meaning': each is true iff B is true or A is false. However, in the new logic they cannot be identified since their assertions represent quite different commitments. Indeed, if I assert $A \rightarrow B$ we shall (according to whether you challenge or not) reach one of the final positions $\emptyset | \emptyset$ or $A | B$. ($\emptyset$ denotes the *empty tenet*, in which nothing is asserted.) Thus in asserting $A \rightarrow B$ I commit myself to be placed in one of these final positions, at your choice. On the other hand, if I assert $B \vee \neg A$ I am obliged to assert either B, thus reaching the final position $\emptyset | B$, or $\neg A$ which (if you challenge) yields the final position $A | F$: i.e. I commit myself to be placed in one of the positions $\emptyset | B$ or $A | F$, at *my* choice.

In order to decide which, if either, of the sentences $A \rightarrow B$ and $A \vee \neg B$ I would willingly assert I must form an opinion regarding the status of the four final positions just mentioned. In general, a particular final position will result in different payments on different occasions. However, depending on my beliefs, there will be a certain set $\mathscr{A}$ of final positions which seem *acceptable* to me (presumably because I expect on average a non-negative gain). To reach one of these is, as far as I am concerned, a 'win'. Thus I will be willing to assert a given proposition P if I am able to conduct the ensuing debate in such a way that an acceptable final position is assured. When this is the case we shall say P *is true* (*for me*) and write $\models P$.[38] (If several speakers are under consideration we can express 'P is true for S' by writing $S \models P$.) P *is false* will mean $\models \neg P$. In general a proposition is neither true nor false. (Note that in the case of an atomic proposition these definitions agree with those of Section 9.)

Let's illustrate this with the above example. $A \rightarrow B$ will be true (for me) if both the positions $\emptyset|\emptyset$ and $A|B$ are acceptable. The former is surely acceptable for any rational speaker, since it involves no commitment at all; the latter may or may not be. (Presumably it will be if I believe that A is at least as likely as B to yield the outcome 'no'.) Similarly, $B \vee \neg A$ will be true if either of the positions $\emptyset|B$ or $A|F$ is acceptable. (The first will be iff I am *sure* that B will yield 'yes'; the second iff I am *sure* that A will yield 'no'.)

The remarks in parentheses show that $A \rightarrow B$ is certainly true if $B \vee \neg A$ is true, *but not conversely*. This may also be deduced in a more abstract way as follows. First note that the final position $F|B$ is surely acceptable to any rational observer, for in this position he cannot lose. Now the 'sum' of the position $F|B$ and $A|F$ is $F, A|B, F$, which is precisely equivalent in commitment to $A|B$. (Here we mean by the *sum* of two final positions that for which each speaker's commitment is obtained by adjoining his commitments in the two given positions.) Surely, for any rational speaker, the sum of two acceptable final positions will also be acceptable. It follows that if $A|F$ is acceptable then so is $A|B$. Similarly, if $\emptyset|B$ is acceptable then so is $A|B$ (since surely $A|\emptyset$ is acceptable, for it cannot result in loss).

11. CLASSICAL SPEAKERS

As we saw in Section 10, the beliefs of any speaker – i.e. the sentences that are true for him – are determined completely by the set $\mathscr{A}$ of final positions acceptable to him;[39] two speakers may therefore be considered equal if they correspond to the same set $\mathscr{A}$. The nature of this set reflects various aspects of the beliefs of a speaker: First, it represents the special information he has concerning particular objects in his vicinity. Secondly, it reflects the physical theory or theories to which he adheres. Thirdly, it may reflect the 'philosophical attitude' of the speaker; what I mean by this will be explained below. Lastly, it reflects the 'rationality' or otherwise of the speaker.

In connection with this last point we must decide under what circumstances a speaker should be considered to be rational. Two conditions suggest themselves: First, he should not find acceptable any final position (such as $\emptyset|F$) in which he is certain to lose; nor indeed any final position in which – although he may not always lose – he expects *on average* to lose. Secondly, he should not consider unacceptable any final position in which he expects on average not to lose. These thoughts lead to the following proposal:

11.1. DEFINITION.[40] A speaker S is *rational* iff, for him,

    (a) for every atomic sentence A, the positions $A|\varnothing$, $F|A$, and $A|A$ are acceptable;

    (b) the position $\varnothing|F$ is not acceptable;

    (c) the sum of any two acceptable final positions is acceptable;

and further, if α is any final position,

    (d) if $\alpha + A|A$ is acceptable then so is α; and

    (e) if $n\alpha + F|\varnothing$ is acceptable for arbitrarily large positive integers n then so is α.

The justification for (b), (c), and the first three parts of (a) is fairly clear; these items have also been illustrated at the end of Section 10, where the *sum* of two final positions is explained. That $A|A$ should be acceptable is plausible: a rational speaker might expect to gain as much from his opponent's assertion (on average) as he loses from his own. The justification of (d) is similar [' + ' here denotes the 'sum' of positions mentioned in (c)]. (e) is justified as follows [here $n\alpha$ denotes $\alpha + \cdots + \alpha$ (n terms)]: If S's average loss in α is strictly positive then his loss in $n\alpha$ would (for sufficiently large n) exceed \$1, which would make the position $n\alpha + F|\varnothing$ unacceptable. The 'Archimedian' axiom (e) is technically convenient but of no great practical import.

We shall henceforth be concerned only with rational speakers, and 'speaker' will mean 'rational speaker'.

In view of the remark at the beginning of this section we may consider any speaker to be characterized by the corresponding set of acceptable final positions, and it will be convenient to regard any such set that satisfies the conditions of Definition 11.1 as corresponding to a 'possible' (rational) speaker.

To illustrate what I mean by the 'philosophical attitude' of a speaker let's take an extreme case, which is also the most familiar, that of an *omniscient classical* speaker. By a *classical* speaker I mean one who believes that every elementary experiment is dispersion-free; for instance, if the elementary experiments under consideration are mathematical calculations (e.g. the example in Sec. 6) most mathematicians would fall in this category; similarly, physicists concerned with certain classical theories such as mechanics and thermodynamics are (in respect of the elementary experiments they choose to consider) classical speakers. Naturally, it is necessary to pin down this notion more closely in a formal definition. We require that, given any atomic sentence A, a classical speaker S must hold one of three opinions: either (a) it is true for him, or (b) it is false for him, or (c) he does not know whether it is true or false. It is to be understood that Case (c) represents *total ignorance*: i.e. in this condition S recognizes that A might be true but on the other hand might be false, and therefore considers acceptable no position that would be unacceptable if he believed A to be true and, equally, none that would be unacceptable if he believed A to be false. [Thus a mathematician who, without computation, is prepared to gamble (at stated odds) that the millionth digit in the decimal expansion of π is less than 8 is not a classical speaker in our sense.]

I call a classical speaker S omniscient[41] if two conditions hold:

(a) S is *omniscient for atomic sentences*: i.e., for every atomic sentence A, $S \models A \lor \neg A$; in other words S is willing, for each such A, to assert either A or $\neg A$. This means he does not merely believe A is dispersion-free but is prepared to say whether it is true or false. [This is not as strong a condition as it seems at first sight. We need not assume that S has amassed an infinite amount of experimental information; we need merely suppose that, in each particular case, he is prepared to carry out, before answering, a trial of the relevant elementary experiment.]

(b) S is *omniscient for quantifiers*: i.e. given any set of sentences, each of which is either true or false for him, he is able to scan the whole set and tell whether or not it contains any true (or any false) sentence.

In the case of an omniscient classical speaker S it is easy to deduce from the rules of debate 10.2–10.7 that *every* sentence is either true or false (for him); moreover, the properties 9.3, used to define the logical connectives in classical logic, hold (when we identify 'true' with 'true for S'). It follows that the sentences true for S are exactly those which – given the true atomic sentences – would be true according to classical logic. (In particular, all classical logical identities are true for S.) Consequently S may if he wishes use classical logic (instead of considerations based on the rules of debate) in deciding which sentences he can safely assert. In this way, and to this extent, classical logic emerges as a special case.

From the definition above, the notion of an omniscient speaker might seem to be rather arbitrary. This is not so; on the contrary, this concept has a natural place in the present approach to logic. It arises from the consideration of a simple and important relation between speakers, defined as follows. Suppose S and S' are two speakers, and the set $\mathscr{A}$ of final positions acceptable to S is contained in the corresponding set $\mathscr{A}'$ for S': $\mathscr{A} \subset \mathscr{A}'$. Then any final position acceptable to S is also acceptable to S', whence one can deduce that every sentence true for S is true for S':

i.e. S' is willing to make any assertion that S is willing to make. We shall then say that S' is *greater than or equal to* S and write S' ≥ S; if, in addition, S' ≠ S (so $\mathscr{A}' ≠ \mathscr{A}$) we say S' is *greater than* S: S' > S. This last relation might be read as 'S' is *more knowledgeable* than S, for S' ≥ S means roughly 'If S (thinks he) knows anything then S' (thinks he) knows it too.' The words in parentheses reflect the absence in this approach of the classical notion of absolute truth. One might emphasize this aspect by saying instead, 'S' is *more confident* than S', the terminology used in [13].

The relation ≥ establishes a *partial ordering* among the (rational) speakers.[42] I don't want to enter into a mathematical discussion of partial orderings here, but one aspect concerns us now: the notion of a *maximal element*. An element in a partially ordered set is *maximal* if there is no greater element. (Do not confuse with a *maximum* element, which is one greater than every other element.) Thus a speaker S is maximal if there is no (possible) 'more knowledgeable' speaker. Are there any maximal speakers? Thanks to the precise form of Definition 11.1 there are plenty. One can prove (essentially as in [13]):

11.2. THEOREM. Given any non-maximal speaker S there is a maximal speaker S' greater than S. Moreover, if S is classical then S' may be chosen to be classical.

Such an S' (which need not be unique) will be called a *maximal extension* of S. What has this to do with omniscient speakers? Consider the classical case. It seems fairly clear intuitively that a classical omniscient speaker is maximal. (After all, to such a speaker *every* sentence is either true or false, so one could not enlarge the set of true sentences without contradiction and hence irrationality.) In fact, this can easily be proved. With a little more difficulty we can show the converse: a maximal classical speaker is omniscient.[43] Thus, for classical speakers at least, 'maximal' and 'omniscient' are equivalent. We shall see later that this is true also for nonclassical speakers. This is the fundamental reason for the importance of omniscient speakers.

We have seen that classical logic may be used freely by any omniscient classical speaker. However, the classical speakers that exist in practice (mathematicians, classical physicists) are not omniscient: there are plenty of sentences K, even in elementary arithmetic, for which such a speaker, S say, can assert without risk of loss neither K nor ¬K. (For example, K might be a formal expression of Fermat's Last Theorem.) But then the sentence $K \vee \neg K$ is not true for this speaker in the dialogue logic. Since this is an instance of a classical 'logical identity' (see Sec. 13) it is clear that classical logic is not appropriate for use (in connection with the dialogue) by a non-omniscient classical speaker.

Nevertheless, there seems to be some sense in which such a speaker can usefully employ classical logic. Indeed, we know very well that the typical classical speaker does in practice use classical logic in his arguments; he will claim that $K \vee \neg K$ is 'really true' (or 'true in reality', or 'objectively true', or 'absolutely true') and explain that he does not choose to formally assert it only because he himself "doesn't know whether K is ('really') true or false".

Although the behaviour shown by S in speaking in this way is (from the present point of view) somewhat curious, it does often succeed in conveying information about S's beliefs, and it should admit an interpretation consistent with the present approach to logic. There is no obvious way in which the reference to reality can be given a meaning in the sense of 9.5, but we can perhaps interpret it as follows. Let us suppose that S envisages a deity S' who Knows All. When S says a sentence is 'really true' he means it is *true for* S'. Concerning S', S believes only that (a) S' is omniscient, (b) S' $\geq$ S (for S is confident that his present state of ignorance involves no actual *errors*) and (c) S' is classical (this simply reflects S's confidence that every elementary experiment is dispersion-free): in other words, S' is a *maximal classical extension* of S. Clearly, if S wishes to introduce such an S' no harm can result, for by Theorem 11.2 such an S' exists. In fact, in general S' could be any one of many possible speakers. Since S has no way of telling which is the 'real' S' he will claim as 'really true' only those sentences which are true for every such S'. This suggests that we adopt the following interpretation:

11.3. DEFINITION. Let P be any sentence and S a classical speaker. Then P *is really true for* S (or *objectively true for* S, etc.) means P *is true for every omniscient classical speaker* S' *such that* S' $\geq$ S. Similarly, P *may be really true for* S means P *is true for some such* S'. (Note that, for S, any *true* sentence is certainly *really true*, but the converse does not generally apply.)

Naturally, statements of this sort give information, not about any 'objective reality', but only about S. In fact, the set of all sentences which are really true for S determines completely the set $\mathscr{A}$ of acceptable final positions for S. It is easy to see that this set contains all classical logical identities and is closed under the usual rules of inference of classical logic.[44] This shows how a non-omniscient classical speaker can, by ignoring the dialogue interpretation of formal sentences and introducing instead the notion of absolute truth, convey useful information about his beliefs and at the same time avoid the need to reject classical logic. This procedure is convenient, since the logic arising from the rules of debate in the case of a nonomniscient speaker is much more complex (see [21]).

Finally in this section, let us consider–in the classical case—how a physical theory is reflected in the beliefs of one who subscribes or *adheres* to the theory. We saw in Section 1 that for any complete theory there must be a *primary language* which serves to express in unambiguous terms the experimental significance of the assertions of the theory. In Sections 6 and 7 we saw how this purpose could be achieved by using a *standard interpreted language*. The only essential function of the *secondary language*, in which the deductive part of theory is expressed, is to distinguish a certain set $\mathscr{T}$ of sentences in the primary language as 'true according to the theory' or, briefly, *theoretically true*. Now, it is too much to require that any speaker who believes the theory–any *adherent* of the theory, as we shall say–should be willing to assert any theoretically true sentence. He might decline, on grounds of his own ignorance, while still claiming that the sentence was 'really true' (cf. the example of $K \vee \neg K$ above). This leads us to the following definition of an adherent of a theory:

11.4. DEFINITION. A speaker is an *adherent* of a given theory iff every theoretically true sentence is 'really true' for him.

The function of a physical theory is thus to lay down certain canons which govern the beliefs of any adherent.

So far, the set $\mathscr{T}$ of theoretically true sentences has been regarded as arbitrary. However, in view of the last definition there is no loss in generality in assuming that $\mathscr{T}$ has a certain 'closure property'. Suppose P is a sentence which is really true for every (classical) adherent of the theory. Then–insofar as classical speakers are concerned–we may as well regard P as itself belonging to $\mathscr{T}$. (This applies in particular to any (classical) *logical consequence* of the sentences in $\mathscr{T}$.) With this understanding $\mathscr{T}$ is a set which contains all classical logical identities and is closed under the rules of inference of classical logic: i.e. it is a (classical) *theory* in the usual sense of mathematical logic [26].

The appropriateness of the terminology is, of course, no accident. In fact, the set $\mathscr{T}$ adequately represents the physical theory: even if there is a secondary language 'behind the scenes', as it were, it can be ignored if the set $\mathscr{T}$ is known. In particular, it may be possible to obtain an axiomatiźation of the theory by selecting a suitable set of sentences in $\mathscr{T}$ as proper axioms. This will then constitute a (in general new) formulation of the theory expressed entirely in the primary language; I shall call it a *primary formulation*. Since the axioms are sentences in the primary language they may have a direct physical interpretation, so we have the possibility of constructing in this way a 'satisfactory formulation' of the theory, with the advantages described in Section 3. A concrete example of this procedure is the case of classical thermodynamics [8], described in Section 4.

12. NONCLASSICAL SPEAKERS

In Section 11 we introduced the notion of a classical speaker as one who is characterized by a certain philosophical attitude–the belief that every elementary experiment is dispersion-free–which prevailed in physics up to roughly the end of the last century. Once established, such an attitude tends to persist. When any experimental procedure was encountered which did not always result in the same outcome it was not treated as an elementary experiment *per se* but rather as an instance of a class of dispersion-free elementary experiments which differed from each other in respect of 'errors', 'statistical effects', 'thermal fluctuations', etc. In this way the classical philosophical attitude could be saved, albeit at the cost of a substantial divergence between the viewpoint implicit in the language used and the actual experimental procedure. (In particular, the modified language can no longer be used to give a 'satisfactory formulation' of a theory.) With the advent of quantum mechanics the difficulties associated with the maintenance of the classical philosophical attitude became formidable, and at the same time the motivation for doing so became less, particularly since the standard theory assigned probabilities to elementary experiments directly, rather than through the medium of some dispersion-free theory such as classical mechanics. As a result, it is not now considered entirely irrational to treat a dispersive elementary experiment directly as such rather than to regard it as a sort of random superposition of dispersion-free experiments.

In any case, we have seen in Section 4, 6, and 7 that we are led to adopt this nonclassical attitude in order to obtain a primary language capable of providing a satisfactory formulation of a dispersive physical theory. Now, we saw in Section 11 that in the classical case the simplest situation arises in the case of an omniscient speaker. There classical logic could be used; moreover, even a nonomniscient speaker could use classical logic in a sense: namely, in speaking of the sentences which would be asserted by any of his 'maximal classical extensions'. This suggests that in dropping the classical attitude we should at first retain the property of omniscience. Of course, this concept must be redefined so as to apply in the general case. Now, in the case of a classical speaker it is possible to express the condition (a) for omniscience (see Sec. 11) as follows: *A classical speaker is omniscient for atomic sentences iff, for every atomic sentence A, one of the positions $F|2A$ and $2A|F$ is acceptable.* (Indeed, the first is acceptable iff A is true and the second iff A is false.) Now, the second of these positions is the *negative* of the first, in that it is obtained by exchanging the roles of the speakers. This suggests the first part of the following definition, where we denote the negative of any position α by $-\alpha$; the second part is an immediate generalization of the corresponding part in the classical case.

12.1. DEFINITION. A speaker is *omniscient* iff the following two conditions hold:

> (a) He is *omniscient for final positions* in that, for any final position α, at least one of the positions α and $-\alpha$ is acceptable to him.
> (b) He is *omniscient for quantifiers* in that given any set of positions he is able to scan the whole set and tell whether it contains any acceptable (or any unacceptable) position.

It is easy to show that this is equivalent in the classical case to the previous definition. Moreover, (a) is a reasonable requirement for omniscience even in the nonclassical case, for if a speaker considers both α and $-\alpha$ unacceptable he has evidently not yet been able to decide what, on average, will happen when the necessary trials are carried out: if he really expected to lose in the position α then surely he would expect to gain equally in $-\alpha$. Thus to this extent he is not omniscient.

The consequences of Definitions 12.1 and 11.1 are striking. It can be shown (see [13]) that any omniscient speaker S *behaves* (in respect of the final positions he considers acceptable) *as if* he had assigned to each atomic proposition A a certain subjective probability $\langle A \rangle$ of yielding the outcome 'no', and then considered a final position acceptable iff his expected gain in it is nonnegative.

For instance, the final position $A|B,C$ would be acceptable iff $\langle B \rangle +

$\langle C \rangle - \langle A \rangle \leq 0$; for $\langle B \rangle$ is his expected loss due to the assertion of B and similarly for C, and $\langle A \rangle$ is the average gain he expects due to his opponent's assertion of A. (Notice that the opponent's beliefs are irrelevant insofar as acceptability to S is concerned.) In particular, A is true for S iff $\langle A \rangle = 0$, and false for S iff $\langle A \rangle = 1$. Always $0 \leq \langle A \rangle \leq 1$, and for a classical speaker $\langle A \rangle$ can only be 0 or 1. We call $\langle A \rangle$ the *risk value* of A, since it represents the expected loss due to an assertion of A.

The results just described are summarized in the following definition and theorem:

12.2 DEFINITION. A *risk function* $\langle \ \rangle$ is any function that associates with each atomic sentence A a *risk value* $\langle A \rangle$ in the closed interval [0, 1], with $\langle F \rangle = 1$. (We also call a risk function a *probability assignment* and refer to $\langle A \rangle$ as the *probability of outcome 'no'* for the elementary experiment denoted by A.) Given any risk function we assign also to each final position $\alpha = A_1, ..., A_m | B_1, ..., B_n$ (in which the proponent has asserted $B_1, ..., B_n$ and the opponent has asserted $A_1, ..., A_m$) the risk value

$$\langle \alpha \rangle = \sum_{j=1}^{n} \langle B_j \rangle - \sum_{i=1}^{m} \langle A_i \rangle$$

and call the final position α *acceptable* (with respect to the risk function $\langle \ \rangle$) iff $\langle \alpha \rangle \leq 0$.

12.3 THEOREM. There exists a one-to-one correspondence between (possible) omniscient speakers and probability assignments (or risk functions):

> (a) Given any omniscient speaker S there is a unique probability assignment $\langle \ \rangle$ for which the acceptable final positions coincide with those for S.
> (b) Given any probability assignment $\langle \ \rangle$ the corresponding set of acceptable final positions coincides with those associated with some (unique) omniscient speaker.

It is fairly easy to prove (b); one need merely verify that the set of acceptable final positions for $\langle \ \rangle$ satisfies the conditions given in Definitions 11.1 and 12.1. To establish (a) one has to construct, for any omniscient speaker S, a corresponding risk function $\langle \ \rangle$. The following paragraph indicates the sort of way in which this is done (see [13]).

For an arbitrary atomic sentence A consider the set of all pairs of positive integers (r, n) such that the final position $rF|nA$ is acceptable to S. For each such pair form the rational number r/n. Let $\mathcal{U}$ denote the set of rational numbers so obtained. Similarly, let $\mathscr{L}$ denote the set of rational numbers r/n obtained from pairs (r, n) for which $nA|rF$ is acceptable. Then it can be shown that $(\mathscr{L}, \mathscr{U})$ forms a *Dedekind cut* of the positive rationals: that is, there is a unique nonnegative real number k such that $\mathscr{U}$ consists of all rational numbers not less than k and $\mathscr{L}$ consists of all positive rational numbers not exceeding k. We set $\langle A \rangle = k$. It can then be shown that the function $\langle \ \rangle$ is a risk function which 'represents' the speaker S in the manner described above: S behaves as though $\langle A \rangle$ were (for each A) his subjective estimate of the probability that a trial of A will yield 'no'. Moreover, by its construction, the risk function $\langle \ \rangle$ is uniquely determined by S.

The new approach to propositions (Definition 9.4) is beginning to pay off. It has now led to the result that to any omniscient speaker there corresponds a mapping which assigns a subjective probability to each atomic sentence; moreover, this probability assignment is determined by the set $\mathscr{A}$ of acceptable final positions. This is a first step towards the fulfillment of the hope that in part motivated the new approach: that for a theory such as quantum mechanics, whose theorems are concerned with relations between probabilities, each theorem should be expressible as a formal sentence in the primary language.

The next step in this direction is to investigate more closely the relation between the set of true sentences and the set $\mathscr{A}$ of acceptable final positions. This can conveniently be done by extending the notion of risk value to all compound sentences and to all positions, final or not. I state the (recursive) definition first and then explain the motivation for it:

12.4. DEFINITION. Let $\langle \ \rangle$ be the risk function of an omniscient speaker S; $\langle A \rangle$ is defined for each atomic sentence A. We extend the domain of the function $\langle \ \rangle$ first to all compound sentences by setting:[45,29]

(a) $\langle P \wedge Q \rangle = \sup\{\langle P \rangle, \langle Q \rangle\}$,
(b) $\langle P \vee Q \rangle = \inf\{\langle P \rangle, \langle Q \rangle\}$,
(c) $\langle P \to Q \rangle = \sup\{0, \langle Q \rangle - \langle P \rangle\}$,
(d) $\langle \neg P \rangle = 1 - \langle P \rangle$,
(e) $\langle \forall x P(x) \rangle = \sup\{\langle P(t) \rangle : t \text{ a constant term}\}$,
(f) $\langle \exists x P(x) \rangle = \inf\{\langle P(t) \rangle : t \text{ a constant term}\}$,

where P and Q are any sentences, and $P(x)$ is a formula in which no variable other than x occurs free (so that $P(t)$ is a sentence for each constant term t). Then we extend it further to all positions by:

(g)   Let α be the position $Q_1, ..., Q_n | P_1, ..., P_m$, where the P's and Q's are arbitrary sentences (cf. Def. 12.2). Then the risk value of α is $\langle \alpha \rangle = \Sigma \langle P_i \rangle - \Sigma \langle Q_j \rangle$.

To see the motivation for this definition consider the case when P, Q, and each $P(t)$ are atomic sentences. He who asserts $P \wedge Q$ is obliged to assert P or Q at his opponent's choice. Since his opponent may consistently choose the 'more risky' of these two sentences he can be *sure* only that his average loss will not exceed the greater of $\langle P \rangle$ and $\langle Q \rangle$. A similar argument applies to (e). For (b) and (f) the situation is similar, except that S can choose, and he will naturally select an option of low risk. If S asserts $P \to Q$ his opponent has the option of admitting the assertion, in which case S's loss is zero, or challenging, which yields the final position $P|Q$, where S's expected loss is $\langle Q \rangle - \langle P \rangle$; as in the case of $P \wedge Q$, S must be prepared for the worst. The motivation for (g) should be clear; compare the example of $A|B$, C discussed above.

Since any sentence is formed from atomic sentences by means of logical symbols, repeated use of Definition 12.4 allows the risk value $\langle P \rangle$ to be computed for any compound sentence P, given the risk values of the atomic sentences. Notice that, from the form of 12.4(a)–(f), the fact that the latter all lie in the closed interval [0, 1] implies that $\langle P \rangle$ will also always lie in this interval.

The *justification* of Definition 12.4 is given by the next two results:

12.5. LEMMA. (a) Whenever–in the course of a dialogue–S makes a choice the risk value of the position never decreases, but the choice can always be made in such a way that it remains constant or at worst (and only when a quantifier is involved) increases arbitrarily little.

(b) Similarly, whenever his opponent makes a choice the risk value never increases, but the choice can always be made in such a way that it remains constant or at best decreases arbitrarily little.

Indeed, if S makes a choice he is either dealing with an assertion of his opponent of the form $P \wedge Q$, $P \to Q$, $\neg P$, or $\forall x P(x)$; or with one of his own of the form $P \vee Q$ or $\exists x P(x)$. In each particular case the assertions in 12.5 follow from the appropriate part of Definitions 12.4 and 10.2–10.7, the 'arbitrarily small increase' in risk value being required only when the sup in 12.4(e), or inf in 12.4(f), is not attained. (b) is proved in a similar way.

The terminology of *2-person game theory* can be applied to dialogues.

If S asserts some initial sentence the ensuing dialogue constitutes a *game* in the technical sense. For any game a *strategy* for a player is a plan which assigns a particular choice to each situation in which he has to choose. From 12.5 we get:[46]

12.6. THEOREM. Suppose S asserts an arbitrary compound proposition P. Then for the ensuing debate, given any number $\varepsilon > 0$, S has a strategy which will guarantee a final position α for which $\langle \alpha \rangle < \langle P \rangle + \varepsilon$ but none for which $\langle \alpha \rangle < \langle P \rangle$.

Indeed, by 12.5(a), S can ensure that the risk value of the position (which is initially $\langle \varnothing | P \rangle = \langle P \rangle$) increases by so little at each of his moves that the net increase is less than ε, and by 12.5(b) it never increases at moves of his opponent. On the other hand, he can never decrease it, nor can he be sure of any given (positive) decrease as a result of his opponent's moves.

From 12.6, together with the definition of truth (for S) of a proposition (see Section 10 and footnote 38) we get the following important result:

12.7. COROLLARY. A proposition P is true for S iff $\langle P \rangle = 0$.

This result, together with Definition 12.4 and the remarks preceding 12.2, shows how an assertion about probability can be expressed by a formal sentence in the primary language. For example, by 12.4 one finds that $\langle A \rightarrow \neg A \rangle = 0$ iff $\langle A \rangle \geq \frac{1}{2}$; thus he who is willing to assert $A \rightarrow \neg A$ believes that the probability that a trial of A will yield outcome 'no' is at least $\frac{1}{2}$. Other sentences can be used to express more complicated assertions: for instance, $A \vee \neg A$ is true iff $\langle A \rangle$ is 0 or 1, and $\neg A \rightarrow (\neg A \rightarrow A)$ is true iff $\langle A \rangle \leq 2/3$.

How does this work out for a physical theory? Take, for example, the language L' for light polarization experiments described in Section 7. Suppose a physical theory yields, as the formal expression of a theorem, the sentence $(SQ)D \leftrightarrow \neg(SQ)D$. [Here $\leftrightarrow$ is a *derived* connective: for any sentences P and Q, $P \leftrightarrow Q$ is an abbreviation for $(P \rightarrow Q) \wedge (Q \rightarrow P)$.] To an omniscient speaker who adheres to the theory this sentence is *ipso facto* true. (Cf. Def. 11.4 and Sec. 12.) Its risk value is therefore zero, which means (see the last paragraph) that $\langle (SQ)D \rangle = 1/2$. The sentence thus asserts that the probability of outcome 'no' for the experiment $(SQ)D$ is 1/2, so it constitutes a formal expression of the assertion in Example 7.1. In a similar way the formal sentences of L', $\forall x \forall f \forall a((xf)a \leftrightarrow x(fa))$ and $\forall a \exists b \forall x(xa \leftrightarrow \neg xb)$, express the content of Examples 7.2 and 7.3.

I offer one more illustration of the power of the language L'. Let us introduce the expression $x = y$ as an abbreviation for the sentence $\forall a(xa \leftrightarrow ya)$. In the theory given by the standard representation in terms of Hilbert space (see the beginning of Section 7) the sentence $x = y$ is true iff the rays representing the (pure) states x and y coincide. But this sentence is richer than that suggests, for its risk value $\langle x = y \rangle$ determines the angle θ between these rays; indeed, a short calculation gives $\langle x = y \rangle = \sin\theta$. [Note in passing that 2θ is the angle between the radii that locate these states on the Poincaré sphere.] To illustrate the significance of this let P denote the formal sentence $\forall x \forall y(x = y \leftrightarrow xQ = yQ)$. At first sight (i.e. interpreted in terms of classical logic) P seems to say simply that Q is a one to one mapping of the set $\mathbf{S}$ of all (pure) states onto itself. But computing its risk value we get $\langle P \rangle = \sup_x \sup_y |\langle x = y \rangle - \langle xQ = yQ \rangle|$, so that P is true iff, for all x, y, $\langle x = y \rangle = \langle xQ = yQ \rangle$: i.e. iff Q is an *isometry* of $\mathbf{S}$ (the 'distance' between two states being the angle θ above) – of course, P is a theorem of the Hilbert space theory. In this way P provides a formal expression for the statement 'Q is an isometry'. [Of course, in case Q is not an isometry, P is not simply *false*; rather, $\langle P \rangle$ then provides a measure of the extent to which Q deviates from being an isometry.] Notice how the interpretation of P in terms of classical logic may be read as an assertion about classical mechanics – that Q is a one-to-one transformation of phase space ($= \mathbf{S}$) – while the interpretation of the *same* formal sentence in terms of the new logic yields the corresponding (but mathematically quite distinct) assertion about quantum mechanics.

These few examples provide only a glimpse of what can be done with the language L' using the nonclassical interpretation of the logical connectives given in Section 10. Whether it is possible to express a whole theory entirely in this language is a question we shall discuss further in the next section.

13. SPEAKERS, LOGICS, AND THEORIES

We have now distinguished two basic classes of speakers: the *classical speakers* and the *omniscient speakers*. The intersection of these forms a third class, the *classical omniscient speakers*. Lastly we have the trivial class of *all* (rational) *speakers*. I shall sometimes, for safety, refer to these classes as those of *arbitrary classical, arbitrary omniscient, classical omniscient*, and *arbitrary* speakers, respectively.

For a classical omniscient speaker any sentence of the form $P \vee \neg P$

(where P is an arbitrary sentence) is true. On the other hand we noted in Section 11 that such a sentence is not necessarily true for a nonomniscient classical speaker and at the end of Section 12 that the same applies to a nonclassical omniscient speaker. The sentences which are true for every speaker of a given class constitute the *logical identities* for that class of speaker and determine a corresponding logic. [Of course, these logics are not unrelated: for instance, any logical identity for arbitrary omniscient speakers is *ipso facto* a logical identity for classical omniscient speakers.] For the class consisting of all classical omniscient speakers we have already (Sec. 11) identified this logic; it is classical logic. For the two classes containing nonomniscient speakers the question of rules of order must be settled before the logic can be determined.[36].

There remains the class of arbitrary omniscient speakers. To identify the logic in this case it is convenient to define the *truth value* (for a particular speaker) of any sentence P to be $1 - \langle P \rangle$. Each sentence then has a truth value in the interval [0, 1] and is true iff its truth value is 1, and Definition 12.4 yields a recursive definition giving the truth values of every compound sentence in terms of those of atomic sentences. Now, it turns out that this definition coincides with that used by Łukasiewicz in the early 1920's to define a now famous infinite-valued logic $\mathcal{L}_\infty$;[47] more accurately, the structure given by this definition in conjunction with the other features just mentioned coincides with the 'infinite valued predicate calculus' which is the natural extension of $\mathcal{L}_\infty$ to include quantifiers. Both $\mathcal{L}_\infty$ and this extension–which we may as well also denote by $\mathcal{L}_\infty$–have received a great deal of attention in the mathematical literature. It is unnecessary to review this work here.[48] The important fact–which has been clearly brought out by the above discussion–is that $\mathcal{L}_\infty$ *plays the same role for arbitrary speakers as classical logic does for classical speakers*. In particular, any omniscient speaker may use the logical identities and rules of inference of $\mathcal{L}_\infty$ in deciding whether a sentence in the primary language is true for him. (He may also use 12.4 with 12.7 or a procedure based on tableaux that represent dialogues.) Space does not permit a detailed discussion of $\mathcal{L}_\infty$ here; the $\mathcal{L}_\infty$ logical identities are of course all valid in classical logic, but there are many classical logical identities (e.g. $P \vee \neg P$ that do not hold in $\mathcal{L}_\infty$.

As in the classical case so also in general, a speaker occurring in practice is not omniscient: A physicist S, for instance, who is concerned with dispersive elementary experiments, is not in practice prepared to lay down (and bet on) a precise probability for every such experiment. How does he behave? Just as the typical mathematician (or 'classical' physicist) claims that each atomic sentence is 'really' true or false, though he may not know which (see Sec. 11), so S talks as though every elementary experiment had some definite 'objective probability' of yielding the outcome 'no', although he has at most partial information about these probabilities.

We can explain this behaviour in the same way as in the classical case (Sec. 11). Indeed, S's conduct suggests that he envisages a deity S', 'more knowledgeable' (see Sec. 11) than S himself, who knows everything. (There is no contradiction in this, since such an S' exists by 11.2.) The said 'objective probabilities' may then be interpreted as those arising from the probability assignment which (by 12.3) corresponds to the omniscient speaker S'. In this way we can regard the locution, "objective probability", as a device which allows S to express information about his beliefs; for instance, if he says that --- are possible probabilities for certain experiments, we would understand him to mean that for some S' with S' ≥ S the corresponding probability assignment yields the probabilities ---. This argument may be regarded as providing a justification for the use of objective probabilities; it does not show that there is any such thing as objective probability, but it does show why it is useful to *pretend* that there is.

It is now natural to extend Definition 11.3 to the nonclassical case:

13.1. DEFINITION.[49] Let P be a sentence in the primary language and S a nonclassical speaker. Then P *is really true for* S means P *is true for every omniscient speaker* S' *with* S' ≥ S.

The set of all sentences which are really true for a speaker contains all L_∞ logical identities and is closed under the rules of inference for $Ł_\infty$.[44] As in the classical case, this set completely determines the set $\mathscr{A}$ of all acceptable final positions. There are thus two distinct ways in which S can convey information about his beliefs: He may either say something about the objective probability assignments he considers possible, or he may name some primary sentences that he considers 'really true'.

If, as is claimed above, $Ł_\infty$ plays the same role for dispersive theories as classical logic does for dispersion-free theories, why is it that in current formulations of theories like statistical mechanics and quantum mechanics Łukasiewicz logic plays no part? The reason is that there are two kinds of formulation in the nonclassical case, corresponding to the two ways mentioned above in which a nonclassical speaker can indicate his belief. To understand this let us first see how the discussion at the end of Section 11 is affected in the case of a nonclassical speaker. We saw that any theory, however formulated, has the effect of distinguishing as *theoretically true* a certain set $\mathscr{T}$ of sentences in the primary language. By 11.4 an *adherent* of the theory is one for whom these sentences are really true, where 'really true' is now defined as in 13.1. As in the paragraphs following 11.4 we can, without loss of generality, assume that $\mathscr{T}$ contains also every sentence that is really true for every adherent of the theory; but this time, since we must include nonclassical adherents, we cannot deduce that $\mathscr{T}$ is a *classical* theory but only that it contains all L_∞ logical identities and is closed under the rules of inference of L_∞; one might call it an 'L_∞-theory'.[44] The notion of a *primary formulation* can now be introduced just as in the classical case: it consists in distinguishing certain sentences in $\mathscr{T}$ as *proper axioms* in such a way that every theoretically true sentence is a logical consequence, *using the logic L_∞*, of these axioms.

As in the case of a classical theory, a primary formulation can be a 'satisfactory' one (Section 3), since the axioms are expressed in the primary language. However, no concrete example of a primary formulation of a dispersive theory has yet been obtained; to produce one, if possible for quantum mechanics, is a major outstanding problem. Apart from the choice of axioms, which always requires some cogitation, the main difficulty is the greater complexity and unfamiliarity of Łukasiewicz logic. We shall return to this problem at the end of this section.

Current formulations of quantum mechanics avoid these difficulties essentially by adopting the second way mentioned above in which a nonclassical speaker can indicate his beliefs: namely, by reference to objective probabilities. The concept of probability is adopted (along with other concepts, for instance *state* and *observable*) as a primitive concept for the deductive part of the theory. Insofar as *objective probability*–like *absolute truth* (or indeed *truth*), from which it is derived–is a 'metaconcept' of the primary language, the secondary language (in which the deductive theory is expressed) may here be regarded as a *metalanguage*: It makes assertions *about* the primary language, in particular laying down conditions on possible probability assignments (or risk values) and hence indirectly determining which primary sentences are theoretically true. Since the secondary language is therefore concerned only with *expressions* in the primary language, and since it is assumed that everyone can read (so that any 'elementary experiment' involving only syntactic operations on expressions is dispersion-free) it follows that the use of classical logic is admissible in connection with the secondary language. For this reason a formulation using probability is technically simpler to construct than a primary formulation. On the other hand, since the axioms are no longer expressed in the primary language they are unlikely to have a simple physical meaning, so one would not expect to attain a 'satisfactory' formulation in this way.

As was mentioned above, the main difficulty in constructing a primary formulation in the dispersive case is the fact that $Ł_\infty$, rather than classical logic, is involved. Now, the principal way in which logic enters into a formulation of a physical theory is via mathematics; mathematics is used in the development of the theory and mathematics depends on classical logic. This is true even in the case of a primary formulation of a classical theory; for instance, the formulation of thermodynamics in [8] involves nontrivial mathematics.

It seems clear that for the effective construction of primary formulations for dispersive theories it will be necessary to develop an '$Ł_\infty$-mathematics', so to speak–a structure which is related to $Ł_\infty$ in the same way as classical mathematics is related to classical logic.

It is not difficult to see how this should be done [15]. Ordinary mathematics is based on the concept of a set, and the notion of a proposition is a derived one. In fact the basic atomic proposition is the assertion $t \in A$ that a given element t belongs to a given set A. Of course, this proposition depends on t; it might be denoted $P(t)$. Before we can change the logic we must reorganize the subject with 'proposition' as a primitive concept, 'set' being derived. We assume given an expression $P(x)$ which becomes a proposition $P(t)$ whenever the variable x is replaced by a constant term t. (P is essentially a *property*, and $P(t)$ asserts that the element t has the property P). Then we take as a typical basic set the set $A = \{x: P(x)\}$ of all objects that have property P and *define* $t \in A$ to mean $P(t)$. In this way each set is defined in logical terms.

If we now replace classical propositions by propositions in the sense of $Ł_\infty$–i.e. having, for any speaker, truth values in the interval $[0, 1]$–we obtain the beginnings of the new mathematics. A characteristic feature is that sets are now 'fuzzy': an element may 'partially belong' to a set; for instance, if $t \in A$ has truth value p we might say that the element t belongs to the set A 'with probability p' (or 'to the degree p').

At this point the work connects with two other independent enterprises: first the purely mathematical work of Chang [6] and Klaua [20] on many-valued set theory; secondly, the stream of papers on 'fuzzy set theory' initiated by Zadeh [34], which have the practical aim of providing a framework for the discussion of problems–for instance in the 'inexact sciences'–which are concerned with vaguely defined concepts. (That the present work should have application in the latter field is not surprising, in view of the everyday examples of dispersive elementary experiments mentioned in Section 8.)

It seems to be possible to generalize many of the structures of classical mathematics to this 'fuzzy mathematics' based on L_∞. A start on this work has been made in [15], but a vast amount remains to be done. The main difficulty is that – since formal sentences that are logically equivalent in classical logic are not usually equivalent in L_∞ – each classical structure has many L_∞-generalizations, very few of which are viable.

Queen's University,
Kingston, Ontario.

NOTES

[1] The partial ordering of physical concepts described here corresponds closely to the order in which concepts appear in consciousness during development from birth. The infant experiences only direct concepts and learns, or rather *constructs*, indirect concepts from these as his mind develops. This process, proceeding initially at a subconscious level, may approach consciousness around school age, but becomes fully conscious only for the esoteric concepts created in scientific research.

[2] The notion of ostensive definition is discussed in some detail by Przełecki in [28]. See also [31].

[3] I use 'expression' here in a semi-technical sense which includes both the 'constant terms' and the 'atomic sentences' of mathematical logic. 'Term' would be a more convenient word, but to use it would conflict with current practice.

[4] It is in this way that a child learns the meaning of, for instance, *apposition* of a noun and a verb: by learning ostensively the meaning of simple sentences like 'John walks', 'Bob walks', 'John kicks', and so on, as well as those of the constituent terms 'John', 'Bob', 'walks', 'kicks', etc.

[5] This account does not cover the use of variables, but it is adequate for our immediate purpose.

[6] Essentially this claim seems to underlie one of the arguments (see, for instance, Braithwaite [4], Chapter 3) against completely interpreted theories – i.e. those in which every primitive concept has a direct experimental meaning.

[7] It will generally be a sublanguage, for the primary language must of necessity contain a wealth of constant terms – enough to name every individial object to which the theory may be applied – whereas in the theoretical language constant terms are often entirely absent. In elementary geometry, for instance, letters A, B, C, ... stand for *arbitrary* points, not for *particular* points: i.e. they are *variables*, not constants.

[8] It should not be inferred that every state is to be given an independent ostensive definition; this would be impossible since there are an infinite number of states (see

Section 2). How it is nevertheless possible to assign meanings to all these states was indicated in Section 2 and will be shown in detail in Section 7.

[9] The concept of a *mixed state* plays a fundamental role in von Neumann's book [27]. Like most physicists since, he uses neither the operational interpretation nor the objective interpretation but what might be called the *ensemble interpretation*, in which a mixed state represents the statistical properties of an imaginary ensemble of systems. Although it is unconventional, there seems to be no reason why a mixed state should not be assigned to an individual system even in the objective interpretation. This would allow a terminology in close correspondence with the operational interpretation.

[10] If x and y are states, the mixed state $\frac{1}{2}(x + y)$ may be realized as follows (operational interpretation): *spin a coin; if 'heads' carry out the procedure x, if 'tails' carry out y.* Exactly the same applies with 'test' replacing 'state'. Mixed tests were introduced in [10], and independently (with a different terminology) in [23]. They have a simple mathematical representation (explained below) in quantum mechanics, but are ignored by the quantum logicians. They cannot be accommodated in the lattice on which 'quantum logic' is based: this lattice consists of the pure tests; the set of all (pure or mixed) tests is a partially ordered set[42] but does not form a lattice.

[11] Not in the sense that $\|P - P'\|$ is small, but rather in that P and P' yield approximately the same expectation values for 'most' physically accessible states.

[12] A defect of the quantum logic approach, as a *foundation* for quantum mechanics, is that it is based on *pure* tests. It is not possible to extend it to include mixed tests for the reason explained in footnote [10].

[13] For this reason it seems to me that orthodox model theory is not directly applicable to the problem of interpretation of physical theories.

[14] An atomic sentence is usually defined as an expression without logical symbols to which a truth value (*true* or *false*) may be attached. In this example, and in mathematics generally, the two definitions conform: the sentence is taken to be true if the outcome of the experiment is 'yes'. But in other applications, and in particular in physics, the conventional definition fails since the experiment may not always give the same outcome. The definition given in the text, on the other hand, can be taken over unchanged.

[15] An arbitrary weighted mean of two mixed states (or tests) x and y can be given a direct empirical interpretation in the same way as is done for $\frac{1}{2}(x + y)$ in footnote 10. See also [10].

[16] We are not using the pure state philosophy! See Section 5.

[17] This is unrealistic and contrary to the principles developed in Section 3. We should instead accept the situation for what it is. Indeed, in testing the assertions of the quantum mechanical theory one would in practice use a source that emits a photon not with certainty but with a probability ε sufficiently small that the probability of two photons being emitted on the same occasion is negligible. However, a realistic interpretation along these lines does not directly provide *any* interpretation of the notion of a single-photon state. It can therefore be implemented only after the theory has been modified, this notion being replaced by some more direct concept, such as that of a weak beam of light. Naturally, the theory itself will be significantly altered in this process. Since our present purpose is not to present a conclusive formulation of quantum mechanics but simply to illustrate in a practical case the construction and interpretation of a formal primary language it seems better to make the assumption given in the text. Most of the considerations which arise here may be expected to recur, *mutatis mutandis*, in any improved treatment.

[18] Another unrealistic assumption made for the sake of simplicity. In fact, no pure state can be precisely realized (see Sec. 5). In the more realistic treatment indicated in the previous footnote there is no need to make this assumption. Indeed, one can then conveniently take as the sole primitive 'state' that of no polarization, and this *is* realistic since such a 'state' is directly available in the form of black body radiation.

[19] Imagine, for instance, a black box containing (a) an ordinary filter that rotates the plane of polarization of light by 90° and (b) a random number generator. The box is equipped with a button, to be pressed on each occasion of use, that causes a random number to be (secretly) selected and (secretly) interposes the filter in the path of the next photon iff the number is even. Using this box in conjunction with the source S we have a device which 'produces vertically or horizontally polarized photons with equal probability'. The output is represented by the centre of the Poincaré sphere: i.e. the device provides a method of preparation for the unpolarized state. Though in principle admissible, such 'boxes' are complicated and unnatural. Note that they are unnecessary in connection with the realistic approach mentioned in the last two footnotes. There the primitive 'state' is an unpolarized beam, and a (weaker) beam of any desired degree of polarization is easily produced by a suitable (absorbing) filter, e.g. a Nicol prism or a polarizing film.

[20] This is another idealization that should be avoided in a more thorough treatment (see the previous footnotes[17,18]). In fact, a typical laboratory detector (e.g. a photomultiplier) responds with an efficiency that is, for each state of polarization, somewhere between 0 and 1 and is usually relatively insensitive to polarization. Such a detector realizes a mixed test (see Section 5) represented by a point lying in the *interior* of the region T (in polarization space) that represents all tests. Indeed, in principle, every test that is represented by a point in the interior of this region can be realized in practice, although the difficulty in doing so increases as the boundary of the region is approached.

[21] Of course, this is simplistic; there are bound to be borderline cases. But we suppose that until they encounter such a case the participants behave as though none were to be expected.

[22] We will frequently denote an atomic sentence and the corresponding elementary experiment by the same symbol.

[23] To be more precise we should recognize that – since the future is not available – the property of being dispersion-free is a relative one. An experiment E may be dispersion free 'for' (i.e. in the opinion of) one speaker and not for another. Similarly, an elementary experiment may be dispersion-free 'for' a particular theory. The statement that *an elementary experiment is dispersion-free* should be understood to mean that it is so for every speaker. (There is no practical value in assigning this statement an objective meaning since it could then never be verified.)

[24] At least, this is clear provided that the chance of an 'error' occurring in the calculation is negligible, and in this example this condition can easily be satisfied. But it should be noted that this condition can never hold *uniformly* for all atomic sentences. No matter how precisely the interpretation of the primitive symbols in the language is described there will always be atomic sentences so long that the cumulative chance of 'making an error' in carrying out the indicated experiment (here computation) is greater than (say) 1/3. Thus dispersion will always manifest itself in the limit of very long atomic sentences. The investigation of this situation is an open problem, not contemplated in orthodox mathematical logic.

[25] Naturally, in this and other examples the account given should not be regarded as

definitive, but only as indicating an experiment whose complete description would require much more elaboration.

[26] For similar reasons to those given in footnote 23 this term should be regarded as denoting a subjective concept. Its value or truth-value may be different for different speakers. An unqualified value (truth-value) will be given only when all speakers concur in their opinions.

[27] This example is taken from [12] by kind permission of the editor.

[28] I use 'proposition' and 'sentence' essentially as synonyms; 'proposition' refers to the semantic aspects, while 'sentence' relates to the syntax.

[29] For simplicity I assume that (as in the case of a standard interpreted language) there is a constant term for every object to which the language refers. The syntax of variables and quantifiers may be found in any standard text on mathematical logic, for instance [26]. 'Iff' means 'if and only if'.

[30] Of course, the conventional response here is to regard each trial as corresponding to a distinct proposition; after the first trial one can assert only that the proposition corresponding to *that* trial is true. Naturally, it is possible to take this view; but unless one abandons the correspondence between sentences and propositions this means that an atomic sentence must specify *one particular trial*. This, in turn, involves cluttering the language with references to absolute time (epoch) and position – references which are extraneous, in that nothing similar occurs in the theorems of the theory in question,[31] and which also involve additional difficulties of interpretation since they must refer to some standard reference frame, which must (presumably) be defined ostensively. Moreover, this approach does not begin in any way to solve the problem of the formalization and interpretation of probability statements; indeed, from this viewpoint it becomes unnatural to associate *any* probability with a sentence.

[31] The situation regarding trials here is not the same as that regarding numerals in the example of Section 6. There, although the theory makes no explicit reference to particular numerals, implicit reference is made in that all occurrences of a variable (in the scope of a given quantifier) refer to the "same" numeral.

[32] The dollar here is really being used as a unit of utility, which suggests that the definition of the logic depends on the concept of utility. Since utility is normally defined in terms of probability and since one of the products of the new logic is a definition of probability, it seems that a vicious circle is involved. This is not really the case, however, since everything we assume about the commitment incurred in asserting an atomic sentence is incorporated in the definition of rationality given below (Def. 11.1). The present informal introduction and the notation, $1, merely provides motivation for that definition.

[33] The term 'opponent' used here should be understood as follows. When a speaker makes an initial assertion it is at first to be regarded as an offer to assume the indicated commitment. Any other speaker may respond to this offer; he says, perhaps, "I doubt that.' He thereupon becomes the *opponent* for this particular transaction, while the first speaker is the *proponent*. If the initial assertion is atomic the transaction concludes with a trial of the appropriate elementary experiment; in other cases (see Sec. 10) a debate intervenes between it and the necessary trials.

[34] I.e. as an *initial* assertion (cf. footnote 33); atomic sentences are frequently asserted in the course of debate, as we shall see later.

[35] However, it will sometimes be convenient to describe a proposition as *true* (without

qualification) if it is true for every speaker that we propose to consider. (Cf. footnote 26.)

[36] The question of the most appropriate choice of rules of trial has yet to be settled; when this has been done it may be possible to relax Assumption 10.8. At the time of writing (summer 1975) an effective set of rules of order has been selected [17] and the structure of the resulting debates and the corresponding logic (see Sec. 13) has been analysed in depth by Liddell [21].

[37] One should not assume that this notation will be suitable for the discussion of rules of order, since such rules – which become necessary in the absence of Assumption 10.8 – may well reflect not only the currently asserted propositions but also the circumstances (in the debate) under which they were asserted.

[38] In the case when the proposition P contains quantifiers a slight modification in the definition is necessary: We say P *is true for* S (written $S \models P$) iff S is willing to assert P *for an arbitrarily small fee*; or, equivalently, iff S can conduct the ensuing debate in such a way as to reach a final position which is 'arbitrarily nearly acceptable'; or (again equivalently) iff, for \$1, S is willing to assert P arbitrarily often: i.e. iff, for an arbitrarily large integer n, S can always reach an acceptable final position from the initial position $F \mid nP$. This modification is technically convenient and makes no practical difference.

[39] Subject to the qualifications in footnote 37 this holds also when rules of order have been laid down.

[40] This was given as an axiom in [13].

[41] Cf. the *principle of omniscience* of Bishop [3].

[42] $\geq$ is a *partial ordering* on a set iff, for all elements a, b, c, (a) $a \geq b$ and $b \geq c$ implies $a \geq c$, (b) $a \geq b$ and $b \geq a$ implies $a = b$, and (c) $a \geq a$.

[43] Strictly speaking, the condition (b) for omniscience cannot be established for *all* sets but (roughly) only for those to which quantifiers can be formally applied.

[44] Such a set is called a *theory* in mathematical logic; see e.g. [26]. Of course, the notion depends on the underlying logic; in Section 11 and in the textbooks it is classical logic, but in Section 13 it is L_∞.

[45] sup $\{...\}$, for any set of numbers $\{...\}$, denotes the *greatest element* of the set when the set is finite, and its *least upper bound* when it is infinite. Similarly, inf denotes the *least element* or the *greatest lower bound*.

[46] Note that the method of proof of this theorem allows the question of rules of order to be bypassed; the theorem continues to hold no matter what rules of order may be adjoined. Thus rules of order are irrelevant for any omniscient speaker.

[47] In fact, Łukasiewicz's definition yields an n-valued logic L_n for each $n = 2, 3, ...$ and ∞. L_2 is classical logic. L_3 would correspond to the class of all omniscient speakers who (behave as though they) assign to each elementary experiment a subjective probability whose value is precisely 0, $\frac{1}{2}$, or 1. There seems little point in considering this rather strange 'philosophical attitude'. Similarly for L_n with $n = 4, 5, ...$ L_∞, on the other hand, implies no restriction on the probability assignments.

[48] For reviews of the work on the propositional calculus L_∞ see [1] and [29]. For the corresponding predicate calculus see, for instance, [2].

[49] In the case when S is classical this definition does *not* reduce to Definition 11.3. This is because the fact that a speaker is classical is not necessarily revealed by his set of acceptable final positions – not, at least, without further elaboration of the language.

Postscript. Since preparing this paper I have noticed that there is a one to one correspondence between the 'classical speakers' as defined here and the 'supervaluations' of Van

Fraassen [32]: if S is any classical speaker there is a supervaluation s such that, for every sentence A, A is 'really true' for S iff $s(A) = T$, and A is 'really false' for S iff $s(A) = F$; and every supervaluation arises in this way from some classical speaker. Thus classical speakers and supervaluations are really the same thing. Similarly, the 'arbitrary speakers' of Section 13 may be identified with '$Ł_\infty$-supervaluations', the latter being defined by the following specialization of Van Fraassen's definition of supervaluation ([32] page 95): *An $Ł_\infty$-supervaluation s is a valuation* (in Van Fraassen's sense ([32] p. 31): i.e. a mapping that assigns T to some sentences and/or F to some sentences) *such that there is a nonempty set K of risk functions (or, equivalently, L_∞-valuations) such that, for any sentence A, A is true for s iff $v(A) = 0$ for all v in K.* In fact, this is the essential content of [13] Theorem 5.17.

REFERENCES

[1] Ackermann, R., *An introduction to many-valued logics*, Dover, New York, 1967.
[2] Belluce, L. P., and Chang, C. C., 'A weak completeness theorem for infinite-valued first order logic', *J. Symbolic Logic* **28** (1963) 43–50.
[3] Bishop, E., *Foundations of constructive analysis*, McGraw-Hill, New York, 1969.
[4] Braithwaite, R. B., *Scientific explanation*, Cambridge University Press, Cambridge, 1953.
[5] Caratheodory, C., 'Grundlagen der Thermodynamik', *Mathematische Annalen* **67** (1909) 355–386.
[6] Chang, C. C., 'Infinite-valued logic as a basis for set theory', *Proc. 1964 Int. Congress for Logic, Methodology, and Philosophy of Science*, North-Holland, Amsterdam, 1965.
[7] Dixmier, J., *Les algèbres d'opérateurs dans l'espace Hilbertien* (second ed.), Gauthier-Villars, Paris. 1969.
[8] Giles, R., *Mathematical foundations of thermodynamics*, Pergamon, Oxford, 1964.
[9] Giles, R., 'Foundations for quantum statistics', *Journal of Mathematical Physics* **9** (1968) 359–371.
[10] Giles, R., 'Foundations for quantum mechanics', *Journal of Mathematical Physics* **11** (1970) 2139–2160.
[11] Giles, R., 'A nonclassical logic for physics', *Studia Logica* **33** (1974) 397–415. (This is a shortened version of [12].)
[12] Giles, R., 'A nonclassical logic for physics', in R. Wójcicki (ed.), *Selected papers on Łukasiewicz sentential calculi*, Ossolineum, Warsaw, 1977, 13–51.
[13] Giles, R., 'A logic for subjective belief', in Harper, W. L. and Hooker, C. A. (eds.), *Foundations of probability theory, statistical inference, and statistical theories of science*, Vol. 1, Reidel, Dordrecht, 1976, 41–70.
[14] Giles, R., 'A pragmatic approach to the formalization of empirical theories', in *Proc. conf. on formal methods in the methodology of empirical sciences, Warsaw, June 1974*, Ossolineum, Warsaw, and Reidel, Dordrecht, 1976, 113–135.
[15] Giles, R., 'Łukasiewicz logic and fuzzy set theory', *Proceedings of the 1975 international symposium on multiple-valued logic, Bloomington, Indiana*, pp. 197–211, I.E.E.E. 1975. Reprinted with minor changes in *International Journal of Man-Machine Studies* **8** (1976), 313–327.

[16] Giles, R., Comment on the paper by Dana Scott, in S. Körner (ed.), *Philosophy of Logic*, Blackwell, Oxford, 1976, 92–95.

[17] Giles, R., unpublished.

[18] Hintikka, J., *Logic, language-games, and information*, Oxford University Press, Oxford, 1973.

[19] Jauch, J. M., *Foundations of quantum mechanics*, Addison-Wesley, Reading, Mass., 1968.

[20] Klaua, D., 'Grundbegriffe einer mehrwertigen mengenlehre', *Monatsb. Deutsch. Akad. Wiss. Berlin* 8 (1966) 782–802 (and other papers).

[21] Liddell, G. F., 'A logic based on game theory' (Ph.D. thesis) Queen's University, Kingston, Canada, 1975.

[22] Lorenzen, P., *Metamathematik*, Bibliographisches Institut, Mannheim, 1962.

[23] Lubkin, E., 'Theory of multibin tests', *Journal of Mathematical Physics* 15 (1974) 663–672.

[24] Mackey, G. W., *Mathematical foundations of quantum mechanics*, Benjamin, New York, 1963.

[25] McKinsey, J. C., Sugar, A. C., and Suppes, P., 'Axiomatic foundations of classical particle mechanics', *Journal of Rational Mechanics and Analysis* 2 (1953) 253–272.

[26] Mendelson, E., *Introduction to mathematical logic*, Van Nostrand, Princeton, N. J., 1964.

[27] Merzbacher, E., *Quantum mechanics* (second ed.), Wiley, New York 1970.

[28] Przelecki, M., *The logic of empirical theories*, Routledge and Kegan Paul, London, 1969.

[29] Rescher, N., *Many-valued logic*, McGraw-Hill, New York, 1969.

[30] Robson, B. A., *The theory of polarization phenomena*, Clarendon Press, Oxford, 1974.

[31] Russell, B., *An enquiry into meaning and truth*, Norton, New York, 1940.

[32] Van Fraassen, B. C., *Formal semantics and logic*, Macmillan, New York, 1971.

[33] Varadarajan, V. S., *Geometry of quantum theory*, Vol. 1, Van Nostrand, Princeton, N.J., 1968.

[34] Von Neumann, J., *Mathematical foundations of quantum mechanics*, Princeton University Press, Princeton, N.J., 1955.

[35] Zadeh, L. A., 'Fuzzy sets', *Information and Control* 8 (1965) 338–353.

M. KUPCZYŃSKI*

IS THE HILBERT SPACE LANGUAGE TOO RICH?

ABSTRACT In order to answer this question, we analyse different phenomena occurring
in general experimental set-ups arranged to analyse the properties of some unknown
beams of particles. We arrive at the conclusion that sometimes the Hilbert space language appears to be too rich and also that there are some phenomena where the notion
of transition probability disappears and any attempt to introduce it leads to the possibility of infinitely many inequivalent descriptions. Our analysis encouraged us to ask
the question whether the Hilbert space language is not too rich in the more realistic
situations, for example to deal with high-energy elementary particle scattering phenomena. A programme of investigations in that direction is formulated.

In the polemic with axiomatic quantum mechanics it is shown that the pure state concept can be formulated independently of the existence of any maximal filter.

1. INTRODUCTION

Axiomatic quantum mechanics aimed to prove the uniqueness of the quantum mechanical Hilbert space description for all future phenomena. The
efforts were concentrated on a search for such a set of axioms, concerning
the general structure of the propositions which can be said about the
physical systems, which would imply the usual Hilbert space or algebraic
representation.

The investigations started by Birkhoff & von Neumann (1936), and continued in many other papers (Dähn, 1968, Finkelstein, 1963; Finkelstein,
Jauch, Schimonovich and Speiser, 1962, 1963; Gunson, 1967; Jauch,
1964, 1968; Jauch and Piron, 1963; Ludwig, 1967; Mackey, 1963; Mielnik,
1968, 1969; Piron, 1964; Pool, 1968a, b) led to different axiomatisation
schemes with the required properties. Though some of the accepted axioms
did not seem to be natural, the general belief is that the problem is solved
and that we can freely use the usual Hilbert space language in future. One
can claim that it is not true, because we have to deal with the rigged Hilbert
spaces and because for the continuous spectrum the eigenvectors are only
the distributions acting on some nuclear space ϕ, but in all practical cases

---

*On leave of absence from Institute for Theoretical Physics, Warsaw University,
Poland.

89

Hooker (ed.). Physical Theory as Logico-Operational Structure, 89–113.
All Rights Reserved.
Copyright © 1978 by D. Reidel Publishing Company, Dordrecht, Holland.

we can use wave packets and regularised fields to obtain measurable results in the framework of some Hilbert space.

Axiomatic quantum mechanics was vigorously attacked by Mielnik (1968, 1969), who claimed to prove that the Hilbert space description is only one degenerate case of the infinitely many non-Hilbertian quantum worlds to be observed in the future. In this paper we show that such conclusions are not well. Mielnik's analysis is based on the quite unrealistic assumption that the physical transition probabilities between some pure states are equal to the static transmission probabilities between two maximal filters used for the preparation of these states.

In our opinion, the approach of axiomatic quantum mechanics is too general to give insight into some specific physical phenomena which can appear in different experimental set-ups. For this reason, we analyse the general experimental set-up E which can be used to investigate the phenomena characterising the ensembles of particle-beams. We assume that our experimental set-up E can consist of the following devices:

(i)   the sources S which produce beams B;
(ii)   the filters F which allow the division of beams into sub-beams having some common properties;
(iii)   the transmitters T which change a beam b into a beam b_T;
(iv)   the detectors D_P which register the intensity of the beams having a property p;
(v)   the instruments I. A beam b enters into an instrument, the instrument measures some property, and a beam b goes out.

However, two observers investigating the same beam, but equipped with a different set of devices, can observe different phenomena and discover different mathematical schemes to describe them. Keeping this possibility in mind, we have been trying to analyse the different cases of the experimental set-up E, differing by the richness of the beams and the devices.

A careful analysis leads us to new definitions of filters, pure ensembles and to the important conclusion that in any considered case the Hilbert space description turns out to be possible. However, sometimes the data do not allow the extraction of the transition probabilities in a unique way, so it is more reasonable to abandon the Hilbert space description and to try to explain a causal evolution of the whole ensembles. Another feature which appears in our analysis is the fact that only some vectors and some scalar products in the Hilbert space description of the phenomena have a physical meaning; so, in some way, the Hilbert space language is too rich.

The 'too rich' language makes possible that using more or less phenomenological models we can always (in no unique way) explain the data without really broadening the understanding of them. The above-mentioned successes in the explanation of the data deepen the belief in the basic and unchangeable character of the language used and build a psychological barrier, making the discovery of a new, more economic and less ambiguous, language much more difficult.

All these considerations encouraged us to raise the important question whether the Hilbert space language is not too rich to explain the observed physical phenomena, for example in high-energy elementary-particle physics. A natural question arises: how could we find out whether this is the case? Although it is evident now that we cannot assign to all vectors in the Hilbert Fock space the physically realised states of the elementary particles system and that not all scalar products can be practically measured, yet it does not mean that the Hilbert space language is too rich. Similarly, in classical mechanics not every solution of an arbitrary Newton equation has a practical meaning and this does not mean that the language of classical physics is inappropriate.

To show that the Hilbert space language is too rich to deal with the scattering phenomena of elementary particles, we would have to show that, for example, the assumption of the unitary S matrix (which is derived using the assumption that any vector in the Hilbert state can be taken as an initial state) is violated. For example, we would have to find two such initial realisable states $|i_1\rangle$ and $|i_2\rangle$, which in our formalism must be represented by the orthogonal vectors, and show that the states $|Si_1\rangle$ and $|Si_2\rangle$ cannot be represented by the orthogonal vectors in the Hilbert space.

A careful analysis of these problems will be continued in the subsequent paper.

2. GENERAL EXPERIMENTAL SET-UPS

At first we shall try to be as general as possible, so we shall consider two sets of objects: a set of sources and a set of devices. The interactions between beams, produced by the sources, and the devices give us the information about them which can enable us to make physics. In general, the information obtained is not unambiguous, so one has to accept some additional interpretational assumptions.

Usually one acts in a different way. Wanting to investigate beams, one constructs some devices based on the knowledge of classical and quantum physics. Such devices make possible the description of the unknown beams in terms of the quantities known before (like mass, momentum, energy, charge, spin, etc.). Such an approach is very reasonable, since it assumes the continuity of the science which worked so well before. However, let us cite Bohr (1961): 'The main point to realise is that a knowledge presents itself within a conceptual framework adapted to account for previous experience and that any such frame may prove too narrow to comprehend new experiences . . .' and '. . . when speaking of a conceptual framework we refer merely to the *unambiguous* logical representation of relations between experiences . . .'.

Keeping this in mind, we now forget about our science and we assume that we know nearly nothing about the sources and the devices. We want to investigate the problem of how we should deal with that case and what kind of language could be used to describe the observed phenomena. Let us start with some statements:

STATEMENT 1. The sources S and all the devices used in the experimental set-up are given *a priori*. They should have the very important feature of reconstructability, by which we understand that identical set-ups can be constructed in any other laboratory at any time.

STATEMENT 2. Among all the devices, we must have a counter g of quanta which must be used to find the intensity of the beams. This counter is at least one 'classical' device which is necessary to make quantitative 'quantum' physics. The problem is to construct such a counter for unknown beams; but let us assume that we have it. It need not be an absolute counter, like in Piron (1964), but it should be the most sensitive one available.

STATEMENT 3. Using the counter g, we can observe the changes of the beam intensities after their interactions with the devices. Those interactions give us the information about the beams and the devices we have. This information allows us to classify the beams and to find among the devices such objects as filters, transmitters, etc. Of course, the information about the beams depends essentially on the devices used, and vice versa the information about the devices depends essentially on the beams which we have at our disposal. So everything we know is to a large extent relative, and we can never be sure that in the future we shall not discover other beams and devices which will change the interpretation of some old phenomena or which will force us to find a new theoretical language to describe the new ones. Similar views are contained in Mielnik (1968). In many cases it seems to be improbable, but it cannot be excluded.

STATEMENT 4. Having some knowledge of the beams and devices, we have to choose some of them for further analysis. The chosen devices can be divided into two groups:

(a)   preparatory and analysing devices,
(b)   transmitters.

Such a division corresponds to three stages of the experiment in which they will be used:

(i)   preparation,
(ii)   transmission,
(iii)   detection.

STATEMENT 5. In the preparation stage we classify and prepare the beams. We introduce the concept of pure and mixed beams. Knowing the properties of the pure beams, we can ascribe states to them. Thus the preparatory stage enables us to find a set of initial states whose change in the transmission stage we should try to explain.

STATEMENT 6. In the transmission stage we let our beams go through some chosen devices called transmitters. By transmitters we can understand also the action of the external fields. If the beam was under the influence of the external field for the time Δt before detection, we can say that it went through the transmitter $T(\Delta t)$. Thus we can have the approximately instantaneous change of the beam or we can observe its more or less continuous evolution.

STATEMENT 7. After the transmission stage we classify obtained beams and we try to find some mathematical language and a model allowing the interpretation of the observed regularities.

Now we have to find out the meaning of some terms which appeared in these statements. It turns out that the definition of filters and pure states is not obvious. We cannot see the details of the transmission process as it looks inside a device. Therefore what we know is the change of intensity of the incoming beam.

Before starting a more detailed discussion, we must add some assumptions in the spirit of Statement 1. We have to work with an approximately stable source, since in order to make predictions we have to know the intensities of the beams relying on previous measurements. We must also assume that our devices have no memory and act in the 'same' way on the 'same' beams. So in fact we are always dealing with ensembles $\tilde{b}$ of the identically prepared beams b. Performing many experiments we find the properties of the ensembles $\tilde{b}$ and often we can ascribe them to every beam of the ensembles $\tilde{b}$. If it is possible, we shall talk about the beams and their properties instead of talking about the ensembles. In our case we cannot always prepare arbitrary mixtures of the produced beams. First we have to check whether the beams behave in a 'classical' or in a 'quantum' way. We now make a short digression about such behaviour, which will be short repetition of well-known things.

If the beams and devices behave classically, then each quantum of the beam can be characterised by some properties, possessed in an attributive way, which can be found with the help of measurements. These measurements can by no means change the properties of the quanta. A device d transparent only to the quanta having a property 'd' is called a classical filter. Such a device is of course idempotent, which means that it is neutral to all quanta to which it is transparent. If the quanta also have some other properties, we can construct, in principle, maximal filters–transparent only to the quanta having all properties the same. Pure beams are those which go through the maximal filters without change.

In quantum physics, a quantum can have a property with certainty, but only up to the moment of the measurement of another property which is incompatible with the first one. To discover the quantum behaviour one must show the incompatibility of some properties. For this purpose we must find at least two idempotent and incompatible devices d and l to perform the following experiment with a beam b. We transmit the beam b through the device d and obtain a beam b_d, for which a device d is transparent. Now we transmit the beam b_d through the device l and we obtain a beam b_{dl} of smaller intensity. Now the device l is transparent to the beam b_{dl}. Finally, we transmit the beam b_{dl} through the device d, obtaining a beam b_{dld}. If the beam $b_{dld} \neq b_{dl}$, then we can say that the beams and devices do not behave in a classical way.

If, on the other hand, $b_{dld} = b_{dl} = b_{ld} = b_{dldl}$ and other devices incompatible with d and l cannot be found in our experimental set-up, then we can say that the beam b_{dl} is pure, the devices l, d and $l \cdot d$ are classical filters and the device $l \cdot d$ is a maximal filter in our set (for simplicity we exclude the existence of other compatible more restrictive filters). Each quantum of the beam b_{dl} has two properties 'd' and 'l'–the filters d and l are transparent to it.

In the quantum case our classical picture of a filter has to be completely changed. We cannot say, as in Piron (1964), ' . . . quantum mechanical filter selects single particle properties'. All the quanta of a beam b_d have the same property 'd' (the device d is to them transparent), but after going through the device l not all of them can still have a property 'd'. So the idempotent device l does not select the quanta having a property 'l' but it only transforms with a probability $p(d, l)$ some quanta having a property 'd' into quanta having a property 'l' and absorbs ones which are not transformed.

We cannot explain this probabilistic approach assuming that the beam b_d is a mixture of the quanta labelled by hidden parameters ξ *constant in time* and that the device l is a classical transmitter which can change the property 'd' and the parameters ξ in a well-defined causal way depending on their initial values. The non-existence of hidden parameters of this kind was shown in a different formalised lange by Jauch and Piron (1963). So we have to assume that the devices l and d act in an intrinsically probabilistic way, but now we can ask whether it is possible to check that the beam b_d is a pure beam.

Let us, for example, assume that the beam b_d of average intensity I is a mixture of two beams of intensities I_1 and I_2 consisting of quanta A and B, respectively. Let the device l act in the following way:

> it transmits each quantum A with a probability a and changes it into a quantum C;
> it transmits each quantum B with a probability b and changes it into a quantum D;
> it transmits all the quanta D and C without change.

Let the device d act in a similar way:

> it transmits each quantum C with a probability c and changes it into a quantum A;
> it transmits each quantum D with a probability k and changes it into a quantum B;
> it transmits without change all quanta A and B.

The transmission probabilities $p(d, l)$ are strictly defined in the following way

(2.1) $p(d, l) = Ss(r)r(d, l) \, dr$

where S denotes a sum or an integral over all values of $r(d, l)$; $s(r)$ is nor-
malised to the unity probability distribution of the ratios $r(d, l)$; the ratios
$r(d, l) = I_l/I_d$ where I_d and I_l are the intensities of the beams b_d and b_{dl}
for all beams $b \in \tilde{b}$. All other probabilities met later are defined in a similar
way. The probabilities $p(d, l)$ and $p(l, d)$ must be the same for all pairs of
$d - l$ and $l - d$, respectively, occurring in the chain $d - l - d - l - d - l$ of
the experiments with the ensemble $\tilde{b}$. It gives us the constraints on the
possible values of a, b, c, k. Other constraints are obvious: $0 < I_1, I_2 < I$,
$I_1 + I_2 = I, 0 < a, b, c, k < 1$. If we analyse these constraints we come
to the following corollary.

COROLLARY. The probabilities a, b, c, k must satisfy the following con-
dition $a \cdot c = b \cdot k = w$; then $p_\lambda(d, l) = (a + b \cdot \lambda)/(1 + \lambda)$ and $p_\lambda(l, d) = w/p_\lambda(d, l)$ where $\lambda = I_2/I_1$. So for every two experimental numbers $p_\lambda(d, l)$
and $p_\lambda(l, d)$ we can adjust two parameters to make the above interpre-
tation possible.

However, as we see $p_\lambda(d, l)$ depends on the relative intensity λ of the
two hypothetical sub-beams; so we can verify our hypothesis by trying
to change λ in the beam b_d. We do not have any other way than to cause
the decrease of the intensity of the beams $b_d \in \tilde{b}_d$ by different methods.
If we do not obtain different values for $p_\lambda(d, l)$ and $p_\lambda(l, d)$, then we must
reject out hypothesis and state that it is not legitimate to assume that the
beams b_d consist of sub-beams, so they can be called pure. But by the expression 'a pure beam' we should not understand a beam consisting of
identical quanta, since the term 'identical' is classical and means 'behaving
in the same manner in all situations'. The devices l and d do not treat all
the quanta from the beams b_d and b in the same way. We cannot understand the mechanism of this differentiation and also usually we do not
observe separate stages of the transition. We just observe the behaviour
of the beam b_d as a whole and find the statistical regularities. Therefore,
in the theoretical analysis of the process we should not represent the
transmission of the beam b_d by a set of yes-no experiments with each quantum, but more properly we should talk about the properties of the beams
as a whole and about states of the beam instead of talking about the
states of the single quanta. In some situations it can happen that only the
states of ensembles $\tilde{b}$ have a precise meaning. We shall discuss such situations later. This wholeness of the physical phenomenon in the microworld
was wisely pointed out by Bohr (1961) many times.

We have spent so much time discussing the devices l and d because they behave in a way analogous to the behaviour of the tourmaline plates which are usually called filters. We wanted to show to what extent they are not classical, if discussed in terms of corpuscular language. Their filtering properties can be understood in the language associating a wave to each beam. We also wanted to get an intuition enabling us to define filters and pure beams in our poor information system.

DEFINITION 1. Filters are devices which:

(i)   are idempotent;

(ii)   for all beams $b \in B$ entering an arbitrary chain consisting of the filters $f_1, f_2, \ldots$, the transmission probabilities for each pair $f_i - f_j$ are constant; $p_b(f_i, f_j) = \text{const}_1(b)$, $p_b(f_j, f_i) = \text{const}_2(b)$;

(iii)   from all devices D satisfying the conditions (i) and (ii), for each device d we can find a set 0_d consisting of all devices l_j which have the same transmission probabilities to all other devices from D as a device d has, namely

$$0_d = \{l_j \in D; p_b(l_j, d_i) = p_b(d, d_i) \text{ for all } b \in B \text{ and all } d_i \in D\}$$

A filter is a minimally transparent element in the set 0_d. This means that $I_{b_f} \leq I_{bl_j}$ for all $l_j \in 0_d$ and all $b \in B$. If the minimally transparent element in 0_d cannot be found, then we call all the devices 0_d relative filters and to further analysis of the beams we choose one of them.

The long property (iii) enables us to differentiate between classical filters and some similar classical transmitters. Our definition of the filters is different from that given, for example, by Mielnik (1969), and many relative filters from his work are treated like normal filters, as they should be.

DEFINITION 2. Provisory maximal filters are the minimally transparent elements in maximal sets of the compatible filters.

DEFINITION 3. A provisory pure beam is one for which the provisory maximal filter is transparent. We use the term 'provisory' since we are not sure whether the set of filters which we have in E is a maximal one. In the transmission stage some provisory pure beams can behave in a way suggesting that they consist of the two sub-beams not separated in the preparation stage.

Usually it is assumed that two filters d and l are characterised by the transmission probability independent on b. In this paper we assume that the beams can be characterised by many properties and the same filters can be sensitive on different properties in a different way. For example, the filter d can be transparent to all beams having a property 'p_1' but reduce the intensity of the beams having a property 'p_2'. If b_1 denotes a beam with the property 'p_1' only, and b_2 denotes a beam with the property 'p_2' only, then it can happen that $p_{b_1}(d, l) \neq p_{b_2}(d, l)$.

Besides the transmission probabilities $p_b(f_i, f_j)$, which we shall denote in all practical cases $p(f_i, f_j)$, we also need the filtration probabilities

$$(2.2) \qquad p(b, f_i) = Ss(r)r(b, f_i)\, dr$$

where a filtration ratio $r(b, f_i) = I_{fi}/I_b$ with I_{fi} and I_b denote the intensities of the beams b_{fi} and b respectively.

Concluding, if we have the filter in E then we can find provisory pure beams and investigate only their properties. If we do not have the filters, we must have some other devices for the determination of the initial states. Such devices are the detectors D_p mentioned in the introduction. From all counters of quanta which we have in E besides the counter g from Statement 2 we eliminate all those which overlook some quanta independent of their properties. They can be recognised by the proportional decrease of the registered intensities of all the beams. We also eliminate all non-linear counters. All those remaining are called D_p. Now with each beam b we can associate registration probabilities by all the detectors D_p. A registration probability $p(b, d)$ is defined by the detector d as

$$(2.3) \qquad p(b, d) = Ss(r)r(b, d)\, dr$$

where registration ratios $r(b, d) = I_d/I_b$ with I_b and I_d denoting the intensities of the beam b measured by the detectors g and d, respectively. (From this moment the detectors will be denoted solely by d_i and the filters by f_i to differentiate between the two kinds of probabilities $p(b, f_i)$ and $p(b, d_i)$.)

If the probability $p(b, d) = K$, we can say that an average quantum of the beam b has the property 'd' (to be registered by the detector d) with the probability K. As usual, we must check the character of the observed probabilities using different intensity reduction procedures.

To visualise what kind of effects can appear in the case discussed above, we shall consider a simple example.

Example. Let us consider four beams of classical objects produced by four sources, i.e., the beams of balls in three colours: pink, green and blue.

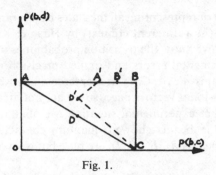

Fig. 1.

All the balls behave in an identical way in all macroscopic experiments. So from the point of view of a colour-blind observer they are identical. However, the observer has three additional detectors: g, d and c; g registers all balls, d registers all pink and green balls, and c registers all green and blue balls.

After repeated experiments, the observer notices that each beam b is characterised by the two registration probabilities $p(b, d)$ and $p(b, c)$, defined as before. The observer checks the stability of the values of $p(b, c)$ and $p(b, d)$ by stopping some of the balls before they arrive at the detectors. Of course, he discovers that the beams behave like classical mixtures, but being unable to select the pure beams he can only represent the beams by some points in two dimensional vector space. If all possible mixtures of the initial beams can be experimentally realised, then all these mixtures can be represented by a convex set on a plane. The specific shape of this set depends on the initial beams. One can say that each set is a convex envelope of the set of the points corresponding to the initial beams. Let us visualise this in a simple picture (see Fig. 1).

The triangle ABC in Fig. 1 is a classical symplectic cone (Mielnik, 1969); the observer notices that the beams A, B, C are pure and the beam D is a mixture of them.

> Every ball of A has a feature 'd' and does not have a feature 'c'.
> Every ball of B has both features 'd' and 'c'.
> Every ball of C has a feature 'c' and does not have a feature 'd'.

The quadrangle C, D', A', B' is another set of the initial beams; now only the beam C is pure and the other beams are mixed, but since we cannot separate pure beams we can investigate the behaviour of all C, D', A', B' beams in the transmission stage.

The possibility of representing all the states by all transition probabilities was pointed out (in a different context) by Haag & Kastler (1964). They also stressed that we know the transition probabilities only approximately, due to the experimental errors and limited precision of the instruments. However, the beams A', B', C, D' are represented by well-separated points in the two-dimensional vector space, so we are not afraid of ambiguities.

If, in some other experimental situation, we obtained the same set A', B', C, D' and the beams showed quantum character, then we would assume that the beams A', B', C, D' are pure but we would represent them in the same way.

Now we want to investigate the behaviour of our beams in the transmission state.

DEFINITION 4. A classical transmitter T is a device which changes the beam b in a unique way into a beam b_T and which is not a filter. A quantum transmitter T is a device which changes the pure beam b into the beams b_s with fixed transition probabilities $p_T(b, b_s)$ and which is not a filter.

Coming back to our example (classical case), we take as a transmitter T the device which changes the colours of the balls in a well-prescribed way. For example, it can change the beam b consisting of green and pink balls into the beam b_T consisting of balls in one or two other colours. The classical transmitter of this type has a characteristic feature of repeatability: in the chains b, b_T, b_{TT}, ... cycles must appear. The experimental values of the registration ratios $r(b_T, c)$ and $r(b_T, d)$ form sharp one-peaked probability distributions $s(r)$, enabling the easy calculation of the registration probabilities $p(b_T, c)$ and $p(b_T, d)$.

If we have a quantum transmitter T' and a quantum beam b, there is no reason for the above-mentioned repeatability. Also, if the transmitter T' transforms the beam b into a set of well-separated beams b_s with the transition probabilities $p_T(b, b_s)$, then the observed experimental values of the registration ratios $r(b_{T'}, c)$ and $r(b_{T'}, d)$ (at least one of them) should form many-peaked probability distributions $s(r)$ with sharp peak values around $r(b_{T'}, c) = p(b_s, c)$ and $r(b_{T'}, d) = p(b_s, d)$, respectively. Therefore, analysing the distributions $s(r)$, one can (in principle in this case) determine the beams b_s and the transition probabilities $p_T(b, b_s)$ uniquely (at least if all $p_T(b, b_s)$ are different).

Wanting to represent mathematically the transmitters T and T', we easily find that T can be represented by a matrix whose range of the domain $A'B'CD'$ must be a convex subset of the square $OABC$. The beam b_T can be represented by a vector $\mathbf{b}_T = (p(b_T, c), p(b_T, d))$ The registration probabilities can be obtained as scalar products with the vectors $e_1 = (1, 0)$ and $e_2 = (0, 1)$, respectively. In the quantum case, to each beam one can only associate a probability measure on the square $OABC$. Then T' transforms measures μ_b (which are nearly 1 on the vectors $\mathbf{b}$ and go quickly to zero outside) into measures $\mu_{bT} = \sum_s p_T(b, b_s)\mu_{b_s}$.

Remark 1. However, it can happen that the distributions $s(r)$ cannot be interpreted in a unique way by means of the transition probabilities $p_T(b, b_s)$. It can even turn out that they can be interpreted in infinitely many ways. This leads us to a serious revision of the definitions of pure beams and transmitters. As we know only the probability distributions $s(r)$ characterise the ensemble $\tilde{b}_{T'}$ completely, the values of the registration probabilities $p(b_{T'}, c)$ and $p(b_{T'}, d)$ can characterise the ensemble $\tilde{b}_{T'}$ well only if the distributions $s(r)$ are sharp, symmetric and one-peaked distributions. In such a case, having a two-number classification of the ensemble $\tilde{b}_T$, we can ascribe the same numbers (with the experimental errors) to all member beams $b_{T'} \in \tilde{b}_{T'}$, and even to the average quanta from each beam $b_{T'}$. If the probability distributions $s(r)$ have a weak structure, then they only characterise the ensemble $\tilde{b}_{T'}$ adequately and we can only discuss well-defined states of the ensemble $\tilde{b}_T$. This consideration leads us to the following definitions.

DEFINITION 5. A state of the ensemble $\tilde{b}$ can be completely characterised only by the probability distributions $s(r)$ of the filtration or the registration ratios for all beams $b \in \tilde{b}$.

DEFINITION 6. A pure ensemble $\tilde{b}$ of pure beams b is characterised by such probability distributions $s(r)$ which remain approximately unchanged:

(i)   for the new ensembles $\tilde{b}_i$ obtained from the ensemble $\tilde{b}$ by the application of the ith intensity reduction procedure on each beam $b \in \tilde{b}$;

(ii)   for all rich sub-ensembles of $\tilde{b}$ chosen in a random way.

DEFINITION 7. A transmitter T is a device which transforms each ensemble $\tilde{b}$ into a well-defined ensemble $\tilde{b}_T$ and which is not a filter.

Now we come to the general conclusions of this long section.

Conclusions

We have considered the experimental set-ups with and without provisory maximal filters (p.m. filters). We also divide into two parts the discussion of mathematical schemes useful for the representation of the results.

A. We have n-p.m. filters $f_1, \ldots, f_n$, which we use not only for the preparation of p.p. beams but also to analyse the final beams obtained in the transmission stage. Now we can have two cases:

(a) In the detection stage we always observe the p.p. beams, but starting from the same p.p. beam b and using the same transmitter T we observe different outgoing pure beams b_i. However, they appear with more or less fixed transition probabilities $p_T(b, b_i)$. In such a situation, we can represent the beams b_i and b_T by the vectors in the n-dimensional euclidean space E_1. The initial beams b_i can be represented by the basic vectors e_i and the beams b_T by the vector $\mathbf{b}_T$ in this space. The probabilities of finding the beams b_i as the outgoing beam b_T can be written in a scalar product form $p_T(b, b_i) = \mathbf{b}_T \cdot \mathbf{e}_i$. Each transmitter T can be represented by a linear operator acting in this space.

(b) The beam b_T never turns out to be one of the p.p. beams b_i. We obtain only probability distributions $s(r)$ of the filtration ratios $r(b_T, f_i)$, which generally have rich peak structure. As we noted in the example, it is possible to describe the ensemble $\tilde{b}_T$ of the beams b_T by probability measures μ_T on the n-dimensional vector space E_2. The vectors $\mathbf{b}$ in E_2, corresponding to beams b, have as their components the filtration probabilities $p(b, f_j)$. The measures μ_b, corresponding to the initial p.p. beams b_i, are nearly equal to 1 on the corresponding vectors $\mathbf{b}_i$ and go quickly to 0 outside. The measures μ_{bT} are only characterised by all $s(r)$. Only in some cases of the $s(r)$ we can extract the transition probabilities $p_T(b, b_s)$ in a more or less unique way. The beams b_s are different from the initial beams b_i. The transition probabilities satisfy a condition $\sum_s p_T(b, b_s) = 1$, contrary to the fact that for any chosen beam b_T from the ensemble $\tilde{b}_T$ in general $\sum_i r(b_T, f_i) \neq 1$. In other cases we can fit our experimental data on $s(r)$ with the different $p_T(b, b_s)$ in many different ways. To have uniqueness we have to accept that the ensemble $\tilde{b}_T$ is only characterised by all $s(r)$, as was already stated in Remark 1 and in Definition 5. In such an approach there is no place for the notion of transition probabilities.

B. The case when we have only n-detectors $d_1, ..., d_n$ is mathematically equivalent to the case A(b). We have only to replace the filtration ratios and probabilities by the registration ratios and probabilities, respectively.

In all cases we should check the purity of the ensemble b_T according to Definition 6. Sometimes we can interpret the values of $p_T(b, b_s)$ found, due to the mixed character of the initial ensemble $\tilde{b}$ with respect to the property analysed by the transmitter T. Such a possibility explains the term 'provisory' occurring in the definitions of the maximal filter and the pure beam. A contrary situation is also possible. Mixed ensembles with respect to some properties can behave like pure ones with respect to the properties analysed by the transmitter T.

The careful differentiation between filtration or registration ratios and the transition probabilities leads us to the conclusion that, in experiments of the type considered, if we succeed with the extraction of the transition probabilities from the experimental data, then in all cases A and B we can represent the beams by unit vectors in the n-dimensional Hilbert space for A(b) and B and the transition probabilities by the appropriate scalar products between them. It stems from the fact that for each ensemble $\tilde{b}_T$ we have a finite number of beams b_s and the transition probabilities $p(b_T, b_s)$, which can be embedded in the Hibert space. However, the following new features appear:

(i) The set of physically meaningful vectors is restricted to those corresponding to the b_i (initial beams), to all beams $b_{iT_i}, b_{iT_iT_j}, ...$ (where T_i are all available transmitters), and also to all beams b_s extracted in the analysis of the data. If we have two vectors $\mathbf{b}_{T_1}$, and $\mathbf{b}_{T_2}$ we do not know whether $\mathbf{b}_{T_1} + \mathbf{b}_{T_2}$ corresponds to a physically realisable beam b_{T_3}.

(ii) The physical interpretations have only some scalar products of the vectors of the type $|\langle \mathbf{b}_T | \mathbf{b}_s \rangle|^2 = p_T(b, b_s)$ where b_s are found in the analysis of all $s(r)$ describing $\tilde{b}_T$ and $\mathbf{b}_s$ are vectors corresponding to them. All the $\mathbf{b}_s$ found for all transmission processes form an orthonormal basis of the Hilbert space considered.

If we accept the philosophy of Definitions 5–7, then the Hilbert space description loses its sense, since the transition probabilities $p_T(b, b_s)$ do not appear.

However, in all the cases discussed above, having many experimental results we can try to find a quasi-theory enabling us to interpret the data and to predict new results.

So we find that the usual quantum mechanical description can turn out to be not too poor, as was suggested by Mielnik (1968, 1969), but too rich or not very appropriate.

The experimental scheme discussed so far is structurally similar to that used for the investigation of the scattering phenomena of elementary particles. Instead of simple filters and detectors we use many complicated instruments which, based on our previous knowledge, enable us to prepare and classify the initial and final beams. The different kinds of chambers and emulsion layers enable us to observe one-particle beams; but, in fact, we observe only statistical regularities characterising ensembles of such beams. The scattering process can be understood as specific transmission processes in two ways. One interpretation is that one-particle beams, example proton beams b_p, are transmitted by the transmitters = protons T_p leading to a many-particle beam $(b_p)_{T_p} = b^f_{p-p}$. Another interpretation is that initial beams b^i_{p-p} consisting of the two free protons are transmitted by the transmitters = strong $p - p$ interaction into the final beams b^f_{p-p}. We still have an additional interpretational freedom; by a transformed ensemble b^f_{p-p} we can understand a set of all final free particle beams b^f_{p-p} or a set of strongly interacting proton-proton beams b^s_{p-p}, which are visualised by the interaction points in the emulsion layers or the photo-pictures from the different chambers. In this second case, the final free particle beams b^f_{p-p} can be interpreted as arising in some kind of measurement process performed on the beams b^s_{p-p}.

One could argue that the formalism discussed in this paper is not very applicable, since in elementary-particle physics we deal with the continuous variables characterising the beams. However, in all preliminary experimental data we characterise our initial and final beams by intervals of, in fact, continuous variables. All these analogies, and the fact that a set of the different scattering processes which can be observed between elementary particles is very limited, show the need for careful investigation of whether the quantum mechanical unitary S-matrix language is not too rich for the description of the observed phenomena. We shall investigate this problem in detail in the subsequent paper.

Now we give for completeness a definition of instruments, leaving the analysis of the practically realised instruments to the subsequent paper.

DEFINITION 8. An instrument I is a device allowing the description of each beam b in terms of earlier known categories (parameters). The ascription of these parameters involves the assumption of the applicability of some earlier known theories. The instrument, by its interaction with the beam b, changes it into a beam b_I. The repeated application of the same instruments usually leads to slightly different values of parameters ascribed to our beams. The measurement made by the instrument for all beams b characterises the ensemble $\tilde{b}$. An ideal instrument is such that we can assume that $b_I = b$.

Generally, one could consider the instrument which can change the ensemble $\tilde{b}$ into the different ensemble $\tilde{b}_I \neq \tilde{b}$. Analysing the results of many measurements, one could find the characteristic features of the instruments used. In spite of the fact that $\tilde{b}_I$ is different from $\tilde{b}$, the values of parameters ascribed to $\tilde{b}$ in the measurements can be used in some way for the labelling of the initial ensembles $\tilde{b}$ or $\tilde{b}_I$. However, in practice we try to use the ideal instruments. A good example of such instruments are the filters f_i as applied on the beams b_{f_i}, different kinds of chambers and so on.

At the end of this section we should like to point out that, in spite of the fact that we have been talking about ensembles of quanta beams, our results can be generalised for experiments with ensembles of identical physical systems. Instead of filters and detectors, we should have other more complicated instruments to determine the states of the initial and final ensembles.

We should also like to remark that the particle character of our beams, implying measurement of the beam intensities by counting the quanta, is not necessary. When discussing states of ensembles $\tilde{b}$, we can measure the intensities and the appropriate ratios for some non-particle beams. The particle character of the beams was essential in the discussion of the pure beams and of the non-classical character of the probabilities.

Now we pass to the polemic with some views presented in papers on axiomatic quantum mechanics.

3. POLEMIC WITH AXIOMATIC QUANTUM MECHANICS

The great success of quantum mechanics in describing many atomic and sub-atomic phenomena, and the fact that classical physics is a limiting case of quantum physics, encouraged people to think that a general framework to describe all physical phenomena had been discovered.

To prove this statement one should find such a set of the natural assumptions on S, B, F, D, T and I characterising a physical process of a measurement in general, which would imply uniquely the usual quantum description.

The more general attitude was accepted in the papers of the Birkhoff and von Neumann (1936), Mackey (1963), Jauch (1964, 1968), Jauch and Piron (1963), Piron (1964), Finkelstein (1963), Finkelstein, Jauch, Schimonovich and Speiser (1962, 1963), Gunson (1967), and others, where the so-called quantum logic of the propositions concerning a physical system was studied. In the papers of Ludwig (1967, 1968) and Dähn (1968) the state-effect structures were investigated; in the papers of Pool (1968a, b) state-event structures.

All these studies aimed to find such a set of natural axioms which would imply uniquely the use of the complex Hilbert space language or the algebraic Haag-Kastler (1964) language for the description of the states and transition probabilities. The required set of assumptions was found in many axiomatic approaches; however, the naturality of some of the accepted assumptions is questionable. They were all chosen by analogy to the experiments performed on the optical bench with the use of colour filters, Nicol prisms, and other devices. The states of the differently polarised light can be represented by all rays in the complex two-dimensional Hilbert space $\mathcal{H}$ $(2, C)$ and each state can be realised in the laboratory. To each linearly polarised beam there corresponds in a one-to-one way an appropriate filter – the Nicol prism or polarisation filter – which is transparent to it. So one has, in principle, an uncountable amount of filters in the laboratory, since in that case nearly each rotation performed on the Nicol prism enables its interpretation as a different filter.

This special case strongly supports the commonly accepted philosophy in quantum mechanics, according to which each pure beam state is prepared by an appropriate maximal filter. Also the filtration probabilities of the pure beams (2.2) in that case are equal to the transmission probabilities (2.1) between the corresponding filters and can be expressed by the scalar products of the unit vectors in $\mathcal{H}$ $(2, C)$ representing those filters. So in fact, instead of talking about the states, one can talk about the filters and the transmission probabilities between them.

This observation gave Mielnik the force to attack the usual quantum logic approach. His main starting point (Mielnik, 1968) was the assumption

that the set of filters with the geometry implied by the transmission probabilities is the main characteristic of all quantum phenomena (however,
Mielnik, instead of saying 'transmission,' says 'transition'. Therefore, to
investigate the problem of the universality of the orthodox Hilbert space
representation, one must study whether the above-menitoned geometry
allows the representation of the filters by the unit vectors and the transmission probabilities as scalar products. In his two clear and provocative
papers (1968, 1969), he realises his programme and comes to the following
conclusions (1969):

> '... It now becomes clear that the orthodox classical and orthodox quantum systems
> do not represent a unique alternative for quantum theories, but they are only partic-
> ularly degenerate members of a vast family of 'quantum worlds' which are mathema-
> tically possible...'
> '... We thus conclude that the concepts reviewed in this article represent the missing
> element necessary to convert non-linear wave mechanics into 'mechanics of non-linear
> quanta... '.

Though there is no mathematical fault in the papers (Mielnik, 1968,
1969), in our opinion the above statements are not well justified. A
simple misunderstanding is due to the interpretation of the transmission
probabilities as the transition probabilities. These latter are a basic notion
measured in all our experiments and depending on the dynamics of the
phenomena. The transition probabilities can be directly connected with
the cross-section, branching-ratios, life-times of the excited levels, and so
on. The value of quantum mechanics consists in its ability to predict those
probabilities in agreement with the experimental data. On the other
hand, the transmission probabilities are the static properties of the filters
and the beams and can only be used (if the filters exist) to characterise
the initial and final beams. A careful analysis of the general experimental
set-ups made in the previous section allowed a clean differentiation between all kinds of probabilities [(2,1), (2.2),(2.3)] and Definition 6 divorced
the concept of a pure state with a concept of a maximal filter. So the
Mielnik statement has to be changed into the following statement:

There can be many sets of filters whose transmission probabilities do
not allow the representation of them by the unit vectors in the Hilbert
space with the transmission probabilities being equal to the appropriate
scalar products.

Besides this main criticism, we have other critical remarks concerning
the paper (Mielnik, 1969). In this paper the maximal transmission systems are considered. Those systems have so rich a class of transmitters
that each physical state can be transformed in any other by means of the

appropriate transmitter. They also have, in general, an uncountably rich set of maximal filters. To each pure beam there correspond two filters; one is completely transparent to the beam, the second is completely non-transparent. In our opinion, dealing with such rich classes of filters is rather unrealistic. Therefore, we cannot accept the second-cited conclusion concerning the quantisation of non-linear theories. The procedure proposed in the paper (Mielnik, 1969) can be devoid of any physical meaning.

To illustrate our arguments we shall discuss a nice example of the drop of non-Hibertian quantum liquid from Mielnik's (1968) paper:

'. . . Someone looked at a small spherical glass bubble: inside there was a drop of liquid. The drop occupied exactly half of the bubble in the shape of a hemisphere. He was able to introduce inside a thin, flat partition dividing the interior of the bubble into two equal volumes. He tried to do this so that the drop would become split, However, the drop exhibited a quantum behavior: instead of being divided into two parts, the drop jumped and occupied the space only one side of the partition. He repeated the attempt, obtaining a similar result. He began to observe this phenomenon and discovered that each time the partition is introduced the drop chooses a certain side with a definite probability. This probability depends upon the angle between the partition and the initial surface of the drop. If the drop occupied a hemisphere s and the partition forces it to choose between the two hemispheres r and r', the probabilities of transition into r and r' are proportional to volumes of $s \cap r$ and $s \cap r'$. He was struck by the analogy between positions of the drop and quantum states and between the partition and the macroscopic measuring apparatus. He wanted to formulate the quantum theory of this phenomenon, but he realised that he could not use Hilbert spaces because the space of states of the drop was not Hilbertian. . . . '.

We disagree with that conclusion and we analyse the behaviour of the observer. To make some predictions he considers an ensemble $\tilde{b}$ of the above-mentioned bubbles b. Before starting to divide the liquid drop he fixes the positions of all the bubbles to make the surfaces of all drops horizontal. He chooses a well-separated set of N partitions t_i labelled by the angles α_i between partitions and the surface of the drops. After many partitions have been made, he observes $2N$ possible positions of the drops after partitions. Repeating the experiments with fixed partition t of the ensemble $\tilde{b}$ he finds out the probabilistic behaviour of the drops with some fixed transition probabilities to the final states. Before accepting a usual quantum interpretation of the probabilities he investigates the purity of the ensemble $\tilde{b}$ as described in Definition 6. If the ensemble $\tilde{b}$ turns out to be pure, he accepts the following interpretation. The partition t is some kind of interaction exerted on the drops so the partition t is some kind of quantum transmitter transforming the ensemble $\tilde{b}$ into pure ensembles $\tilde{b}_1$ and $\tilde{b}_2$ with fixed transition probabilities, $p_t(b, b_1)$ and $p_t(b, b_2)$. Of course, those probabilities for all the partitions t_i we can represent as a scalar products in $2N$ dimensional real Hilbert space of the vectors $\mathbf{b}_{t_i}$ corresponding to the b_{t_i} with appropriate vectors $\mathbf{b}_i$'s. Naturally, the scalar products $\mathbf{b}_{t_1} \cdot \mathbf{b}_{t_2}$ have no physical meaning. Therefore, the Hilbert space description of this phenomenon is in some sense too rich and not too poor, as was claimed in Mielnik (1968).

Returning to the discussion of axiomatic quantum mechanics, we state that in our opinion the problem of Birkhoff and von Neumann, although skillfully solved in the different axiomatisation schemes, was stated in too general a way. In our opinion, it is not very economic to talk about all possible propositions concerning the physical systems in general. In practice we have to perform the experiments and the analysis of the results gives us a set of physically meaningful propositions about the system. This set depends on the particular experimental set-up and its richness depends on the richness of the observed phenomena. The careful analysis of the particular experimental set-ups can lead us to the discovery of new, more economical and fruitful descriptions, though the old language of Hilbert spaces could be used. Being too general, we cannot get insight into such problems and we cannot hope to arrive at the conclusive new statements to be verified in the experiments.

Finally, we should like to question some axioms of Gunson (1967) and Pool (1968a). Gunson considers a set of propositions P and a set of states S. States are the probability measures on the propositions, taking the real values from 0 to 1. The axiom A4. is: 'For every $a, b \in P$ we have $a \leq b$ if and only if $f(a) \leq f(b)$ for all $f \in S$. For the propositions, the relation $a \leq b$ is equivalent to the usual implication relation a implies b. Gunson also uses the following definition of the orthogonality $a \perp b$ $\leftrightarrow a \leq b'$, where b' is the logical negation of the proposition b.

Counter example. Let us consider the following situation. We have only two pure ensembles $\tilde{f}$ and $\tilde{g}$ and two detectors d and l. The only things we can measure are the registration probabilities (2.3) by those detectors. With each detector we can associate two propositions. For example: the propostion 'd' – 'the physical system from the ensemble is registered by the detector d', and the proposition 'd'''–'the physical system from the ensemble is not registered by the detector d'. As we see for $f \in \tilde{f}$ we have $p(f, d) = f('d')$ in Gunson's notation.

Now let us assume that we observe the following values of $p(f, d)$ and $p(f, l)$. $f('d') = 1/4$, $g('d') = 1/3$, $f('l') = 1/8$, $g('l') = 1/7$, $f('d'') = 3/4$, $g('d'') = 2/3$, $f('l'') = 7/8$, $g('l'') = 6/7$.

As we see, the propositions 'd' and 'l' satisfy the axiom A. 4 so 'd' $\leq$ 'l' which is equivalent to 'd' implies 'l', but such implication is physically completely unjustified. Now, using the definition of the orthogonality, we find that $d \leq d'$; therefore, $d \perp d$.

Pool in his papers accepts the following definitions and axioms:

DEFINITION I. 1. An event-state structure is a triple (E, S, P):

(i) E is a set called the logic of the even-state structure and an element of E is called an event:

(ii) S is a set and an element of S is called a state;

(iii) P is a function $P: E \times S \to [0, 1]$ called the probability function and if $p \in E$ and $\alpha \in S$, then $P(p, \alpha)$ is called the probability of the occurrence of the event p in the state α;

(iv)   if $p \in E$, then the subsets $S_1(p)$ and $S_0(p)$ of S are defined by

$$S_1(p) = \{\alpha \in S: P(p, \alpha) = 1\}$$
$$S_0(p) = \{\alpha \in S: P(p, \alpha) = 0\}.$$

AXIOM I. 3. If $p, q \in E$ and $S_1(p) \subset S_1(q)$, then $S_0(q) \subset S_0(p)$.

AXIOM I. 4. If $p \in E$, then there exists an event $p' \in E$ such that $S_1(p') = S_0(p)$ and $S_0(p') = S_1(p)$.

In our opinion, these axioms are not general enough. For example, if α are the states of ensembles consisting of beams, and p are the events of the type (transmission through the filter p), then the $P(p, \alpha)$ can be the transmission probabilities. In this case, the properties and the richness of the sets $S_1(p)$ and $S_0(p)$ depend on the beams and the filters in the particular experimental set-up and it is easy to give an example for which the Axioms 1.3 and 1.4 are not satisfied.

The above two examples support our thesis that it is extremely difficult, if not impossible, to axiomatise all possible experimental set-ups in the natural way.

Now we pass to the last section, where we formulate a programme of future investigations which could enable the answer of the title question of our paper.

4. A PROGRAMME OF INVESTIGATIONS

We could not answer the title question of this paper, since we have been analysing some hypothetical general experimental set-ups. To answer the question whether the Hilbert space language is too rich to describe some physical phenomena, we should carefully analyse all real physical set-ups and observed phenomena, starting from solid-state physics and ending with high-energy elementary-particle physics. Such analysis should be done by physicists who really work in the specific branch of physics and who know all the subtleties of the experimental set-ups and of the theoretical analysis used to explain the data (to obtain the curves).

It is clear that it is quite difficult to find out that the language used is too rich; moreover, with the help of computers a beautiful agreement with the data can be obtained in most cases. However, one feature of the too rich language is the possibility of obtaining the same predictions using quite different models, which is equivalent to the lack of the unique theoretical explanation. The observation of such a situation can be a first hint for future investigations. In our opinion, one more or less sure method is to find such rigorously derived experimental predictions of a general nature, which can be verified in experiment, and to test them with full objectivity. In elementary-particle physics it can be the unitarity of the S matrix. The other method is to try to invent more economic language. In the discussion of the general experimental set-up, such possibilities were indicated. Especially interesting was that of Remark 1 where the notion of the transition probability disappears.

The other interesting problem is an operational status of quantum mechanics in its applications to many new phenomena. The operational status of quantum mechanics was discussed on the basis of experiments with polarised light and Stern–Gerlach experiments. Quantum mechanics as applied to high-energy elementary-particle scattering was not discussed in that context.

Another important problem is to investigate to what extent the good results which we obtain depend on all our particular assumptions and on the basic assumptions of the theory we used. Many models in elementary particle physics are believed to be checked by the agreement of their predictions with experiment and are supposed to have a deeper physical meaning (not only to be a convenient parameterisation of the data). However, sometimes a careful analysis of the results shows that they are not deduced from the assumptions and they can only be rigorously derived from another set of assumptions which can have nothing in common with the physical ideas involved in the initial assumptions. To give an example, a careful anlaysis performed in the papers (Kielanowski and Kupczyński,

1971; Kupczyński, 1971) showed that the additivity assumption in the quark model applied with success for high-energy elementary-particle scattering can have nothing to do with the physical picture of a static quark model where the quarks are treated like hypothetical constituents of the elementary particles.

The programme which we have presented can be summarised as follows. Let us be more critical of the models we propose, of the conclusions we obtain, and let us check the operational status of the language we use to deal with data.

The investigation in this direction will be continued in the subsequent paper.

International Centre for Theoretical Physics
Trieste, Italy

ACKNOWLEDGEMENTS

The author is grateful to Prof. Abdus Salam, the International Atomic Energy Agency and UNESCO for hospitality at the International Centre for Theoretical Physics, Trieste.

REFERENCES

Birkoff, G. and von Neumann, J. (1936). 'The logic of quantum mechanics, *Annals* of *Mathematics* **37**, 823.

Bohr, N. (1961). *Atomic Physics and Human Knowledge*. Science Editions Inc., New York.

Dähn, G. (1968). Attempts of an axiomatic foundation of quantum mechanics and more general theories–IV, *Communications of Mathematical Physics*, **9**, 192.

Finkelstein, D., Jauch, J. M., Schimonovich, S. and Speiser, D. (1962). Foundations of quaternionic quantum mechanics, *Journal of Mathematical Physics*, **3**, 207.

Finkelstein, D., Jauch, J. M., Schimonovich, S. and Speiser D. (1963). Some physical consequences of general Q-covariance, *Journal of Mathematical Physics*, **4**, 788.

Finkelstein, D. (1963). *The Logic of Quantum Physics*, p. 621. The New York Academy of Science.

Finkelstein D. (1968). *The Physics of Logic*, ICTP, Trieste, preprint IC/68/35.

Gunson, J. (1967). Structure of quantum mechanics, *Communications of Mathematical Physics*, **6**, 262.

Haag, R. and Kastler, D. (1964). An algebraic approach to quantum field theory, *Journal of Mathematical Physics*, **5**, 848.

Jauch, J. M. (1964). The problem of measurement in quantum mechanics, *Helvitica Physica Acta*, **37**, 293.

Jauch, J. M. (1968). *Foundations of Quantum Mechanics*. Addison-Wesley, Reading, Mass.

Jauch, J. M. and Piron, C. (1963). *Helvetica Physica Acta*, **36**, 827.

Kielanowski, P. and Kupczyński, M. (1971). Relativistic quark model predictions for the transversity amplitudes, *Nuclear Physics*, **B29**, 504.

Kupczyński, M. (1971). *On the Additivity Assumptions*, Warsaw University preprint IFT/71/15.

Ludwig, G. (1967). Attempt of an axiomatic foundation of quantum mechanics and more general theories–II, *Communications of Mathematical Physics*, **4**, 331; for Part III see (1968). *Communications of Mathematical Physics*, 9, 1.

Mackey, G. W. (1963). *Mathematical Foundations of Quantum Mechanics*. W. A. Benjamin Inc., New York.

Mielnik, B. (1968). Geometry of quantum states, *Communcations of Mathematical Physics*, **9**, 55.

Mielnik, B. (1969). Theory of filters, *Communications of Mathematical Physics*, **15**, 1.

Piron, C. (1964). Axiomatique quantique, *Helvetica Physica Acta*, **37**, 439.

Pool, J. C. T. (1968a). Baer*–semigroup and the logic of quantum mechanics, *Communications of Mathematical Physics*, **9**, 118.

Pool, J. C. T. (1968b). Semimodularity and the logic of quantum mechanics, *Communication of Mathematical Physics*, **9**, 212.

BOGDAN MIELNIK

GENERALIZED QUANTUM MECHANICS

ABSTRACT. A convex scheme of quantum theory is outlined where the states are not necessarily the density matrices in a Hilbert space. The physical interpretation of the scheme is given in terms of generalized "impossibility principles". The geometry of the convex set of all pure and mixed states (called a statistical figure) is conditioned by the dynamics of the system. This provides a method of constructing the statistical figures for non-linear variants of quantum mechanics where the superposition principle is no longer valid. Examples of that construction are given and its possible significance for the inter-relation between quantum theory and general relativity is discussed.

1. INTRODUCTION

In turn of development of quantum theory efforts were made to present a geometric description of quantum mechanics independent of "wave functions" and "complex amplitudes". The best known such description was originated by Birkhoff and von Neumann [2] and completed by Piron [18]. It explores a partial order relation in an idealized set of "yes-no measurements" called a "quantum logic". The resulting approach though mathematically profound is not physically complete. In the last ten years two other approaches have been developed. One is the algebraic approach reflecting the physics of operations which can be performed on statistical ensembles. This aspect has been introduced to axiomatic quantum field theory by Haag and Kastler [10] and it reappears as the main motif in the present day quantum statistics. The other approach, originated already in the fourties (Segal [1]) might be called "convex". It explores the convex structures of quantum mechanics with a special attention concentrated on the convex set of all states (pure and mixed) of a quantum system. The description of quantum mechanics from that point of view was most systematically explored by Ludwig [14] and further developed in [3–6, 11, 15–17, 19, 21, 22]; it now becomes one of main currents in the foundation of quantum theory. The synthesis of the convex and the algebraic approaches has been gradually achieved [3, 5, 6, 9–11, 16, 19]. It brought the complete geometrization of quantum mechanics including the descrip-

Commun. math. Phys. 37, 221–256 (1974)

Hooker (ed.), Physical Theory as Logico-Operational Structure, 115–152.
All Rights Reserved.
Copyright © 1978 by D. Reidel Publishing Company, Dordrecht, Holland.

tion of the present day formalism of Hilbert spaces in terms of physically meaningful axioms [9, 14, 15, 19, 21], the general classification of the operation [6, 11], the definition of filters as endomorphism of a convex *cone of beams* [3, 16] and finally, the construction of the transition probabilities as affine geometric invariants of a convex set [16]. As a result of that development a generalized convex scheme of quantum mechanics has emerged from the point of view of which the scheme of the present day theory is not unique but is a particular member of a vast family of "quantum worlds" mathematically admissible. The conjecture was also raised that the convex set theory might play a similar role in quantum physics as the Riemannian geometry in general relativity [16]. The aim of the present paper is to take the next step by showing that the "convex scheme" is flexible enough to comprise nonlinear versions of quantum mechanics in which a non-linear wave equation would play the role of the Schrödinger equation. With this aim the geometric description of quantum mechanics based on the convex set theory is outlined in §2. In §3 and §4 the geometry of a system is related with the dynamics which allows the construction of the convex manifolds of quantum states for systems obeying a generalized wave mechanics. Some applications of the resulting scheme are indicated in §4 and its relation to other physical theories is discussed in §5.

2. CONVEX SCHEME (OUTLINE)[1]

The elements of convex set theory are rooted in primitive concepts of quantum mechanics. The most fundamental such concept is that of a quantum state. Given a statistical ensemble of objects of any nature, the *state* is the collection of the physical properties of an average ensemble individual. For the above notion of a state the following concept of a *mixture* becomes natural. Given certain ensembles $\mathscr{E}_1, \ldots, \mathscr{E}_n$ corresponding to states $x_1, \ldots, x_n$ and given numbers $p_n, \ldots, p_1 \geq 0$, $p_1 + \cdots + p_n = 1$, one can form a new ensemble $\mathscr{E}$ of which a fraction p_j comprises randomly chosen objects of $\mathscr{E}_j (j = 1, \ldots, n)$: the ensemble $\mathscr{E}$ defines a new state x which will be denoted $x = p\,x_1 + \cdots + p_n x_n$ and called a *mixture* of $x_1, \ldots, x_n$. The concept of a mixture induces that of a *pure state*: a state is called *pure* if it cannot be represented as a mixture (with all coefficients non-vanishing) of any physically distinct states. These definitions suggest that the set of all states of a physical object should possess the structure of a convex set. Some definitions concerning that structure are given below.

Definitions. A *convex set* is a subset of an affine space containing together with any two points the interval joining them. Here, an *affine space* is any set E of elements called *points* with a linear combination operation assigning to each finite system of points $x_1, x_n \in E(n = 1, 2, ...)$ and any system of numbers $\lambda_1, ..., \lambda_n \in \mathbb{R}$, $\lambda_1 + \cdots + \lambda_n = 1$ a new point $\lambda_1 x_1 + \cdots + \lambda_n x_n$; the linear operation has properties which allow one to represent E as a plane in a real linear space. An *affine topological* space is an affine space E with a topology in which the linear operation is continuous. Given an affine space E and a system of points $x_1, ..., x_n \in E$, any linear combination $p_1 x_1 + ... + p_n x_n$ with $p_1, ..., p_n \geq 0$, $p_1 + \cdots + p_n = 1$ is called a *convex combination* of $x_1, ..., x_n$. For E an affine topological space, a continuous analogue of that operation can be introduced. Given a subset $X \subset E$ with the topology induced by that of E and given a positive measure μ defined on Borel subsets of X such that $\mu(X) = 1$ (a *probability measure* on X) the integral $\int_X x d\mu(x)$, if it exists, is called a *convex integral* of points $x \in X$ over the measure μ. The convex combination is a special case of a convex integral obtained by taking the measure μ to vanish outside of a finite set of points. Given an affine space E and two points $x_1, x_2 \in E$, $x_1 \neq x_2$, the set of all linear combinations $E(x_1, x_2) = \{\lambda_1 x_1 + \lambda_2 x_2 : \lambda_1, \lambda_2 \in \mathbb{R} \ \lambda_1 + \lambda_2 = 1\}$ is called the *straight line* determined by x_1 and x_2 while the set of all convex combinations $I(x_1, x_2) = \{p_1 x_1 + p_2 x_2 : p_1, p_2 \geq 0, p_1 + p_2 = 1\}$ is called the *straight line interval* joining x_1 and x_2. Any point of $I(x_1, x_2)$ different from the end points x_1 and x_2 is called an *internal point* of $I(x_1, x_2)$. Given a convex set $S \subset E$ an element $x \in S$ is called an *extremal point* of S if it cannot be represented as a convex combination with both coefficients positive of any two distinct points of S. Thus, x is extremal if it is not an internal point of any interval $I \subset S$.

The most general axiom reflecting the phenomenology of mixtures in quantum mechanics can be now formulated as follows. *For any quantum system the set of all states can be represented as a closed convex set S in a certain affine topological space E. The convex combinations in S correspond to the state mixtures while the extremal points of S represent the pure states of the system. The topology on S reflects the observable properties of quantum states.*

Here, physical significance may be attributed to the set S alone: the surrounding affine space E is introduced only as an auxiliary construct[2].

For reasons of economy it will be assumed that S spans E. For quantum mechanical systems which are the objective of this paper it will be assumed in addition that S contains a set of extremal points rich enough to represent any point of S as a convex integral of extremal points. The convex set S plays a fundamental role in quantum statistics; it will be further called a *statistical figure* (see also [16]). Two simple examples of that structure are given below.

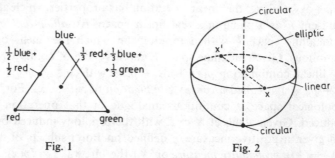

Fig. 1   Fig. 2

Figure 1 represents the mixtures which can be formed of classical objects of three types (for example, red, green and blue balls). The statistical figure here is a triangle in a 2-dimensional affine space: the vertices are pure states corresponding to one-colour ensembles, while the remaining points represent the mixed states with one mixed state (the center) distinguished (completely random mixture). A similar structure can be obtained by considering mixtures of classical objects of n possibles types: in that case the statistical figure is a simplex with n vertices in an $(n - 1)$-dimensional affine space. The case of $n = \infty$ is essential for realistic models of classical mechanics. Here, the pure states correspond to points of a classical phase space P (endowed with a certain natural topology) and the statistical figure S is the convex set of all probability measures on P with the topology induced by that of P. The convex set of all probability measures on P with the topology induced by that of P. The convex set of all probability measures on a certain topological space is a *generalized simplex* whose vertices are all point-concentrated measures. It is an important property of the simplexes that each point of a simplex can be uniquely represented as a convex integral of extremal points. This fact reflects the classical nature of the corresponding objects: it is a crucial feature of classical objects that their statistical ensembles can be uniquely decomposed into the pure components. Thus, the simplexes have to be considered the statistical figures of classical theories.

A different example is shown in Fig. 2, which represents the polarization states of a photon. Here, the statistical figure is an ellipsoid in a 3-dimensional affine space. The surface of the ellipsoid represents the pure polarization states: the equator comprises the linear polarizations, the poles are the circular polarization states and the remaining points of the surface correspond to the elliptic polarizations. Each two antipodes of the ellipsoid correspond to the "opposite" polarizations (for instance, each two antipodes on the equator represent two mutually orthogonal linear polarizations). The internal points of the ellipsoid stand for the mixed states with one mixed state θ (the centre) distinguished (polarization chaos). For the above statistical figure the decomposition of mixtures into their pure components is no longer unique: each mixed state can be represented in many ways as a combination of pure states. Thus, the chaos state can be represented as $\theta = \frac{1}{2}x + \frac{1}{2}x'$ where x and x' are any two antipodes of the ellipsoid. Physically, this means that having a light beam in the polarization state θ one cannot say whether the beam has been prepared by mixing two linearly polarized beams or by mixing two circularly polarized beams or in any other way. This fact illustrates a certain general "impotence law" coded in the geometry of S.

Principle of Impossibility

The law which emerges from the example in Fig. 2 might be given the following form. *Having a mixed statistical ensemble of non-classical objects one cannot determine uniquely its pure components and find out how the mixture has been prepared. Two mixtures created in two distinct ways by taking different collections of pure states may be physically indistinguishable.* This statement is one of the most general negative laws limiting the perception of quantum ensembles: it might be called the *first principle of im-*

"First impossibility"

$$x = p_1 x_1 + p_2 x_2$$

$$x = q_1 y_1 + q_2 y_2$$

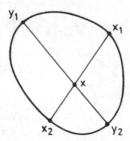

Fig. 3

possibility of quantum theory and considered the main manifestation of the non-classical nature of microobjects. The above law is not exclusive for the orthodox theory but it can be read from the geometry of any statistical figure S which is not a simplex (see Fig. 3). A non-simplicial shape of S is a geometric expression of non-classical character of the corresponding objects.

Counters

Though it was known for a long time that the shape of S reflects the physics of the corresponding quanta, it was only recently discovered that this shape contains the complete information concerning the properties of quantum states. In reading that information the following concept of a normal functional is essential.

Definition. Given an affine space E, a functional $\phi: E \to \mathbb{R}$ is called *linear* if $\phi(\lambda_1 x_1 + \cdots + \lambda_n x_n) = \lambda_1 \phi x_1 + \cdots + \lambda_n \phi x_n$ for any $x_1, \ldots, x_n \in E$, $\lambda_1, \ldots, \lambda_n \in \mathbb{R}$, $\lambda_1 + \cdots + \lambda_n = 1$. Given a convex set S in an affine topological space E, a linear continuous functional $\phi: E \to \mathbb{R}$ is called *normal* on S if $0 \le \phi x \le 1$ for every $x \in S$.

The normal functionals admit a simple geometric representation. Any non-constant linear continuous functional ϕ in an affine topological space is completely determined by a pair of closed parallel hyperplanes on which it takes the value 0 and 1. Now, ϕ is normal on S if the set S lies in between the hyperplanes $\phi = 0$ and $\phi = 1$ (see Fig. 4).

If the convex set S is a statistical figure for a certain physical system, the normal functionals possess a natural physical interpretation. A meaningful theory besides physical objects describes also measuring devices. The

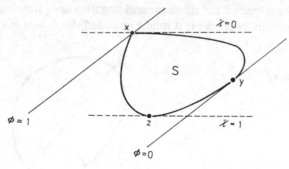

Fig. 4

typical measuring device of quantum theory is a particle counter. Given a counter ϕ and a state $x \in S$ the number ϕx will mean the average fraction of systems in x-state detected by the counter ϕ ($0 \le \phi x \le 1$). Since counters considered here react only to individual systems of the ensemble (without taking into account the interrelations between them), it follows that for any mixed ensemble the total number of particles (systems) detected is the sum of the corresponding numbers for all the components of the mixture. This leads to: $\phi(p_1 x_1 + \cdots + p_n x_n) = p_1 \phi x_1 + \cdots + p_n \phi x_n$ and so, each counter defines a certain normal functional on the statistical figure S (and thus, on the whole of E). The question now arises: how rich is the set of all normal functionals which correspond to certain physical counters? As no counter-example is known, it will be assumed that each normal functional represents a way of detecting a property of the system which, at least in principle, could be realized by constructing an adequate counter. For example, if the convex set in Fig. 4 were a statistical figure for certain physical particles, the pair of planes $\phi = 0$ and $\phi = 1$ would represent a counter registering unmistakenly all particles in state x ("x-particles") and blind to all y-particles, whereas the planes $\chi = 0$ and $\chi = 1$ would represent a counter registering all z-particles, $1/2$ of the y-particles and blind to x-particles. Since for the convex set of quite arbitrary shape the non-trivial normal functionals might not exist, some general assumptions as to the structure of S are still necessary. What will be assumed below is that the shape of S allows the existence of a class of normal functionals rich enough to distinguish the points of S: the convex set with that property is called *bounded*.

The assumptions up to now can be summarized as follows. *For any quantum system the set of all states is a closed and bounded convex set S (called a statistical figure) in an affine topological space E.* (The physically essential structure here is S while the surrounding affine space E is spanned by S as an auxiliary construct.)

The statistical figure S uniquely determines the class of normal functionals represented by pairs of parallel hyperplanes enclosing S: every normal functional corresponds to a quantum mechanical counter which might be used to test the system properties.

On these assumptions, all the physical information contained in the geometry of S may be now decoded. One of the most familiar such information is the "quantum logic" of Birkhoff, von Neumann, and Piron [2, 18, 12].

Quantum Logic

In the geometry of convex sets the following concept of a *wall* is of importance.

DEFINITION. Given a convex set S, a wall of S is any convex subset $S' \subset S$ such that, whenever S' contains an internal point of any straight line interval $I \subset S$, it must also contain the total interval I. Thus, S' is a wall if the following implications hold: 1) $x_1, x_2 \in S'$, $p_1, p_2 \geq 0$, $p_1 + p_2 = 1 \Rightarrow p_1 x_1 + p_2 x_2 \in S'$; 2) $x_1, x_2 \in S$, $p_1 x_1 + p_2 x_2 \in S'$ with $p_1, p_2 > 0$, $p_1 + p_2 = 1 \Rightarrow x_1, x_2 \in S'$.

The above concept of a wall generalizes that of an extremal point: the extremal points are simply one-point walls of S. Geometrically, a wall can be associated with an intuitive idea of a "maximal plane fragment" (of any dimensionality) of the boundary of S. Each convex set has at least two inproper walls the empty set $\emptyset$ and whole of S. For any convex set S the walls form a partially ordered set with the ordering relation $\leq$ meaning set theoretical inclusion. As is immediately seen, the common part of any family of walls is again a wall: hence, the walls form also a lattice (where for any family of walls the intersection of all upper bounds defines the lowest upper bound). If S is a statistical figure this lattice admits a natural physical interpretation. It is an old question whether the formalism of quantum theory is adequate to describe the properties of single systems. What is verified directly in the most general quantum experiment are rather the properties of statistical ensembles. However, the properties of single systems can be introduced by an abstraction process [12, 16]. One can agree, that a property P of quantum ensembles defines a property of single systems provided that it is "additive" and "hereditary"[3]: if any two ensembles have the property P their mixtures must have it too. Conversely, if any mixed ensemble has the property P, so must have each of the mixture components. These conditions mean that the subset of all states with the property P should be a wall of S. One thus guesses that the lattice of walls of S represents the set of all possible physical properties of a single system ordered according to their generality. In axiomatic approaches to quantum mechanics an important role is attributed to the notion of *completely exlcuding* properties. This notion finds a simple geometric description too. Given two walls (properties) $S_1, S_2 \subset S$ and a normal functional ϕ, it will be said that ϕ *completely separates* S_1 and S_2 if either $\phi x_1 = 0$, and $\phi x_2 = 1$ or $\phi x_1 = 1$ and $\phi x_2 = 0$ for every $x_1 \in S_1$ and $x_2 \in S_2$ (i.e. if the corresponding counter is completely blind to the particles with one of these properties and detects all the particles with the other). Now, two properties (walls) are called *excluding* or *orthogonal* ($S_1 \perp S_2$) if there exists at least one counter completely separating them. The set of all physical properties of a microparticle with the relations of inclusion $\leq$ and exclusion $\perp$ is what one traditionally calls a *logic of the particle* or *quantum logic*[4].

Hence, in the present approach the quantum logic is no longer a fundamental structure but one of the particular aspects of the geometry of S.

Second Principle of Impossibility

Besides the structure of the "logic" the geometry of S allows one to read a numerical relation between pure states generalizing the basic invariant $|(\psi, \varphi)|^2$ of the orthodox theory. Let x and y be two extremal points of a statistical figure S and let $Q(y)$ denote the set of all normal functionals taking the value 1 at y. If $x \perp y$, there exists in $Q(y)$ at least one functional vanishing at x. In general, however, such a functional in $Q(y)$ may not exist because of the geometry of S. An example of this situation is shown on Fig. 5.

For the convex set here each point of the arc $z_1 z_2$ is orthogonal to the extremal point y: suitable separating functionals are determined by all possible pairs of parallel support lines P_j, P'_j one of which supports S at y and the other at an arbitrary point of $z_1 z_2$. A similar separating functional, however, does not exist for the pair of points x and y as there is no parallel pair of straight lines supporting S at x and y. As seen on Fig. 5 the smallest value possible at x for the functionals of the family $Q(y)$ is 1/2 and is accepted by the functional ϕ represented by the pair of lines P_1, P'_1. Thus, if the convex set in Fig. 5 were a statistical figure for certain physical quanta, it could be infered that no counter can be constructed registering all y-quanta had less than the average fraction 1/2 of x-quanta. This illustrates a certain general impossibility law coded in the geometry of the statistical figure. In order to formulate it more precisely the following geometric quantity is needed.

Definition. Given a closed convex set S and a pair of extremal points x, $y \in S$ the ratio $x : y$ is the lower limit at x of all normal functionals taking the value 1 at y:

$$(2.1) \qquad x : y = \inf_{\phi \in Q(y)} \phi x.$$

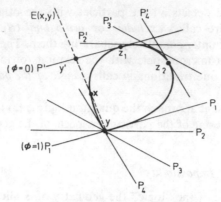

Fig. 5

For any convex set S the ratios of the extremal points are uniquely defined by the shape of S and can be determined by an abstract construction employing support planes [16].

If S is a statistical figure of certain physical systems, the non-vanishing of the quantity (2.1) reflects an inavoidable lack of selectivity of quantum mechanical measurements. *Given two pure states x and y no counter can be constructed which detects all y-systems and less than a fraction $x : y$ of x-systems. More generally, every physical process leading to a certain macroscopic effect for all y-systems must inavoidably lead to the same effect for at least $x : y$ of x-systems.* The above statements form one of essential quantum laws which will be further called the *second principle of impossibility.* Because of their role in that principle, the ratios of the extremal points of any statistical figure will be called the *detection ratios* (see also [16]).

A particular form of the "second impossibility" is observed in orthodox quantum mechanics. Here, the statistical figure is the set of all density operators in a certain Hilbert space $\mathcal{H}$: $S_{\mathcal{H}} = \{x \in \mathcal{L}(\mathcal{H}): x^+ = x \geq 0,$ $\mathrm{Tr}\, x = 1\}$. The extremal points of $S_{\mathcal{H}}$ are the simple projection operators of the form $|\psi \rangle \langle \psi|$ where ψ are the unit vectors in $\mathcal{H}$. For an arbitrary pair of extremal points $|\psi \rangle \langle \psi|$ and $|\varphi \rangle \langle \varphi|$ the detection ratio has been found in [16]:

(2.2) $|\psi \rangle \langle \psi| : |\varphi \rangle \langle \varphi| = |(\psi, \varphi)|^2.$

This fact allows one to deepen the statistical interpretation of the quantities $|(\psi, \varphi)|^2$ in orthodox theory. According to the traditional interpretation these quantities possess the following meaning of the *transition probabilities*. Given any pure state $|\varphi \rangle\langle \varphi|$ the theory assumes the existence of a measuring device testing for microsystems in that state. When this device is applied to an ensemble systems in another state $|\psi \rangle\langle \psi|$ a fraction $|(\psi, \varphi)|^2$ of them, on average, passes the test with the result positive. What may be infered now however, is not merely the existence of this sort of verifying device, but also the basic impossibility of any more selective apparatus. Given two pure states corresponding to unit vectors $\psi, \varphi \in \mathcal{H}$, no particle absorber is possible transparent to all φ-systems and transmitting less than a fraction $|(\psi, \varphi)|^2$ of ψ-systems. Generally, no physical process is possible which tolerates all φ-systems and less than a fraction $|(\psi, \varphi)|^2$ of ψ-systems. The quantities $|(\psi, \varphi)|^2$ establish an absolute selectivity limit for quantum experiments. The above facts, though proved analytically [16] possess a simple geometric meaning. As an illustration consider polarized light beams. If x and y are the states of the circular and linear polarizations of a photon respectively, there exist a device transparent to linear y-photons and transmitting ony half of the circular x-photons (the Nicol prism with its polarization plane coinciding with that of the y-photons). However, no window can exist more selective than the Nicol prism, that is transparent to linearly polarized photons but absorbing more than a fraction 1/2 of circularly polarized photons. The impossibility of such a device follows immediately from the geometry of the ellipsoid in Fig. 2 (construction with the support planes), whereas the analytical representation of the points of that ellipsoid by density operators in a Hilbert space is much less obvious.

It is worth while to notice that the "second impossibility" exhibits an implicit dynamical content of the statistical figure. In fact, the assertion about the impossibility of certain physical processes is a genuine dynamical statement limiting a priori the class of evolution processes admissible. This limitation is present on the apparently pre-dynamical level of the theory, as a necessary condition for the applicability of its mathematical language. Thus, in supplementing the structure of quantum states by dynamical equations the precaution must be taken to respect an "innate" dynamical information contained in the geometry of S. This seems to be a special case of Haag's general idea about the necessary consistency between the primitive measurement axioms of quantum theory and its mature form of a dynamical theory[5].

Problem of Generality

As becomes clear, the convex set theory introduces into quantum mechanics a flexibility similar to that which the Riemannian geometry
achieves in the space-time physics. Indeed, it is significant that one can
describe the structure of quite an arbitrary convex set in terms of typically
quantum mechanical concepts such as the "detection ratios" and the
"quantum logic". When the set S deviates from the traditional manifold
of "density matrices", those concepts do not loose their sense: they only
change the geometry. This allows a consistent description of generalized
systems where the geometry of the "transition probabilities" and the
structure of "logic" would not be originated by the Hilbert space theory.
Thus, one might think about a possibility of non-orthodox "quantum
worlds" where the "logical" axioms of Birkhoff, von Neumann and Piron
[2, 18] would be relaxed and the "operational" axioms of Pool [19] and
Gunson [9] would not hold. One might also construct hypothetical systems with the lattice of "properties" being not orthocomplemented and
the axiom of Ludwig [14] about the "sensitivity increase of effects" broken.
The simplest example embodying all those non-orthodox features is
represented by a hypothetical statistical figure of the form of a square in
a 2-dimensional affine plane:

Here, the pure states are the vertices x, x', y, y' and the remaining
points of the square are mixed states with one mixed state (the centre)
distinguished as the "total chaos". For any pair of pure states there
exists a separating normal functional (thus, e.g., the pair of support lines
P, P' in Fig. 5, defines a normal functional separating x and y from both
x' and y'). Hence, the detection ratios vanish and there is no second im-

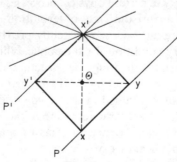

Fig. 6

possibility. Still, the first impossibility is present: the mixtures $\frac{1}{2}x + \frac{1}{2}x'$ and $\frac{1}{2}y + \frac{1}{2}y'$ are indistinguishable as they both define the "total chaos". The logic of properties forms a Dedekind's lattice (there are 4 types of walls: the empty set $\emptyset$, the vertices, the sides, and the whole square). However, it is not orthocomplemented: for any pure state x, among the walls orthogonal to x no greatest one exists, and so, no unique negation in "quantum logic" can be defined.

A question now arises: is it so, that a statistical figure like that on Fig. 6 could indeed reflect the statistics of some real objects, or is it bound to represent a kind of science fiction structure in the framework of axiomatic quantum mechanics? In spite of known arguments [9, 18, 19] the answer to that question would be still premature. Instead of looking for plausible axioms limiting the structure of S to that of the orthodox theory it is better to analyse first what the other structures might bring. A programme for such an investigation has been raised in [16]. In the present paper the "convex scheme" is used in whole generality: it will be shown that only this unrestricted scheme is wide enough to describe the possible non-linearities in quantum mechanics.

3. RELATION TO NON-LINEARITY

The non-linearity may affect various levels of quantum theory. One can deal with certain physical quanta which, when they propagate in vacuum, are well described by linear wave equations: nevertheless their dense clouds (approximately described by c-number fields) obey non-linear propagation laws[6]. This kind of non-linearity will be called *secondary*. One can also think about more basic type of non-linearity affecting not only the macroscopic fields but also the isolated quanta: so, that even a free particle in vacuum would not propagate according to a linear equation. The non-linearity of the kind might be called *primary*. A hypothetical experiment in which such a phenomenon could be detected can be imagined as follows. Suppose, one has a quantum beam and a pair of slits like those in Young's experiment. The beam intensity is low and an analogue of Fabricant's experiment is performed: the quanta pass the slits and drop separately onto a screen behind. Suppose now, that the statistics of quantum hits on the screen reveals the interference fringes. However, unlikely to the traditional quantum mechanical experiments, the intensity of the fringes is not a sinusoidal function of the screen cartesian coordinate: the phenomenon, though typically quantum in spirit, fails to fit the scheme of the orthodox quantum mechanics. The question then arises: what sort of quantum mechanics could be constructed on the basis of such a hypothetical experiment? An interesting variant of this question is obtained by assuming that the shape of the interference fringes can be associated with a certain non-linear wave equation. The question then becomes: how can one construct a non-linear analogue of quantum mechanics with a non-linear wave equation playing the role of the Schrödinger equation?

To give an answer one is led back to the origin of quantum mechanics. The development of the theory involved two basic heuristic steps: 1) the guess of the manifold of the pure states (to be that of Schrödinger's waves), and 2) the guess of the statistical interpretation (stating that the transition probabilities are determined by the scalar products). In attempt of generalizing quantum mechanics the problem of the pure states is not the main difficulty, because some natural imitations of the step 1) are possible. The most obvious of them arises if one starts from a certain generalized "Schrödinger equation" (linear or not) which admits a certain conserved quantity $e(\varphi) \geq 0$, characterizing the solutions φ and interpreted as a "total wave charge". In that case, one can define the set Φ of all "normalized" solutions φ with $e(\varphi) = 1$: one might assume that the solutions $\varphi \in \Phi$ represent the pure states of certain hypothetical quanta.

The choice of the statistical interpretation is much less obvious. For waves obeying non-linear equations the scalar products, in general, are not conserved by the time evolution and so, they are not appropriate to define the transition probabilities. The question thus arises, which functions of the non-linear waves should substitute the orthodox quantities $|(\psi, \varphi)|^2$? One of the advantages of the convex scheme, is that this question does not need to be answered a priori. Indeed, in the approach outlined in §2 the transition probabilities are not fundamental but secondary: it is enough to have the statistical figure in order to reconstruct every detail of the statistical interpretation. The problem which becomes now essential concerns S as a whole: how can one reconstruct the shape of the statistical figure starting from some minimal physical information? In particular: how can one find the structure of the mixed states if it is assumed that the pure states obey a non-linear wave equation? In the axiomatic approaches the structure of S was usually determined by some universal regularity arguments always leading to the Hilbert space form of S. A possibility of a more general construction has been noticed in [16]; it yields S provided that the following two elements are given:

1) A topological manifold Φ of elements φ, ψ, ... which are candidates to

represent the pure states of a hypothetical system. The correspondence between the elements $\varphi \in \Phi$ and the pure states needs not to be one-to-one: each pure state, in general, may be represented by a whole subclass of elements of Φ[7]. The topology on Φ should be compatible with the observable properties of the pure states. In what follows Φ is called a *manifold of pure states*.

2) A class of real continuous functions on Φ representing the observational data and called the *observables*.

The second of the construction elements quoted here differs from the traditional algebra of observables and deserves a special description.

Observables

The concept of an observable in the present day theory has a complex status. It stands for both the measurement operation and its numerical output. The mathematical entities used to unify these two aspects are the self adjoint operators with their spectral decompositions. Below, this "conceptual union" will be split. The operator entities will be reserved to represent the transition processes rather than the observation acts [16]. The idea of observable will be associated with a purely numerical result of a measurement mathematically represented by a c-number function of the pure states.

Definition. Given a manifold of pure states Φ an *observable* or a *statistical quantity* is any function $f: \Phi \to \mathbb{R}$ whose values $f(\varphi)$ are interpretable as the statistical averages on various pure states $\varphi \in \Phi$ of a certain quantum mechanical measurement.

Here, the quantum mechanical measurement stands for any experimental mechanism assigning real numbers to the members of a quantum ensemble. The limiting assumption is that the mechanism should be sensitive only to the properties of the single ensemble individuals and not to their correlations within the ensemble. Since no other limitations are present, the above introduced concept of observable is wider than the orthodox one: it stands not only for an average indication of a "perfect" measuring device composed of ideal filters (as in the majority of papers adopting von Neumann's approach) but can also describe average effects shown by "imperfect" or "mixed" devices [8] or even by quite arbitrary macroscopic bodies sensitive to the presence of quanta and endowed with numerical scales (thus coinciding with a more general concept of *effects* introduced by Ludwig [14]). In spite of that generality it is not at all so that every function on the set of pure states is an observable. To the contrary, the structure of the present day theory is precisely due to the fact that the observables form a relatively narrow subclass of all functions on Φ. Indeed, the manifold Φ here is the unit sphere in a Hilbert space and the statistical averages are all given by the scalar products $(\psi, A\psi)$ $(\psi \in \Phi)$ where A are the self-adjoint operators. Thus, in the orthodox theory only the quadratic forms are observables: the other functions of ψ, though they can be experimentally determined, are not statistical averages of any quantum mechanical experiment. The above distinction is one of intriguing features of orthodox quantum mechanics: there is some mystery in the fact that for a microsystem certain functions of the pure states can and certain cannot be measured as statistical averages.

Since this is so, it is reasonable to assume that a similar distinction should also exist in case of a general theory. This fact is so essential that it might be choosen a new fundamental aspect of quantum theory competitive to "quantum logic" and "algebra of observables". *Given the set of pure states of a certain hypothetical system, the nature of the system should be characterized by indicating which functions on the pure states are the observables.* From now on the class of these functions will be denoted F. Given the pure states, the contents of F may serve to classify the theories: the richer the class of observables the "more classical" the theory. Some general statements concerning the structure of F can be made. Thus, F should be a linear class: given two devices destinated to measure two observables $f_1, f_1 \in F$, a new device can easily be constructed which would measure, as a statistical average, any linear combination $\lambda_1 f_1 + \lambda_2 f_2 (\lambda_1, \lambda_2 \in \mathbb{R})$. Because of the continuity of the macroscopic world some assumptions asserting the closed character of F should also be adopted. It would be too strong to assume that any point-wise limit of functions from F must also belong to F. This could lead to the appearence in F of some discontinuous functions of the pure state corresponding to over-idealized measuring devices inconsistent with the "physical topology" (see also the discussion by Giles [8]). In what follows, a weaker property will be only assumed, namely, that F is closed in the set $C(\Phi)$ of all real continuous function on Φ endowed with the topology of the point-wise convergence. This property will be called the *relative closure* of F.

The possibility of describing quantum systems in terms of observables suggests the following formal definition.

Definition. A quantum system without dynamics is a pair of entities (Φ, F) where Φ is a manifold of pure states and F is a relatively closed, linear class of continuous real functions on Φ called "observables".

Before exhibiting the relation with the convex scheme of §2 it is worth while to notice that this definition is naturally suited to the particular task of constructing non-linear variants of quantum mechanics. In fact, suppose that the manifold of pure states Φ coincides with a certain set of "non-linear waves". The most natural next heuristic step then is not so much to guess the character of quantum logic or the shape of S but rather to postulate the class of observables F. Here, some natural hints exist based on the character of the non-linear wave dynamics. One feels, that a class of functions should be found which would reflect the structure of a non-linear wave equation as naturally as the class of the quadratic forms reflects that of linear equations.

The derivation of the observables from the dynamics of the ψ-wave is to be treated in §4. Now, it will be shown that the description of a quantum system in terms of pure states and observables is complete: given the pair of entities (Φ, F), the statistical figure of the system can be uniquely constructed.

Construction of S

Indeed, suppose that Φ and F are known. Then, consider the set Π of all the probability measures π defined on Borel subsets of Φ for which the integrals $\int_\Phi f(\varphi)d\pi(\varphi)$, $(f \in F)$, are convergent. The set Π has an innate structure of a convex set: given two probability measures $\pi_1, \pi_2 \in \Pi$ and two numbers $p_1, p_2 \geq 0$, $p_1 + p_2 = 1$, the linear combination $p_1\pi_1 + p_2\pi_2$ is again a probability measure and it belongs to Π. The elements of Π have a natural physical interpretation. Every $\pi \in \Pi$ is a prescription for preparing a mixed state: the prescription says that for any Borel subset $\Omega \subset \Phi$ the fraction of the pure components from Ω taken into the mixture is $\pi(\Omega)$. Thus, Π might be used to label the mixed states of the system. However, the correspondence between the elements $\pi \in \Pi$ (prescriptions) and the resulting mixtures is not necessarily one-to-one: this still depends upon the observable properties of the mixed states. For any $f \in F$ and $\pi \in \Pi$ the value of the observable f on the mixed state prepared according to the prescription π may be defined by the integral $f(\pi) = \int_\Phi f(\varphi)d\pi(\varphi)$. The numbers $f(\pi)$ $(f \in F)$ represent the collection of the observable properties of the mixture π. Now, two measures π, π' are called *equivalent* $(\pi \equiv \pi')$ iff $f(\pi) = f(\pi')$ for every $f \in F$. Any two equivalent measures are interpreted as two prescriptions for producing mixtures which lead to physically indistinguishable mixed states. The equivalence relation $\equiv$ is a crucial element of the construction which accounts for the "first impossibility principle". Having given that relation one now defines a *state* (pure or mixed) as a class of equivalent probability measures. Consistently, one constructs the statistical figure S as the convex set of all equivalence classes $S = \Pi/ \equiv$.

While Π is a generalized simplex, the quotient set S, in general, has a distinct structure: its geometry reflects the physics of the hypothetical quanta. Having given S one can reconstruct the whole rest of the scheme as described above. This shows that the class of observables F is indeed the key element of the theory which provides a unique construction of the generalized quantum scheme. A particular example of this construction accounts for the origin of the orthodox quantum mechanics.

Orthodox Theory

That theory borrows some essential features from the linear classical field theory. Its main assumption states that each pure ensemble of quanta can be described by a complex field ψ (a "wave function") which obeys a linear wave equation. Since for the linear equations the densities of the typical conservative quantities are quadratic in the field, it is natural to assume that one such quadratic density (which will be denoted by $\psi^+(x)\psi(x)$) represents the average space density of the ensemble and the corresponding integral $(\psi, \psi) = \int_\infty \psi^\dagger(x)\psi(x)\, d_3x$ is the total expected number of the ensemble particles. Since the total particle number does not enter into the description of a state, one can thus label the pure states by those ψ-field for which $(\psi, \psi) = 1$; this leads to the manifold Φ being the unit sphere in a Hilbert space. At this moment certain elements of the statistical interpretation are already clear: it is decided that the probability of finding an average ensemble particle in an arbitrary space domain Ω is given by the quadratic form $p_\Omega(\psi) = \int_\Omega \psi^\dagger(x)\psi(x)d_3x$. This assumption will be further called the *primitive statistical interpretation* and the forms p_Ω will be called the *primitive observables*. In agreement with the previous consideration the construction of the statistical figure S requires the knowledge of the total class of observables F. The main indication as to the nature of that class follows again from the linearity of the quantum mechanical evolution equation. As seems reasonable to assume, the general quantum mechanical measurement upon ψ-wave can be accomplished in two stages. First, the ψ-wave undergoes a preliminary evolution process which is mathematically described by a certain norm conserving operation $\psi \rightarrow T\psi$. Because of the linearity of·the evolution equation, the operation T too is linear. Second, some primitive observables are measured upon the evolved wave $\psi' = T\psi$. This is done by capturing the particle in one of disjoint reception domains Ω_1, Ω_2, ... characterizing the structure of the measuring apparatus.

If the reception domains Ω_j are labelled by real numbers λ_j (the scale of the apparatus) the statistical quantity measured is $f = \lambda_1 p'_{\Omega_1} + \lambda_2 p'_{\Omega_2} + \cdots$, where p'_Ω are the "evolved" primitive observables $p'_\Omega(\psi) = p_\Omega(T\psi)$. Now, since p_Ω are quadratic and T is linear, the evolved p'_Ω are again quadratic and so is the statistical quantity f. This implies that only the quadratic forms are measured as statistical quantities in quantum mechanical experiments. A question now arises as to how large is the set of quadratic forms which are observables? The simplest assumption is, that each real, continuous quadratic form on Φ is an observable and can be measured, at least in principle, as a statistical average of an adequate experiment. (A motivation of that assumption will be given in §4). With these assumptions made the whole scheme of the orthodox theory can be uniquely deduced. Indeed, let π be a probability measure on the unit sphere Φ and let f be an observable. Since f is quadratic, there exists a linear functional f defined and continuous on a subset of the tensor product space $\bar{\mathcal{H}} \otimes \mathcal{H}$, such that $f(\psi) = f(|\psi \rangle \langle \psi|)$ ($\psi \in \Phi$). Hence, $f(\pi)$ is determined by values of f on $\bar{\mathcal{H}} \otimes \mathcal{H}$:

$$(3.1) \qquad f(\pi) = \int_\Phi f(\psi)\,d\pi(\psi) = \int_\Phi f(|\psi\rangle\langle\psi|)\,d\pi(\psi) = f\left(\int_\Phi |\psi\rangle\langle\psi|\,d\pi(\psi)\right).$$

This formula implies that the physically essential properties of the mixture produced according to the prescription π depend only upon the following hermitean, positive, unit-trace element of $\bar{\mathcal{H}} \otimes \mathcal{H}$ called a "density matrix":

$$(3.2) \qquad x_\pi = \int_\Phi |\psi\rangle\langle\psi|\,d\pi(\psi).$$

This is how the orthodox "density matrices" emerge from the division of Π into equivalence classes. The resulting statistical figure S is isomorphic with the convex set of the entities of form (3.2); $S = \{x \in \mathcal{G}(\mathcal{H}): x = x^\dagger \geq 0, \mathrm{Tr}x = 1\}$. The rest of the orthodox scheme follows. The whole structure is essentially conditioned by the choice of F: one might define the orthodox quantum mechanics as a theory of such a c-number wave for which only the quadratic forms are the observables. If another class of observables were choosen, different from that of the quadratic forms, the same mathematical mechanism would produce a distinct statistical figure corresponding to a different theory. In order to illustrate that dependence, a sequence of hypothetical schemes will be now discussed.

Higher Order Schemes

Similarly as before, let Φ be the unit sphere in $\mathcal{H}$. Now, however, assume that the class of observables F is not the set of the quadratic forms like in orthodox theory but the set F_{2n} of all the continuous $2n$-th order forms f given by

$$(3.3) \qquad f(\psi) = h(\psi, ..., \psi; \psi, ..., \psi),$$

where $h(\varphi_1, ..., \varphi_n: \psi_1, ..., \psi_n)$ are hermitean multiforms in $\mathcal{H}$ linear in the variables ψ_j and antilinear in φ_j's. Since $(\psi, \psi) = 1$ on Φ, each $2n$-order form $f \in F_{2n}$ coincides on Φ with some forms of higher order: $f(\psi) = (\psi, \psi)^k f(\psi)$ ($k = 1, 2, ...$). Hence, $F_2 \subset F_4 \subset F_6, ...$ and so, the classes of observables F_{2n} correspond to hypothetical theories with extending varieties of macroscopic measuring devices. A characteristic property of such a sequence of theories is a step-wise recess of the "first impossibility": the wider the class of observables the more selective the perception of the mixed ensembles and more kinds of mixture become physically distinguishable.

Denote now by $S(\mathcal{H}, 2n)$ the statistical figure constructed for $F = F_{2n}$. An analytic description of $S(\mathcal{H}, 2n)$ is possible similar to that employed by the orthodox theory. Indeed, suppose that f is a $2n$-order form. Then, there exists a linear, hermitean form $\mathbf{f}$, defined and continuous on a subset of $\underbrace{\mathcal{H} \otimes \cdots \otimes \mathcal{H}}_{n} \otimes \underbrace{\mathcal{H} \otimes \cdots \otimes \mathcal{H}}_{n}$ such that $f(\psi) = \mathbf{f}(\bar{\psi} \otimes \cdots \otimes \bar{\psi} \otimes \psi \otimes \cdots \otimes \psi)(\psi \in \Phi)$. Hence, for any $\pi \in \Pi$:

$$(3.4) \qquad f(\pi) = \int_\Phi f(\psi)\, d\pi(\psi) = \int_\Phi \mathbf{f}(\bar{\psi} \otimes \cdots \otimes \bar{\psi} \otimes \psi \otimes \cdots \otimes \psi)\, d\pi(\psi)$$

$$= \mathbf{f}\left(\int_\Phi \bar{\psi} \otimes \cdots \otimes \bar{\psi} \otimes \psi \otimes \cdots \otimes \psi\, d\pi(\psi)\right).$$

This formula implies that the physical properties of the mixture prepared according to the prescription π are fully determined by the following hermitean, positive, unit-trace element of $\mathcal{H} \otimes \cdots \otimes \mathcal{H} \otimes \mathcal{H} \otimes \cdots \otimes \mathcal{H}$ which is a natural generalization of the orthodox "density matrix";

$$(3.5) \qquad x_n \big|_{2n} = \int_\Phi \bar{\psi} \otimes \cdots \otimes \bar{\psi} \otimes \psi \otimes \cdots \otimes \psi\, d\pi(\psi).$$

The entity (3.5) might be called a "density tensor". The $S(\mathscr{H}, 2n)$ is precisely the convex set of all density tensors of form (3.5) (the subset of all the positive, unit-trace elements of $\mathscr{H} \otimes \cdots \otimes \mathscr{H} \otimes \cdots$ which are decomposable into convex integrals of the simple multitensors $\bar{\phi} \otimes \cdots \otimes \bar{\phi} \otimes \phi \otimes \cdots \otimes \phi$). One thus arrives here at a new realisation of the old scheme of quantum states, quantum observables and expectation values; the states are now the decomposable density tensors of an even order, the observables are arbitrary hermitean tensors of that same order and the expectation values are given by the tensor contractions.

The generalization introduced by $S(\mathscr{H}, 2n)$'s is formally similar to that achieved by the introduction of higher order multipole moments to the description of a density distribution. In fact, what the orthodox theory amounts to is the description of the probability measures $\pi \in \Pi$ in terms of their hermitean quadrupole moments with respect to the centre of the unit sphere Φ. In case of $S(\mathscr{H}, 2n)$ $(n > 1)$ higher order multipole moments of the measures are used. Since for the measures distributed over the unit sphere the higher order moments determine the lower order moments, the information contained in the subsequent "density tensors" is increasingly precise. Consistently, the statistical figures $S(\mathscr{H}, 2n)$ are "increasingly classical": each next of them represents, from the point of view of each previous, a kind of a "hidden parameter" scheme with an increased manifold of the mixed states and recessed impossibility principles. This leads to a tempting question: is the impossibility of measuring as statistical average of anything but quadratic forms of ϕ indeed so fundamental as assumed by the present day quantum mechanics or is it only a technical barriere? Are the higher order forms of ϕ basically beyond the reach of quantum statistics or, perhaps they could be measured if only sufficiently subtle experimental techniques were employed?

Forbidden Measurements

Though no indications exist about the incompleteness of the present day theory, it is one of the advantages of the "convex" approach that it exhibits some areas in which that theory, at least in principle, could be broken. Since the quadratic character of the observables is conditioned by the linearity of the evolution processes the most obvious such area consists in hypothetical evolution processes in which the quantum mechanical wave function would undergo a non-linear change. The most formal way of introducing such processes would be to assume some new couplings between the wave function and the external world. Thus, for instance, having a spinorial wave ψ of Dirac electron one might assume the existence of a hypothetical external scalar field φ coupled with ψ according to $L_{int} = G(\bar{\psi}\psi)\varphi(G' \neq$ const) and making ψ to evolve according to the non-linear equation

$$(3.6) \qquad (\gamma^\mu\partial_\mu - im)\,\psi + \varphi G'(\bar{\psi}\psi)\psi = 0.$$

The evolution (3.6), though allowed by the general principles of constructing couplings between c-number fields, would nevertheless break the consistency of the orthodox quantum mechanics. In fact, if an external field φ was created catalyzing a non-linear behaviour of ψ, this field could be used to measure, as statistical averages, some non-quadratic forms of ψ: in order to do that it would be sufficient to let a pure electron beam pass an external field φ and then measure, upon the evolved ψ, some conventional quantum mechanical observables. As a result, the orthodox impossibility principles of quantum mechanics would be broken and the traditional manifold of the "density matrices" would become insufficient to represent the enriched set of the mixed states. Hence, if one assumes the sufficiency of the orthodox scheme, one must also assume the impossibility of non-linear response processes like (3.7). This exhibits a limitation of quantum theory close to Haag's consistency requirement: the orthodox electron cannot enter into arbitrary couplings without loosing its identity.

The above example is of rather theoretical merit: it seems not probable that quantum mechanics will be broken just by inventing new types of external fields. However, the non-linear processes might arise from less artificial sources. The partial differential equations of quantum mechanics (like those of Schrödinger or Dirac) concern, in essence, only such evolution processes in which a microparticle interacts with an infinitely heavy macroscopic surrounding which is not affected by the presence of the particle. The processes of that kind form the proper domain of quantum mechanics and are well described by the known types external potentials none of which violates the traditional linearity. The situation is less obvious if the microparticle interacts with an object which, though macroscopic, is not infinitely inert but "subtle" and can modify its properties under the influence of the approaching particle. In that case one is tempted to consider the possibility of a hypothetical interaction process in which the wave function of the micro-object would undergo a non-linear change. The process may be described as follows.

There are two objects participating in the interaction: a microparticle and a macroscopic system. The microparticle is described by a certain wave function whereas the state of the macrosystem is determined by a set of classical parameters. The state of the whole of micro + macro system is simply the pair of states of the systems components. At the beginning of the process the microparticle and the macroscopic medium do not yet influence each other: the state of the microparticle is given by a certain wave ϕ which, at least approximately, evolves according to a linear wave equation whereas the macrosystem is in a certain standard initial state. Then the mutual interaction starts: the state of the macrosystem (which has the "subtle" ability of reacting to the particle presence) is modified under the influence of the approaching particle. This, in turn, modifies the way how the particle propagates. Thus, the wave function of the micro-particle interacts with itself by modifying its own macroscopic environment. At a certain conventionally choosen final moment the interaction is again insignificant: the microparticle is now in a new state ϕ' which depends upon the initial state ϕ: $\phi' = A(\phi)$. The result of the microparticle self interaction via the macroscopic system is the non-linearity of the operator A.

Note, that such theoretical schemes find some concrete realizations in the framework of the existing theory. Thus in the quasi-classical electrodynamics the electron is represented by a spinorial wave ϕ which is supposed to interact with a classical electromagnetic field. The wave ϕ here is assumed to produce a classical field A_μ which, in turn, influences the propagation of the wave:

$$(3.6) \qquad A_\mu(x) = \int_\infty \Delta^{\text{ret}}(x - x')\,j_\mu(x')\,d_4 x';\, j_\mu = e\,\bar{\phi}\gamma_\mu\phi;$$

$$(3.7) \qquad \left[\gamma^\mu\!\left(\partial_\mu - \frac{e}{c}\,A_\mu\right) - \text{im}\,c\right]\phi = 0.$$

According to the orthodox quantum mechanics the schemes like the quasi-classical electrodynamics or the theory of a "self-consistent wave" cannot describe correctly the propagation of single quanta but can only provide approximate data concerning the average behaviour of clouds of many interacting particles. However, attempts are also made to assign to them an exact meaning [13, 20]. According to that point of view, the electromagnetic field would be classical in nature and the quantal effects of electrodynamics would exclusively follow from the behaviour of fermions involved. Not entering into details of the discussion it is interesting to

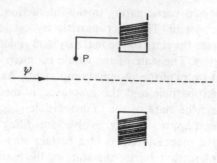

Fig. 7

notice, however, that the serious assumption about the classical nature of the electromagnetic field would imply some deep changes in the quantum mechanics of fermions themselves. Indeed, the classical electromagnetic field interacting with the electron would be an example of a "subtle" macroscopic medium leading to a non-linear behaviour of the electron wave function ψ. This non-linearity, even if quantitatively small, could be arbitrarily amplified by adequate electron transmitters employing the classical character of A_μ. One such hypothetical device is represented in Fig. 7.

The device on Fig. 7 is composed of an electromagnet and a sensor registering the intensity of the electric field at a detection point P. When the electric field at P is above a certain critical value, the electromagnet responds connecting its own magnetic field; otherwise it remains inactive. Obviously, the device reacting in this manner could produce a strongly non-linear transformation of the electron ψ-wave and so, it would allow one to get out of the "enchanted circle" of the quadratic observables[8]. An intriguing question thus returns: are the non-linear response processes of the quantum mechanical wave function indeed impossible? Or, perhaps, they could be produced if the experimental techniques were advanced enough to construct some "subtly reacting" macroscopic devices as that suggested in Fig. 7?

From the point of view of the present day theory the answer is decisively negative. However, it must be replied that the orthodox theory should not be used to prove the unphysical character of non-linear response processes in quantum mechanics. It may not be excluded that the scheme of the modern theory of coupled systems with its formalism of tensor products (which is the alternative of non-linearity) plays a similar role in particle physics as the epicycle structure did in ancient kinematics. In that case the "forbidden processes" would not be impossible but to the contrary, they would form a natural area to look for new quantum phenomena.

4. DYNAMICS AND GEOMETRY

The scheme of the generalized quantum mechanics begins now to emerge. As is clear from the second impossibility principle, there is a link between the dynamics and the geometry of quantum systems (see discussion in §2). As follows from the considerations of §3 an important element of that link is the class of observables F. Now, the link will be completed by constructing the class of observables for systems with given dynamics.

Similarly as in §3 a hypothetical system will be considered with the manifold of pure states Φ being the set of all "normalized waves" $\psi = \{\psi(x)\}$, where the values $\psi(x)(x \in \mathbb{R}^3)$ belong to a certain finite dimensional real or complex vector space $\mathcal{E}$ and the normalization condition is given by $e(\psi) = \int_\infty I(\psi(x))d_3x = 1$, $I(\xi)$ being a positive functional of the vector variable $\xi \in \mathcal{E}$. For so normalized waves the preliminaries of the statistical interpretation will be defined by the quantities $p_\Omega(\psi) = \int_\Omega I(\psi(x))d_3x$ which will be called the *primitive observables* and interpreted as the probabilities of localizing the system in a pure state ψ in various space domains $\Omega \subset \mathbb{R}^3$.

Given the primitive observables, the knowledge of dynamics happens to be the only element necessary to reconstruct the total class of observables F and so, to determine the shape of the statistical figure.

Dynamics

In classical field theory the dynamics is usually introduced by postulating an evolution equation which governs the propagation of the fields. This point of view must be now modified. What, in fact, is the dynamics is not so much one evolution law but rather a whole family of such laws, determining the behaviour of the system in various "external environments". Indeed, what one deals with in case of Schrödinger or Dirac dynamics are whole families of structurally similar partial differential evolution equations with arbitrary potentials representing the "external world"; it would be dynamically empty if one knew only the vacuum versions of the Schrödinger or Dirac equations. This leads to the following notion of dynamics as a complex entity.

Definition. An *evolution law* is any law (equation) which allows one to reconstruct the evolution of a physical system if the initial conditions are given. The *dynamics* is a class of evolution laws defining the behaviour of the system in various external conditions.

The concepts used here are open to further specifications. Thus in 'non-relativistic theories the initial conditions define the state of the system

Table 1

	External Potentials	Dynamics	Normalization
1	V	$i\hbar\dfrac{\partial\psi}{\partial t} = -\dfrac{\hbar^2}{2m}\Delta\psi + V\psi$	$\int_{\infty}|\psi|^2 d_3\boldsymbol{x} = 1$
2	V	$i\hbar\dfrac{\partial\psi}{\partial t} = -\dfrac{\hbar^2}{2m}\Delta\psi + V|\psi|^2\psi$	$\int_{\infty}|\psi|^2 d_3\boldsymbol{x} = 1$
3	V,U	$i\hbar\dfrac{\partial\psi}{\partial t} = -\dfrac{\hbar^2}{2m}\Delta\psi + V\psi + U|\psi|^2\psi$	$\int_{\infty}|\psi|^2 d_3\boldsymbol{x} = 1$
4	V	$i\hbar\dfrac{\partial\psi}{\partial t} = -\dfrac{\hbar^2}{2m}\Delta(|\psi|^2\psi) + V\psi$	$\int_{\infty}|\psi|^4 d_3\boldsymbol{x} = 1$
5	V	$i\hbar\dfrac{\partial\psi}{\partial t} = -\dfrac{\hbar^2}{2m}\Delta(|\psi|^2\psi) + V|\psi|^2\psi$	$\int_{\infty}|\psi|^4 d_3\boldsymbol{x} = 1$

in a certain initial time moment and the knowledge of the evolution in the knowledge of the system states in all other moments. In the relativistic case the initial conditions stand for the Cauchy data defining the properties of the system on a certain space like surface and the knowledge of the evolution is the possibility of reconstructing the system behaviour on the whole rest of the space time. It is also open how rich a variety of data should be substituted for the external conditions. One might be interested in simplified models of dynamics with relatively poor classes of external conditions (as, for instance, the dynamics of a wave diffracting on a rigid, macroscopic body). In practice, more complex models of dynamics are important with the external reality described by at least arbitrary potentials. Some examples are listed in Table 1.[9]

All the examples of Table 1 are the generalizations of the nonrelativistic Schrödinger's dynamics which occupies the Position 1. The Example 2 has the vacuum propagation law identical with the Schrödinger vacuum equation but represents a distinct case of dynamics because the field ψ is coupled distinctly to the external world. The dynamics of Example 3 is essentially richer than those of Example 1, 2 for it assumes an external reality in which two different types of potentials are present. In all three

Examples 1–3 the quadratic form $\int_\infty |\psi|^2 d_3 x$ is the basic conservative quantity used to define the normalization and thus suggesting the choice of the primitive observables. It is no longer so in case of the Examples 4 and 5 which are based on the different vacuum propagation law:

$$(4.1) \qquad i\hbar \frac{\partial \psi}{\partial t} = -\frac{\hbar^2}{2m} \Delta(|\psi|^2 \psi).$$

For that law the 4-th order form $\int_\infty |\psi|^4 d_3 x$ plays an analogous role as $\int_\infty |\psi|^2 d_3 x$ in case of the Schrödinger equation. Hence, if Examples 4, 5 reflected the dynamics of a certain quantum wave, the origins of the statistical interpretation could be based upon $|\psi|^4$ as a fundamental statistical density. Examples of wave dynamics for which the form $|\psi|^k (k > 0)$ would be a basic conservative quantity could be as easily constructed.

Motion Group

In what follows the decisive role belongs not to the dynamics itself but rather to a certain superstructure generated by the class of dynamical laws. In order to define it precisely the relative character of quantum states must be recalled. It is convenient to agree that the waves ψ considered here represent the pure states in Schrödinger's picture. These are relative entities: by telling that a state of a system is given, one has in mind the situation which takes place on a certain space-like hyperplane Σ in Minkowski (or Galileo) space-time as perceived by an inertial observer for whom Σ is a plane of simultaneous events with the time coordinate $t = 0$. Hence, the exact entity to which the concept of a (Schrödinger) state is refered is a space-like hyperplane endowed with a cartesian-frame of three space-like coordinates and with the "past" and the "future" *hyperplane*. In order not to complicate the scheme at this stage, all the framed hyperplanes considered below will be assumed to determine a common direction of the future in space-time.

Let now Σ and Σ' be two framed hyperplanes and let the state of a hypothetical object on Σ be described by a certain wave ψ. Assume, that the dynamics of the ψ-wave is known and that some definite external conditions in the space-time exist. Then, a definite dynamical law is valid which determines a new state on Σ' described by a certain wave ψ'. Given the dynamics, the mapping $\psi \leftarrow \psi'$ depends upon the pair of framed

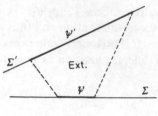

Fig. 8

hyperplanes Σ and Σ' and upon the external conditions involved: $\psi' = T_{\Sigma, \Sigma', \text{Ext.}}(\psi)$[10].

The operators $T_{\Sigma, \Sigma', \text{Ext}}$: $\Phi \to \Phi$ represent the possible evolutions of the ψ-waves (inbetween various pairs of framed hyperplanes and in presence of various external conditions) consistent with the dynamics assumed: they will be further called *motions*. The dynamical information essential for the structure of a quantum system can now be synthetized by introducing the following concept of a *motion group* which gathers the totality of all *could be* motions of the system.

Definition. Given a manifold of pure state Φ, the *motion group M* is the smallest relatively closed group of transformations $\Phi \to \Phi$ containing all the operations $T_{\Sigma, \Sigma', \text{Ext}}$. Here, the *relative closure of M* means that every homeomorphism $\Phi \to \Phi$ which is a point-wise limit of transformations belonging to M must also belong to M. Thus, the group M is closed in the set of homeomorphisms of Φ endowed with its natural topology.

The above *motion group* is essentially wider than the sometimes employed *dynamical group* of a system. It contains all one-parameter transformation groups corresponding to all admissible turns of evolution. Since Σ and Σ' can, in particular, be two distinctly framed versions of the same geometrical plane, the group M, appart of genuine dynamical processes, must also contain purely kinematical transformations of Φ corresponding to isometries on a fixed hyperplane in the Minkowski (or Galileo) space-time: it might be said that M reflects the total "kineto-dynamical mobility" of the system.

The group M provides an essential information concerning the structure of the observables. Indeed, having any measuring device destinated to measure a certain statistical average $f \in F$ one can produce more observables by alternating the measurement process. Instead of measuring straightforwardly the statistical average f on a given wave $\psi \in \Phi$ one can let ϕ undergo first a certain preliminary kineto-dynamical process $T \in M$ and only afterwards measure f on the evolved wave $\psi' = T\phi$ thus obtaining a new statistical quantity: $(fT)(\phi) = f(\psi') = f(T(\phi))$. In this way the existence of "motions" prevents one to assume too poor a class of observables: having any observable $f \in F$ one must also assume the existence in F of an infinity of other observables of form fT generated by all possible evolution processes which the system might perform under the influence of various external forces. This suggests the following formal assumption:

ASSUMPTION. The class F is invariant with respect to the motion group M.

With that assumption made, there is now only one heuristic step necessary to complete the generalized quantum scheme.

Main Assumption

One of simplifications made by the present day theory consists in attributing an equal status to all the observables. Each self-adjoint operator in a Hilbert space is assumed to represent a physical observable: each of them is as well measurable as any other. It is a feature of quantum phenomenology, however, that it fails to reflect this abstract equality: for what one observes in reality is a distinguished role of the position measurements. Indeed, the known quantum experiments seem to follow the same general scheme in which the act of the localization is the ultimate experimentator's tool to extract the physical data[11]. The scheme consists in letting the particle wave function undergo a certain preliminary evolution process as a "preparatory stage" of the measurement and then in detecting the particle in one of spatially separated "reception domains".

As already noticed (compare §3) the statistical quantity measured in so arranged experiment is a linear combination of the "evolved" primitive observables $f = \sum_j \lambda_j p_{\Omega_j} T$ where p_{Ω_j} are the position observables corresponding to the space domains Ω_j ($p_\Omega(\phi)$ means the probability that a particle in a pure state ϕ will be localized in Ω and T denotes the operator of the preliminary evolution process). This indicates that an arbitrary observable $f \in F$ can be either expressed or at least approach by linear combinations of the quantities pT where p are the primitive observables and T are the motions of the system. The universality of this measurement technique does not seem to be conditioned by the particular character of the orthodox quantum mechanical evolution equations (like that of Schrödinger or

Dirac) but it rather forms on immanent feature of the macroscopic measuring techniques. Hence, it is reasonable to believe that the reducibility of the general to the primitive observables should also exist in a general theory, though the operators T are no longer linear there. This leads to the following general assumption which represents the required guess of the observables for a system with a non-linear dynamics.

ASSUMPTION. Given a manifold of pure states Φ, a collection P of real functions on Φ interpreted as the "primitive observables" and a motion group M, the complete class of observables F is the smallest linear and relatively closed class of continuous real functions on Ω containing P and invariant under the group M.

This assumption is the last heuristic step relating the geometry of quantum system to the dynamics assumed. Once this step is taken one has no longer freedom of specifying further the structure of quantum states and the statistical interpretation since those elements are already determined: having given F one uniquely reconstructs the shape of the statistical figure and subsequently, all other geometric aspects of the theory. It is worth while to notice that from the heuristics just completed a certain abstract description of quantum systems emerges which is not even restricted to the particular domain of quantum theories based on wave equations.

Group-Theoretical Model

In fact, for the applicability of the constructions here outlined it is not essential to assume that the pure states correspond to certain c-number waves. Instead, one can take Φ to be a topological manifold of elements $\varphi, \psi, \ldots$ of arbitrary nature. The primitive statistical interpretation, too, does not need to depend upon the possibility of constructing positive conservative densities for some wave functions. Instead, one might just pick up a certain class P of real functions on Φ and decree them to be the "primitive observables" (the problem whether the elements $p \in P$ represent localization experiments and, eventually, how are they related to the space domains $\Omega \subset \mathbb{R}^3$ is to be resolved on the level of concrete theories). Finally, in order to fix the "dynamics" one has to choose a certain group M of transformations $\Phi \to \Phi$ and decree it to cover the totality of the kineto-dynamical processes which the system can undergo in various physical circumstances. This leads to the following group theoretical model of a quantum system.

DEFINITION. A *quantum system with dynamics* is a set of five entities (Φ, M, P, F, S) where:

1) Φ is a topological manifold of points ψ, φ, ... covering the pure states of a hypothetical physical system (the correspondence between the pure states and the elements $\varphi \in \Phi$, in general, is not one-to-one).

2) M is a relatively closed subgroup of homeomorphisms $\Phi \to \Phi$ called the *motion group* and representing the dynamics of the system. (There is a generating subset of elements in M which can be identified with the evolutions of the system in various external conditions.)

3) P is a class of continuous functions on Φ which are called the *primitive observables* and interpreted as statistical averages of some "elementary measurements" which can be performed upon the system.

4) F is the total *class of obserables* constructed as the smallest linear and relatively closed class of real continuous functions on Φ which contains P and is invariant under the transformation group M.

5) S is the *statistical figure* constructed according to the prescription of §3.

The intrinsic structure of quantum mechanics described by this definition might be represented in the following scheme.

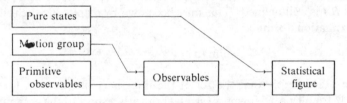

One of the most traditional realizations of that scheme is the non-relativistic quantum mechanics of one particle.

One-Particle Orthodox Quantum Mechanics

For simplicity, a non-relativistic particle in one space dimension will be considered. The pure states then are represented by complex waves $\psi = \{\psi(x)\}$ ($x \in \mathbb{R}^1$) with the normalization $\int_{-\infty}^{+\infty} |\psi|^2 \, dx = 1$ and the manifold Φ is the unit sphere in the complex Hilbert space $\mathcal{H} = L^2(-\infty, +\infty)$. The dynamical laws are of Schrödinger's form

$$(4.2) \qquad i\hbar \frac{\partial \psi}{\partial t} = -\frac{\hbar^2}{2m} \frac{\partial^2 \psi}{\partial x^2} + V(x)\psi$$

where V are arbitrary potentials. The quantity $|\psi|^2$ is a positive conservative density of the evolution laws (4.2); one thus guesses that it has the meaning of the probability density for the localization experiments (Born) and consistently, one chooses the primitive observables in the form:

$$(4.3) \qquad p_\Omega(\psi) = \int_\Omega |\psi|^2 \, dx; \; \Omega \subset \mathbb{R}^1.$$

The motion group of the system is now to be determined. This group should contain all the unitary one-parameter subgroups of form $U = e^{-i/\hbar H\tau}(\tau \in \mathbb{R}^1)$ which are the Hamiltonians with all possible external potentials: $H = -(\hbar^2/2m)(\partial^2/\partial x^2) + V$. The further considerations form only an outline and have to be completed by exact proofs. Since the group M is relatively closed, hence, together with any two families of operators $e^{\tau A}$, $e^{\tau B}$ (where A and B are imaginary operators in $\mathcal{H}$) it should also contain any operator of the form $e^{\alpha A + \beta B}$. This may be seen by applying the Trotter's formula:

$$(4.4) \qquad e^{\alpha A + \beta B} = \lim_{n \to \infty} (e^{\alpha A/n} e^{\beta B/n})^n.$$

Similarly, if $e^{\tau A}$, $e^{\tau B} \in M(\tau \in \mathbb{R}^1)$, the group M should contain the unitary subgroup of the form $e^{\tau[A, B]}$ (provided that the commutator $[A, B] = AB - BA$ is well defined). This may be shown by employing the following approximation formula

$$(4.5) \qquad e^{[A, B]} = \lim_{n \to \infty} (e^{A/n} e^{B/n} e^{-A/n} e^{-B/n})^{n^2}.$$

These facts imply that the set of the imaginary operators K for which $e^{\tau K} \in M$ for all $\tau \in \mathbb{R}^1$ forms a Lie algebra with respect to the commutation $[A, B] = AB - BA$. This algebra can be justly called the "Lie algebra" of the generalized "Lie group" M and denoted ad M. Now, taking $A = -(\hbar^2/2m)(\partial^2/\partial x^2)$ and $B = -(\hbar^2/2m)(\partial^2/\partial x^2) + V$, where V is an arbitrary potential, one infers that $V = B - A \in$ ad$M$. Similarly, the operator $K_1 = (\partial/\partial x)f(x) + f(x)\partial/\partial x = [A, B] \in$ ad M, where $f = -(\hbar^2/2m)V'$ is an arbitrary function. The K_1 happens to be the most general operator of form $K = k(x, p)$ which is linear in $p = -i\hbar(\partial/\partial x)$. In turn, by commuting $\partial^2/\partial x^2$ and K_1 one shows that $K_2 \in$ ad M, where K_2 is the most general imaginary function $k(x, p)$ quadratic in p. By induction one shows that ad M contains any imaginary operator $k(x, p)$ which is analytic in p. Since M is relatively closed, this indicates that ad M coincides with the set of all imaginary operators in $\mathcal{H}$ and so, M coincides with the unitary group acting on the unit sphere

Φ. The reconstruction of the class of observables F now becomes the problem of the representation theory: the required F is the smallest linear and relatively closed class of continuous real functions on Φ invariant under the unitary group and containing the quadratic forms p_Q. The known facts of the representation theory of the finite-dimensional unitary groups indicate that this class is just the set of all bounded quadratic forms on Φ. The rest of the "genesis" of the orthodox theory was already described in §3.

Note, that the construction presented above exhibits some new consistency relations between the known elements of the orthodox quantum mechanics. Indeed, in the traditional presentation of that theory some essential elements of the statistical interpretation (such as the general expression for the transition probabilities by $|(\psi, \varphi)|^2$) are introduced in form of separate assumptions apparently independent of the primitive facts of the theory. This is not so in the group theoretical model just outlined. Here, once it is assumed that the wave ψ obeys the Schrödinger's dynamics and once the Born's interpretation is accepted of $|\psi(x)|^2$ as the probability density, the whole rest of the scheme, together with the general expression for the transition probabilities, uniquely follows via construction of S. Thus, the group theoretical model yields a deepened insight into the "anatomy" of orthodox quantum mechanics.

General Programme

The possibility of more general realizations of the scheme is now open. An obvious thing to do would be to consider the known examples of classical non-linear wave dynamics with some naturally choosen "primitive observables" and then, to use the group theoretical model in order to see, to what kind of quantum systems do they lead. Mathematically, this involves a wide programme of structural investigations of non-linear field theories centered around two problems.

• One has to find out the motion groups for non-linear theories. This task, though it does not present a fundamental difficulty, is nontrivial as far as the effective knowledge of the motion group is concerned. As is already clear, for some non-linear versions of the Schrödinger's dynamics (like Examples 2, 3 in Table 1) the resulting motion group is significantly richer than the unitary group[12]. An essential question arises, as to, when M coincides with the group of all homeomorphisms of the wave manifold Φ? The theory in which this happens would lead to the class of observables

F coinciding with the set of all continuous real functions on Φ and so, the resulting statistical figure would be a generalized simplex. It would be interesting to know what kind of non-linearities in the quantum mechanical wave equation imply, in this way, the return to a classical theory.

● ● Given a motion group M acting on a wave manifold Φ and given a class of primitive observables P one has to reconstruct the total class of observables F. Mathematically, this amounts to looking for a class of continuous real functions on Φ which would be a minimal representation subspace of the group M containing P. The effective solution of that problem requires the development of the representation theory of infinite dimensional Lie groups and the theory of special functions on infinite dimensional differential manifolds. Those branches are already taking their first steps (see e.g. [7]); they form a natural area to look for the way out of the 50 years old formalism of Hilbert spaces in quantum physics.

5. CONSISTENCY PROBLEMS

Though the question about the practical use of the generalized quantum scheme is still open, this scheme already exhibits some essential consistency relations between quantum mechanics and other physical theories. This can be seen by applying consistency ideas of Haag's type to interacting physical systems.

Suppose, there are two generalized quantum systems A and B. The phenomenology shows that the structure of A can be described by a manifold of pure states Φ_A and a class of observables F_A. Similarly, B can be described by a manifold of pure states Φ_B and a class of observables F_B. Suppose now, that a theory is formulated which tells, how A interacts with B. In this moment the problem of consistency arises. Indeed, when A is assumed to interact with B a certain new technique of measuring the statistical averages upon A-states is created. The techniques consists in letting the members of an ensemble of A-systems interact with replicas of B and then in measuring some statistical quantities on the evolved B instead of A.

Obviously, with the help of such a technique any observable of F_B induces a certain statistical average to be measured on the states of A. The question is: are these induced quantities already included in F_A? If so, the theory of interaction is consistent with the originally assumed structure of A. If not, the scheme is inconsistent and must be modified in at least one of three directions. Either one can interpret such a measurement as a sort of "subtle device" which yields a deeper insight into the real structure of A and consistently, requires an extension of F_A. Or, one can infer that the assumptions concerning the structure of B were premature and the consistency must be repaired by removing from F_B some functions which are not indeed observables. Or, finally, one can suspect that there is something to be changed in the assumed interaction mechanism. A typical example of an inconsistent interaction scheme is that of quasi-classical electrodynamics described in §3. It seems, that there is no way how the classical electromagnetic field could be coupled with the orthodox electron without introducing too rich a class of observables, inconsistent with the orthodox structure of the electron states. This causes no difficulty in the present day electrodynamics where the mechanism of the inconsistency is removed by quantizing the electromagnetic field. The situation is much less obvious as far as the interaction between the microparticles and the gravitational field is concerned. Though some schemes with quantized gravitation are already emerging, the opinions also exist that the gravitation is an immanently classical phenomenon, even on the micro-level. If that was so, the consequences of that fact for quantum mechanics would be much deeper than generally recognized. Indeed, the classical gravitational field interacting with the electron would be a 'subtly reacting medium" as described in §3 and the measuring devices based upon gravity detection would provide a method of measuring non-quadratic observables against one of fundamental prohibitions of the present day quantum mechanics. This might indicate, that we are facing the following alternative: *either the gravitation is not classical or quantum mechanics is not orthodox.* It is possible (though there is no mathematical proof at the moment) that, under the assumption about the classical nature of gravitation, the detectors of the gravitational field of a microparticle would allow one to measure, as a statistical average, any continuous functional of the microparticle pure state. This would lead to even stronger alternative: *either the gravity is quantum or the electron is classical.* These alternatives exhibit a somewhat peculiar situation of the present day theories: though it may be very difficult to quantize the gravitation, it is even more difficult not to do it. The incompleteness of the present day science at this point is, perhaps, one more reason why the scheme of quantum mechanics should not be prematurely closed.

Acknowledgements. The author is indebted to Professor I. Bialynicki-Birula, Professor J. Hulanicki, Professor J. Plebański, Dr. R. Raczka, Professor A. Trautman, Professor

J. Werle, Dr. L. Woronowicz and other colleagues for their interest in this paper and helpful discussions. Special thanks are due to Professor F. A. E. Pirani for his hospitality at King's College in London in May and June 1971 where a part of this investigation has been carried. The author is also grateful to his colleagues from the Institute of Theoretical Physics of Chalmers in Göteborg and from the Institute of Theoretical Physics of the Royal Institute of Technology in Stockholm for critical and helpful discussions of non-linearity problem in quantum mechanics.

NOTES

[1] This section is a review of essentially known material.

[2] One could think about an axiomatic approach where the convex set S would be defined without involving the surrounding space. A step in that direction was recently taken by Gudder [22].

[3] Not every property of an ensemble is of such a nature that it can be attributed to each single ensemble individual. An example can be obtained by considering a particle beam and a semi-transparent window: the fact that the average beam particle penetrates the window with probability 1/2 reflects a certain property of the beam as a whole. This property does not necessarily concern each single beam particle: for it may happen that the beam is a mixture of two distinct types of particles one of which is completely transmitted and the other completely absorbed by the window.

[4] The so introduced concept of a logic is wider than in the majority of axiomatic approaches. The notion of orthogonality employed here is more primitive than the usually introduced concept of negation. Given the lattice W of walls with the orthogonality $\perp$, the physically interpretable negation can or cannot be introduced depending on the structure of the relations $\perp$ and $\leq$ on W. If for an $a \in W$ the family $a^{\perp} = \{x \in W : x \perp a\}$ contains its lowest upper bound, this element can be denoted a' and called the "orthogonal complement" (negative) of a. If not, then no unique negative can be assigned to a. For an arbitrary convex set the resulting logic $W(\leq, \perp)$, in general, does not admitt the construction of the orthocomplementation $a \to a'$.

It is an open question whether all the walls of S should be considered the physically essential elements of the "logic" or some regularity requirements should be suplemented (stating, for instance, that only the closed walls of S correspond to "physically verifiable properties"). An extensive discussion of an equivalent problem is due to Giles [8].

[5] The importance of this type of consistency for quantum theory was pointed out to the author by Professor R. Haag during the winter school in Karpacz, February 1968.

[6] This happens in quantum electrodynamics where each single photon in vacuum, although "dressed", propagates according to Maxwell equations, whereas the intense photon beams exhibit non-linearities because of photon-photon interactions.

[7] It is also no harm if Φ covers a submanifold of quantum states wider than just the set of the pure states. This may happen in case of superselection rules, when Φ is taken to be the unit sphere in a Hilbert space, but some elements of Φ, because of the structure of observables, are *a posteriori* identified with mixed states of the system. (For an interesting discussion of that point the author is indebted to Dr. S. Woronowicz.)

[8] By employing the classical character of A^{μ} one would be able to construct a large

family of non-orthodox measuring devices based on the detection of the electromagnetic field. For instance: a counter which clicks only when the value of $\psi^\dagger\psi$ at a certain detection point is great enough; a counter which may not register the electron at a point A if $\psi^\dagger\psi$ at another point B is too big, etc.

[9] For the non-linear equations of Table 1 the problem about the precise shape of the corresponding dynamics is still open. It is to be decided whether the theory should deal with the singular solutions, which are likely to appear in case of non-linear wave equations, or, perhaps, it may be restricted to the regular ψ-waves at the cost of limiting the external potentials and the initial conditions involved.

[10] In case of dynamics like those of Table 1 the dependence $\psi \to \psi'$ should be determined by solving the Cauchy problem for a partial differential equation. Given a wave $\psi = \{\psi(x)\}$ as an initial condition on the framed hyperplane Σ and given the external potentials in the partial differential evolution equation one has to find out the unique space-time extrapolation $\{\psi(x,t)\}$ of the initial wave ψ such that $\psi(x,\ 0) = \psi(x)$ (the coordinates $x,\ t$ are induced by the frame on Σ). The values $\psi(x,\ t)$ at points $(x,t) \in \Sigma'$ after being transformed to the coordinate frame of Σ' (according to their covariant character) determine the required wave ψ'. For the non-linear wave equations the mapping $\psi \to \psi'$ may produce singular solutions on Σ' which might indicate that the theory should be opened toward singular Cauchy data.

[11] The author appreciates a stimulating discussion of that point with Professor J. Werle.

[12] Authors notes. Unpublished.

University of Warsaw
Institute for Theoretical Physics

REFERENCES

1. Segal, I. E.: Ann. Math. **48**, 930 (1947).
2. Birkhoff, G., von Neumann. J.: Ann. Math. **37**, 823 (1936).
3. Dähn,G.: Commun. math. Phys, **9**, 192 (1968); **28**, 109 (1972), **28**; 123 (1972).
4. Davies, E. B., Lewis, J.T.: Commun. math. Phys. **17**, 239 (1970).
5. Davies, E. B.: Commun. math. Phys. **15**, 277 (1969); **19**, 83 (1970); **22**, 51 (1971).
6. Edwards, C. M.: Commun. math. Phys. **16**, 207 (1079); **20**, 26 (1971); **24**, 260 (1972).
7. Flato, M., Sternheimer, D.: Commun. math. Phys. **14**, 5 (1969).
8. Giles, R.: J. Math. Phys, **11**, 2139 (1970).
9. Gunson, J.: Commun. math. Phys, **6**, 262 (1967).
10. Haag, R., Kastler, D.: J. Math. Phys. **5**, 848 (1964).
11. Hellwig, K. E., Kraus, K.: Commun. math. Phys. **11**, z14 (1969); **16**, 142 (1970).
12. Jauch, J. M., Piron, C.: Helv. Phys. Acta **42**, 842 (1969).
13. Leiter, D.: Intern. J. Theor, Phys, **4**, 83 (1971).
14. Ludwig, G.: Z. Naturforsch. **22a**, 1303 (1967); **22a**, 1324 (1967).
15. Ludwig, G.: Commun. math. Phys. **4**, 331 (1967); **9**, 1 (1968); **26**, 78 (1972).

16. Mielnik, B.: Commun. math. Phys. **15**, 1 (1969).
17. Neumann, H.: Commun. math. Phys. **23**, 100 (1971).
18. Piron, C.: Helv. Phys. Acta **37**, 439 (1964).
19. Pool, J. C. T.: Commun, math. Phys. **9**, 118 (1968); **9**, 212 (1968).
20. Sachs, M.: Intern. J. Theor, Phys, **4**, 83 (1971).
21. Stolz, P.: Commun. math. Phys, **11**, 303 (1969); **23**, 117 (1974).
22. Gudder, S.: Commun. math. Phys, **29**, 249 (1973).

PETER MITTELSTAEDT

QUANTUM LOGIC

I. INTRODUCTION

It has been shown by Birkhoff and v. Neumann (1936) and by Jauch and Piron (1963, 1964, 1968) that the subspaces of Hilbert space constitute an orthocomplemented quasi-modular lattice L_q, if one considers between two subspaces (elements) a, b the relation $a \subseteq b$ and the operations $a \cap b$, $a \cup b$, $a^\perp$. Furthermore, since the subspaces can be interpreted as quantum mechanical propositions, and since the operations $\cap$, $\cup$, $\perp$ have some similarity with the logical operations $\wedge$ (and), $\vee$ (or) and $\neg$ (not), the question has been raised already by Birkhoff and v. Neumann, whether the lattice of subspaces of Hilbert space can be interpreted as a prop-ositional calculus, sometimes called *quantum logic*.

There are many kinds of lattices which can be interpreted as a prop-ositional or logical calculus. A Boolean lattice L_B of propositions corre-sponds to the calculus of classical logic and an implicative (Birkhoff, 1961) lattice L_i has as a model the calculus of effective (intuitionistic) logic. Therefore one could ask the question, whether also the lattice L_q of quantum mechanical propositions can be interpreted in some sense as the calculus of quantum logic. A formal condition which must be satisfied, if one wants to interpret a lattice as a propositional calculus, is that the partial ordering relation $\leqslant$ can also be defined as an operation. This condition results from the requirement that the lattice must be the Lindenbaum Tarski algebra of the logical calculus considered. For the lattices L_B, L_i and L_q it is well known that this condition is fulfilled (Mittelstaedt, 1972).

More important is the question of the semantic interpretation of the lattice L_q. A Boolean lattice L_B can be interpreted by a two-valued truth function. It has been shown by Gleason (1957), Kamber (1965), Kochem and Specker (1967), that neither a two-valued truth function nor a con-veniently generalized truth function does exist on the lattice L_q. On the other hand, it is well known that also the intuitionistic logic, i.e. the lattice

Hooker (ed.), Physical Theory as Logico-Operational Structure, 153–166.
All Rights Reserved. Reprinted from PSA 1974.

L_i cannot be interpreted by truth values. However, an implicative lattice can be considered as a logical calculus if one takes into account a more general method – the operational method which makes use of the dialogic technique (Lorenzen, 1962; Lorenz, 1968; Kamlah and Lorenzen, 1967). Although this interpretation cannot directly be applied to the lattice L_q it turns out that a generalisation of the dialogic method can be used as an interpretation of a lattice L_{eq} which is isomorphic to L_q if one adds to its axioms the 'tertium non datur' $I \leqslant a \vee \neg a$ (Mittelstaedt, 1972a; Mittelstaedt and Stachow, 1974; Stachow, 1973).

In order to demonstrate the possibility of a logical interpretation of the lattice L_q by means of the dialogic technique we proceed in the following way: (Figure 1). We start with the discussion of some obvious modification of the operational method, which come from the incommensurability of the quantum mechanical propositions, i.e. from the theory of quantum mechanical measurement. With the aid of the modified dialogic method we derive a propositional calculus Q_{eff}, which is similar to the calculus of the intuitionistic logic but contains a few restrictions which come from the incommensurability of quantum mechanical propositions. Therefore, we call this calculus the calculus of the effective quantum logic. Taking account of the definiteness of truth values for quantum mechanical propositions – which property follows again from the theory of measuring process – we arrive at the calculus Q of full quantum logic, which also contains the 'tertium non datur'. This calculus represents a model for the lattice L_q and can therefore be used as a logical interpretation of this lattice.

II. THE DIALOGIC INTERPRETATION OF LOGIC

We consider elementary statements about a quantum mechanical system, which we will denote by a, b, c. We will assume that it has been established how these elementary propositions can be proved, i.e. that for each proposition a method is known which is to be regarded as evidence for the respective proposition. On the set S_e of elementary propositions we define connected propositions $c(a, b)$, where a, $b \in S_e$. In particular we use the connectives:

Conjunction: $a \wedge b$ (a and b)
Adjunction: $a \vee b$ (a or b)

Subjunction: $a \to b$ (a then b)
Negation: $\neg a$ (not a)

Apart from the colloquial denotation 'a and b', 'a or b', 'a then b' and '$\text{not } a$', we explain these connectives as follows by a dialog (Lorenzen, 1962; Lorenz, 1968; Kamlah and Lorenzen, 1967): One of the participants in the discussion (Proponent P) asserts the connected proposition $c(a, b)$, the other (Opponent O) attempts to refute this assertion. If the proponent asserts, for instance, the proposition $a \to b$, then he assumes the obligation, in the case that a can be proved by the opponent, of justifying b. Hence a dialog in which P successfully defends the proposition $a \to b$ reads:

P	O
$a \to b$	proof of a. Why b
proof of b	

On the basis of explanations of this kind, which are described in detail in the literature (Lorenzen, 1962; Lorenz, 1968; Kamlah and Lorenzen, 1967; Mittelstaedt, 1972a; Mittelstaedt and Stachow, 1974), it is defined how a connected proposition $c(a, b)$, formed by the connectives $\wedge$, $\vee$, $\to$ and $\neg$ can be proved, if a and b are elementary propositions.

It is obvious that the connectives $\wedge$, $\vee$, $\to$ and $\neg$ can be iterated and that the concept of dialogic proof can be generalized to iterated connected propositions. A simple example of such a dialog is

P	O
1. $b \to (a \to b)$	1. proof of b. Why $a \to b$
2. $a \to b$	2. proof of a. Why b
3. proof of b.	

The set of all arbitrarily connected propositions – including the elementary propositions – will be denoted here as S_P. The connectives $\wedge$, $\vee$, $\to$, $\neg$ are then *operations* on this set.

If a proposition $c \in S_P$ can be proved, i.e. defended in a dialog, we write $\vdash c$. It is useful in addition to the *operations* $\wedge$, $\vee$, $\to$, $\neg$, on the set S_P of propositions to define a *relation* $a \leqslant b$ with a, $b \in S_P$ by

$$a \leqslant b \rightleftharpoons \vdash a \to b.$$

This relation is called 'implication' and must be distinguished from the operation $a \to b$, denoted here as 'subjunction'. According to the definition the relation $a \leqslant b$ between the propositions a and b holds if and only if the proposition $a \to b$ can be defended in a dialog. If for two propositions the implications $a \leqslant b$ and $b \leqslant a$ are valid we call them equivalent and write $a = b$.

A proposition which can always be successfully defended in discussion irrespective of the elementary propositions it contains, is denoted as a logical statement or, more precisely, as an *effective logical* statement. A logical statement is one for which there is always a certain strategy of success within the scope of dialog, which does not depend on the content of the elementary propositions contained in it. A certain strategy of success can only exist if the proponent can never be committed by the opponent to prove a proposition explicitly – since the proposition in question could be false and in that case the proponent would lose the dialog. As an illustrative example we use the dialog mentioned above. In the framework of the possibilities of defending a proposition in a dialog so far considered the proponent is committed to prove proposition b in $P3$. However this proposition has been proved previously by the opponent in $O1$. If the proposition b proved in one instance is still valid at a later stage of the discussion, when b is challenged again, then the proponent should be allowed to refer to proposition b in $O1$ instead of proving it again in $P3$.

P	O
1. $b \to (a \to b)$	1. proof of b; why $a \to b$
2. $a \to b$	2. proof of a; why b
3. b (refer to $O1$)	

The assumption that a proposition once proved in the dialog remains valid throughout the discussion, is usually accepted in ordinary logic. Therefore the proponent should always be allowed to cite a proposition which has been proved previously by the opponent. By means of this additional possibility some connected propositions (i.e. logical statements) can be defended in a dialog independent of the particular propositions contained in it.

It is obvious that the totality of logical statements can only be presented if the rules for the dialog are formulated in detail. This has been done extensively in the literature (Lorenzen, 1962; Lorenz, 1968; Kamlah and Lorenzen, 1967) and should therefore not be repeated here. Using these rules for a dialog the calculus of effective logic can easily be derived.

III. THE CALCULUS OF EFFECTIVE QUANTUM LOGIC

If one considers a connected proposition about a quantum mechanical system, for the dialogic proof of this proposition it is of importance whether the elementary propositions contained in it are commensurable or not. The reason is that a proposition b, proved in one instance, may not simply be cited at a later stage of the discussion, when b is challenged once again. For in general in the course of the dialog other propositions not commensurable with b will have been proved and thus proposition b is no longer available. This situation can again be illustrated by the proposition $b \to (a \to b)$. If a and b are not commensurable, b may not simply be cited in the dialog mentioned above, and therefore $b \to (a \to b)$ cannot (in general) be successfully defended.

In order to incorporate this point of view into the detailed rules for a dialog, it must be clarified whether two propositions can be considered as commensurable. Generally one denotes two propositions a and b as commensurable, if the corresponding observables can be measured in arbitrary sequence on the investigated system without thereby influencing the result of the measurement. From this definition it follows that two propositions a and b, between which one of the implications $a \leqslant b$ or $b \leqslant a$ is valid, are always commensurable. Furthermore, it can be shown, for instance, that a proposition a is commensurable with $a \to b$, independently of b, and with $\neg a$.

A reformulation of the rules of dialog taking account of the restrictions important for quantum mechanical propositions has the consequence that not all dialogically provable implications can still be successfully defended in dialog. Propositions which, even after the inclusion of these restrictions, can still successfully be defended, will be denoted as *quantum-dialogically* provable. The detailed formulation of the rules of dialog incorporating the restrictions, which come from the incommensurability of quantum mechanical propositions, has been given elsewhere (Mittelstaedt, 1972a; Stachow, 1973) and shall therefore not be repeated here.

The totality of statements, which are formed with the connectives $\wedge$, $\vee$, $\rightarrow$, $\neg$ and can always be successfully defended quantum dialogically, i.e. independent of the content of the elementary propositions contained in it, will be called *effective quantum logic*.

For a formulation of the effective logic, which is most convenient for a comparison with lattice theory, it is useful to introduce the two special propositions $\bigvee$ (truth) and $\bigwedge$ (falsity). The use of both of these propositions within the framework of the dialogic method shall be established in such a way that $\bigvee$ cannot be questioned by either participant of the dialog and that whoever maintains $\bigwedge$ shall have lost the dialog.

From this definition it follows that the propositions $a \rightarrow \bigvee$ and $\bigwedge \rightarrow a$ can be proved in dialog for all propositions a. Therefore we have the general validity of the implications

$$a \leqslant \bigvee \qquad \bigwedge \leqslant a$$

from which in particular it follows that $\bigvee$ and $\bigwedge$ are commensurable with all propositions. According to the dialogic definition of $\bigvee$ we find that for an arbitrary proposition a the statement $\vdash a$ is equivalent to $\vdash \bigvee \rightarrow a$. On the other hand $\vdash \bigvee \rightarrow a$ is equivalent to the relation $\bigvee \leqslant a$. Therefore a proposition a can be defended in a dialog, i.e. $\vdash a$, if and only if the relation $\bigvee \leqslant a$ holds. Furthermore, since for any proposition a we have $\vdash a \rightarrow \bigvee$ the relation $\bigvee \leqslant a \rightarrow \bigvee$ holds for an arbitrary a. Correspondingly since for any a we have $\vdash \bigwedge \rightarrow a$ the relation $\bigvee \leqslant \bigwedge \rightarrow a$ and consequently the relation $a \leqslant \bigwedge \rightarrow a$ is valid for any a.

We wish to represent the effective quantum logic in the form of a calculus, that is, we will present a system of rules with the aid of which all propositions of effective quantum logic can be derived from a few statements which we will include in the rules as the point of departure of the calculus. For the formulae of the calculus we use combinations of the symbols $\bigvee$, $\bigwedge$, a, b, c,... with $\wedge$, $\vee$, $\rightarrow$, $\neg$ and $\leqslant$ and the bracket symbols. With these formulae we establish the rules for the derivation of implications $a \leqslant b$. For the designation of the rules we use the double arrow $\Rightarrow$ and the double comma „. With the aid of the rules of dialog which take account of the restrictions for quantum mechanical propositions we arrived at the following calculus, which will be denoted by Q_{eff}. ($b * c$ in 4.4 means any one of the connected propositions $b \wedge c$, $b \vee c$, $b \rightarrow c$, $\neg b$).

THE CALCULUS Q_{eff}

1.1	$a \leqslant a$
2	$a \leqslant b \,,\, b \leqslant c \Rightarrow a \leqslant c$
2.1	$a \wedge b \leqslant a$
2	$a \wedge b \leqslant b$
3	$c \leqslant a \,,\, c \leqslant b \Rightarrow c \leqslant a \wedge b$
3.1	$a \leqslant a \vee b$
2	$b \leqslant a \vee b$
3	$a \leqslant c \,,\, b \leqslant c \Rightarrow a \vee b \leqslant c$
4.1	$a \wedge (a \rightarrow b) \leqslant b$
2	$a \wedge c \leqslant b \Rightarrow a \rightarrow c \leqslant a \rightarrow b$
3	$a \leqslant b \rightarrow a \Rightarrow b \leqslant a \rightarrow b$
4	$a \leqslant b \rightarrow a \,,\, a \leqslant c \rightarrow a \Rightarrow a \leqslant (b*c \rightarrow a)$
5.1	$a \wedge \neg a \leqslant \bigwedge$
2	$a \wedge b \leqslant \bigwedge \Rightarrow a \rightarrow b \leqslant \neg a$
3	$a \leqslant \neg a \rightarrow a.$

The rule $\alpha \Rightarrow \beta$ states that if the implication α can be proved quantum dialogically, then the implication β can also be justified quantum dialogically. For the proof of a rule $\alpha \Rightarrow \beta$ the implication α will therefore be presupposed by the opponent as a *hypothesis* before the dialog. The proponent has then to defend the implication β (quantum dialogically) whereby he may refer to the hypothesis α which has been accepted by the opponent. It is obvious that the referability of the hypothesis α is not restricted in any way.

In order to illustrate the meaning of the calculus Q_{eff} a few important consequences, which can be derived from Q_{eff}, will be given here:

THEOREM I. The following implications can be proved in Q_{eff}:

(1)	$a \wedge b \leqslant a \rightarrow b$
(2)	$\neg a \leqslant a \rightarrow b$
(3)	$\neg a = a \rightarrow \bigwedge$
(4)	$\bigvee \leqslant a \rightarrow a$
(5)	$a \leqslant \neg \neg a.$

Note: (3) is well known as the intuitionistic definition of the negation;

(4) shows that the special proposition $\bigvee$ is equivalent to $a \to a$ for any arbitrary a; the inverse of (5) i.e. $\neg\neg a \leqslant a$ cannot be proved within the calculus Q_{eff}, just as in the ordinary effective logic.

THEOREM II. The following rules can be proved in Q_{eff}

(1) $a \leqslant b \Leftrightarrow \bigvee \leqslant a \to b$

(2) $a \leqslant \bigwedge \Leftrightarrow \bigvee \leqslant \neg a$

(3) $a \leqslant b \Rightarrow \neg b \leqslant \neg a$

(4) $b \leqslant a \,,\, c \leqslant \neg a \Rightarrow a \wedge (b \vee c) \leqslant (a \wedge b) \vee (a \wedge c)$

(5) $\bigvee \leqslant a \vee \neg a \Rightarrow \neg\neg a \leqslant a.$

Note: (1) reconfirms in terms of the calculus Q_{eff} the dialogic definition of the relation $a \leqslant b$ by $\vdash a \to b$; the inverse of the rule (3) cannot be proved within the calculus Q_{eff}, just as in the ordinary effective logic. The rule (4) will be called 'weak quasimodularity', since in the framework of full quantum logic II.4 expresses the quasimodularity of the lattice of propositions. (5) is important for the transition from the effective quantum logic to the full quantum logic.

Furthermore, the connected propositions $a \wedge b$, $a \vee b$, $a \to b$ of two propositions a and b and the negation of $\neg a$ of a proposition a, which had been originally explained by a dialog, are now uniquely defined by the calculus Q_{eff}, i.e. we have the

THEOREM III. For two given propositions a and b it follows from Q_{eff} that
   (1) the conjunction $a \wedge b$ is uniquely defined by 2.1, 2.2, 2.3.
   (2) the disjunction $a \vee b$ is uniquely defined by 3.1, 3.2, 3.3.
   (3) the subjunction $a \to b$ is uniquely defined by 4.1, 4.2, 4.3, 4.4.
   (4) the negation $\neg a$ is uniquely defined by 5.1, 5.2, 5.3.
The meaning of these uniqueness theorems is that the calculus Q_{eff} of effective quantum logic can be used without any further reference to the dialogic definitions of the connected propositions. Whereas the uniqueness of $a \wedge b$ and $a \vee b$ is well known from ordinary logic, the uniqueness of $a \to b$ and $\neg a$ is by no means trivial, since the 4.1–4.4 and 5.1–5.3 of Q_{eff} are quite different from the corresponding rules in the calculus L_{eff} of effective logic. In the calculus L_{eff} of effective logic, one has instead of 4.1–4.4 and 5.1–5.3 the rules

4.1* $a \wedge (a \rightarrow b) \leqslant b$
4.2* $a \wedge c \leqslant b \Rightarrow c \leqslant a \rightarrow b$
5.1* $a \wedge \neg a \leqslant \bigwedge$
5.2* $a \wedge b \leqslant \bigwedge \Rightarrow b \leqslant \neg a$

from which the uniqueness of the propositions $a \rightarrow b$ and $\neg a$ follows immediately. The proofs of Theorem I–III have been given elsewhere and should therefore not be repeated here. (Mittelstaedt, 1972a; Mittelstaedt and Stachow, 1974; Stachow, 1973).

IV. CONSISTENCY AND COMPLETENESS

The calculus Q_{eff} of effective quantum logic is *consistent* with regard to the class of quantum dialogically provable implications, i.e. every implication derivable from Q_{eff} can be proved quantum dialogically. The proof of this consistence property is straightforward. However, since we have not formulated in this paper the detailed rules of a quantum dialog, we cannot present this proof here and must refer to the literature (Mittelstaedt, 1972a; Mittelstaedt and Stachow, 1974; Stachow, 1973). Furthermore, the calculus Q_{eff} is also *complete* in respect to the class of quantum dialogically provable implications, i.e. every quantum dialogically provable implication can be derived from the calculus Q_{eff}. Consistency and completeness of Q_{eff} guarantee that the dialogic technique can indeed be completely replaced by the calculus Q_{eff}.

The proof of the completeness property is rather complicated. It is well known from ordinary logic that in order to prove the completeness of a logical calculus one first has to transform the rules for a dialog into an equivalent calculus for so called tableaux (Lorenzen, 1962; Lorenz, 1968). This tableau calculus can be shown to be equivalent to a calculus of sequences which was first introduced by Genzen. In a last step one shows the equivalence of this calculus of sequences with the Brower calculus of effective logic. The calculus Q_{eff} of effective quantum logic, which we have presented here, has the form of a Brower calculus. Therefore, for the completeness proof we have to proceed in the same way just mentioned. At first, one has to construct a tableau calculus T_{eff} of effective quantum logic which is equivalent to the rules of quantum dialog. In a second step, one must transform this calculus T_{eff} in an equivalent cal-

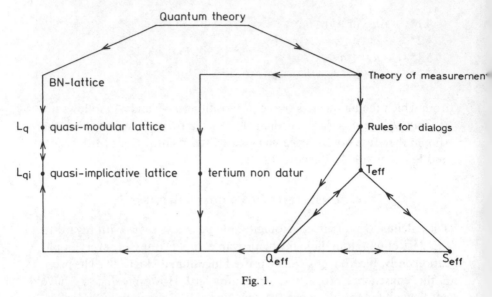

Fig. 1.

culus S_{eff} of quantum logical sequences. Finally, in a third step one has to show the equivalence of this calculus S_{eff} with the Brower like calculus Q_{eff} of effective quantum logic. Alternatively, one can also show directly the equivalence of T_{eff} and Q_{eff} (Figure 1). All these completeness proofs have now been finished by E. W. Stachow and will be published in a forthcoming paper (1975).

V. THE CALCULUS OF FULL QUANTUM LOGIC

In order to construct a quantum logical calculus which can be shown to be equivalent to the lattice L_q of quantum mechanical propositions mentioned above, we have to incorporate into the logical calculus the fact that the propositions considered have well defined truth values, that is, it can be decided whether they are true or false. For quantum mechanical propositions this assumption is always true, since for every quantum mechanical proposition one can decide between the two possibilities by an experiment. This weak assumption, that every proposition can be decided to be true or false, must not be confused with the much stronger postulate of the existence of a two-valued truth value function. It has already been mentioned above, that on the lattice L_q of quantum mechanical propositions such a two-valued truth function does not exist (Kamber, 1965; Gleason, 1957; Kochen and Specker, 1967), and that this fact is one of the essential reasons for the difficulty to interpret the lattice L_q as a logical calculus.

The assumption that the propositions considered have well defined truth values has the important consequence that for any arbitrary proposition a the 'tertium non datur' can successfully be defended in dialog, since either a or $\neg a$ can actually be proved (Kamlah and Lorenzen, 1967). Furthermore, from $\vdash a \vee \neg a$ it follows $\bigvee \leqslant a \vee \neg a$. Consequently, for a quantum mechanical proposition, the relation $\bigvee \leqslant a \vee \neg a$ is always valid. In a dialog which contains the elementary quantum mechanical propositions a, b, c,... the relations

$$\bigvee \leqslant a \vee \neg a, \quad \bigvee \leqslant b \vee \neg b, \quad \bigvee \leqslant c \vee \neg c, \ldots$$

respectively can therefore be presupposed as hypothesis.

In order to incorporate the 'tertium non datur' into the quantum logical calculus one has to add the implication

6.1 $\bigvee \leqslant a \vee \neg a$

to the rules of the calculus Q_{eff}. The resulting system of rules will be called the calculus Q of *full quantum logic*. This calculus differs essentially from Q_{eff}. The important consequences of the incorporation of the 'tertium non datur' into the quantum logical calculus can be demonstrated by

THEOREM IV. The following rules can be proved in the calculus Q but not in Q_{eff}

(1) $\neg \neg a \leqslant a$
(2) $\neg b \leqslant \neg a \Rightarrow a \leqslant b$
(3) $a \rightarrow b = \neg a \vee (a \wedge b)$.

Theorem IV.1 is the inverse of Theorem I.5 and follows from the 'tertium non datur' according to Theorem II.5. Theorem IV.2 is the inverse of Theorem II.3. Theorem IV.1 and IV.2 is important for the orthocomplementarity of the lattice of propositions. A further consequence which can be derived in the calculus Q of full quantum logic is the fact that the subjunction $a \rightarrow b$ of the two propositions a and b can be expressed by the other operations (4) due to Theorem IV.3. In the effective quantum logic the subjunction $a \to b$ is also uniquely defined by a and b (Theorem III.3), but in the framework of the calculus Q_{eff} the subjunction cannot be replaced by the other operations.

With the aid of the rules of the calculus of full quantum logic Q, the set of all quantum mechanical propositions can be considered as a lattice. Since we have already defined the relation $\leqslant$ between two elements, it is not necessary here to construct first the Lindenbaum-Tarski algebra of the propositional calculus. The calculus Q of full quantum logic can immediately be expressed in terms of lattice theory. However, for the lattice theoretical characterisation of Q it is not useful to translate the calculus Q directly into a lattice, but first to reformulate the rules of the calculus by means of Theorems I–IV in a system of more convenient but equivalent rules.

Because of the rules 1.1–1.2 the set of propositions forms with respect to the relation $\leqslant$ a partly ordered set. For every two elements a and b the rules 2.1–2.3 cause the existence of a greatest lower bound $a \wedge b$ and 3.1–3.3 provide the existence of a least upper bound $a \vee b$, so that the structure investigated is a lattice. The assumption that a proposition $\bigwedge$ and $\bigvee$ exists means that the lattice has a null and a unit element respectively. The rules 5.1, 6.1 together with Theorems I.5, II.3, III.4, IV.1, IV.2 imply that the lattice considered is orthocomplemented and that the orthocomplement of an element a is the uniquely defined negation $\neg a$. The rules 4.1, 4.2 and Theorem IV.3 means that the orthocomplemented lattice considered is also quasi-implicative, that is, for any two elements a and b there exists one and only one element $a \to b$, the quasi-implication which satisfies the rules 4.1, 4.2 and Theorem IV.3. The lattice of quantum mechanical propositions is therefore an orthocomplemented quasi-implicative lattice, which we will denote by L_{qi}. It can be characterized by the following axioms.

$$L_{qi} \quad {}^{(1)} \quad a \leqslant a$$
$$L_{qi} \quad {}^{(2)} \quad a \leqslant b, \ b \leqslant c \Rightarrow a \leqslant c$$
$$L_{qi} \quad {}^{(3)} \quad a \wedge b \leqslant a$$
$$L_{qi} \quad {}^{(4)} \quad a \wedge b \leqslant b$$
$$L_{qi} \quad {}^{(5)} \quad c \leqslant a, \ c \leqslant b \Rightarrow c \leqslant a \wedge b$$
$$L_{qi} \quad {}^{(6)} \quad a \leqslant a \vee b$$
$$L_{qi} \quad {}^{(7)} \quad b \leqslant a \vee b$$

L_{qi} (8) $a \leqslant c, \; b \leqslant c \Rightarrow a \vee b \leqslant c$

L_{qi} (9) $a \leqslant \bigvee, \; \bigwedge \leqslant a$ lattice with null and unit element

L_{qi} (10) $a \wedge \neg a \leqslant \bigwedge$

L_{qi} (11) $\bigvee \leqslant a \vee \neg a$

L_{qi} (12) $a = \neg \neg a$ orthocomplemented lattice

L_{qi} (13) $a \leqslant b \Leftrightarrow \neg b \leqslant \neg a$

L_{qi} (14) $a \wedge (a \to b) \leqslant b$

L_{qi} (15) $a \wedge c \leqslant b \Rightarrow \neg a \vee (a \wedge c) \leqslant a \to b$ quasi-implicative lattice

It can be shown (4) that the axioms L_{qi} (14) and L_{qi} (15) are equivalent to the axiom

$$L_{qi} \text{ (16)} \quad b \leqslant a, \; c \leqslant \neg a \Rightarrow a \wedge (b \vee c) \leqslant (a \wedge b) \vee (a \wedge c)$$

i.e. the lattice L_{qi} is quasimodular. Therefore we arrive at the lattice L_q, which we have already mentioned in connection with the subspaces of Hilbert space.

In conclusion, we find that the lattice of subspace of Hilbert space, which has been shown to be an orthocomplemented quasimodular lattice, can be considered as a model for that lattice L_q, which has been obtained on the other hand from the investigation of the quantum mechanical propositional calculus. This result shows that the lattice of subspaces of Hilbert space can indeed be interpreted as a propositional calculus. Our conclusion is that this interpretation can be based on the operational foundation of logic which makes use of the dialogic technique.

Institut für Theoretische Physik
Universität zu Köln

BIBLIOGRAPHY

Birkhoff, G.: 1961, 'Lattice Theory', *Ann. Math. Soc. Coll. Publ.* XXV, Rev. Ed., p. 147, 195.

Birkhoff, G. and Neumann, J. v.: 1936, *Ann. of Math.* 37, 823.

Gleason, A. M.: 1957, *J. of Math. and Mech.* 6, 885.

Jauch, J. M.: 1968, *Foundations of Quantum Mechanics*, Addison-Wesley Publishing Co., Reading, Mass.

Jauch, J. M. and Piron, C.: 1963, *Helv. Phys. Acta* 36, 827.

Kamber, F.: 1965, *Math. Ann.* 158, 158.

Kamlah, W. and Lorenzen, P.: 1967, *Logische Propädeutik*, Bibliographisches Institut, Mannheim.

Kochen, S. and Specker, E. P.: 1967, *J. of Math. and Mech.* 17, 59.

Lorenz, K.: 1968, *Arch. f. Math. Logik und Grundlagenforschung*.
Lorenzen, P.: 1962, *Metamathematik*, Bibliographisches Institut, Mannheim.
Mittelstaedt, P.: 1972, *Z. Naturforsch.* **27a**, 1358.
Mittelstaedt, P.: 1972a, *Philosophische Probleme der modernen Physik*, Bibliographisches Institut, Mannheim, English ed.: *Philosophical Problems of Modern Physics*, D. Reidel Publishing Company, Dordrecht, Holland, 1976.
Mittelstaedt, P. and Stachow, E. W.: 1974, *Found. of Physics* **4**, 335.
Piron, C.: 1964, *Helv. Phys. Acta* **37**, 439.
Stachow, E. W.: 1973, Diplomarbeit, The University of Cologne.
Stachow, E. W.: 1975, Dissertation, The University of Cologne (to be published).

C.H. RANDALL AND D.J. FOULIS

THE OPERATIONAL APPROACH
TO QUANTUM MECHANICS

1. THE OPERATIONAL APPROACH TO QUANTUM MECHANICS VERSUS THE OTHER APPROACHES.

The operational approach in the title of this paper refers to the primitive and fundamental status accorded to the concept of a physical experiment or operation by the authors. This emphasis on the operational approach should not be construed as an adoption of operationalism, logical positivism, or radical empiricism, as may initially appear to be the case. Our approach does not entail the rejection of subjective methods; in particular, it does not deny – and in fact readily accomodates – the unifying and explanatory power of idealized models.

In order to emphasize the novel character of this approach, we begin by comparing it to and contrasting it with some of the more popular alternate approaches. All approaches to quantum mechanics recognize that the basic ingredients of any realistic theory are *events, observables, states*, and *operations*. These approaches are distinguished by which of the four ingredients are taken as primitive. In our approach, operations are regarded as the only primitives – all other notions are considered to be less fundamental.

(a) The Orthodox Approach

(Dirac [10], von Neumann [40]).

In orthodox quantum mechanics, the physical states of a microscopic system are represented by unit vectors in a Hilbert space H. The temporal development of such a system, governed by the Schrödinger equation, produces a trajectory on the unit sphere in H; hence, this unit sphere serves as the quantum mechanical analogue of the phase space of classical mechanics. Here the Hilbert space H is primitive, and the events, observables, states, and operations enter as derived concepts. A (general) state, corresponding to an ensemble of physical systems, is represented by a positive trace class operator W of trace 1. An observable is represented by

167

Hooker (ed.), Physical Theory as Logico-Operational Structure, 167–201.
All Rights Reserved.
Copyright © 1978 by D. Reidel Publishing Company, Dordrecht, Holland.

a (possibly unbounded) self-adjoint operator A, and the expectation value of A in the state W is given by Tr (WA). The spectrum of A is interpreted as the (closure of the) set of possible numerical values produced by measurements of A.

The projection operators are precisely the self-adjoint operators whose spectra consist of at most the two values 1 and 0. Thus, it is reasonable to regard a projection as an observable whose measurement indicates whether or not some event has occurred. Conversely, it is worth noting that all self-adjoint operators (observables) can be recovered from projections (events) according to the spectral theorem.

In orthodox quantum mechanics, the procedure of measuring an observable and securing an outcome is often represented by the mapping which carries a state vector to its renormed projection onto an eigenspace of the observable. This is perhaps as close as one comes in conventional quantum mechanics to an explicit mathematical representation of a physical operation.

(b) The Quantum Logic Approach

(Birkhoff & von Neumann, [4], Jauch [23], Piron [32]).

This approach assigns a primtive role to events (or, if you prefer, propositions or questions). It is supposed that the occurrence or non-occurrence of such an event can be determined by carrying out a suitable physical procedure (or operation). Once again these operations have no explicit mathematical representation. However, the observables and states do enter explicitly as derived concepts.

It is assumed, to begin with, that the events form at least an orthomodular partially ordered set L (sometimes called a logic). An observable is represented by a σ-homomorphism $A: \mathscr{B} \to L$ from the σ-algebra $\mathscr{B}$ of all real Borel sets into L. For $E \in \mathscr{B}$, $A(E)$ is interpreted as the event that a measurement of A produces an outcome in E.

A state (or generalized probability measure) is represented by a σ-orthoadditive map $\alpha: L \to [0, 1]$ from the logic into the closed unit interval. For $E \in \mathscr{B}$, and an observable A, $\alpha(A(E))$ is interpreted as the probability that a measurement of A on a system in state α produces an outcome in E. Thus, the function F defined by $F(x) = \alpha(A(-\infty, x])$ is a cumulative probability distribution function, so that the expectation value of A in state α can be computed by $\int_{-\infty}^{\infty} x dF(x)$.

If one assumes that L is isomorphic to the orthomodular lattice of projection operators on a Hilbert space, then the structure just described is equivalent to orthodox quantum mechanics (Mackey [27]). This has suggested to some authors (e.g., Pool [31]) that operations might still be represented by conditioning maps on the set of states, much as they are in orthodox quantum mechanics. On the other hand, classical probability theory (e.g., Kolmogorov [25]) suggests that an operation might correspond to a maximal orthogonal set of events.

(c) The Algebraic Approach

(Jordan, von Neumann & Wigner [24], Segal [38], Haag & Kastler [19]).

This approach makes the observables primitive and supposes that they either form or generate a linear algebra – for instance, a C^*-algebra, a von Neumann algebra, or a Segal algebra. Here it is assumed that a physical state determines the average values of the observables; moreover, it is imagined that such states can be represented by normed, positive linear functionals on the algebra.

Events are not ordinarily introduced in this approach, although it is clear that idempotent elements of the algebra of observables could play such a role. Moreover, operations usually do not play an explicit part in the algebraic approach. If the notion of an operation appears at all, it is again in the form of a conditioning map on the set of states.

(d) The Ordered Vector Space Approach

(Stone [39], Mielnik [29], Davies & Lewis [9], C. M. Edwards [12], Ludwig [26]).

In this approach, the states of a physical system are taken as primitive – in effect, these states are represented by a convex set B which forms the base of a convex cone K in a real vector space V. It is assumed that (V, B) is a base normed space with a dual order unit space (V^*, e) (Alfsen [1]). The vectors in the cone K are often viewed as describing ensembles or beams of particles and the intensity of such a beam v is presumed to be $e(v)$.

Whereas in the preceding approaches the notion of an operation played at best a minor part, the ordered vector space approach assigns a major role to the conditioning maps corresponding to 'filtering' operations.

Naturally, such operations are represented by positive linear operators $T:V \rightarrow V$ for which the intensity $e(Tv)$ of the filtered beam does not exceed the intensity $e(v)$ of the incident beam. Thus one of the principal desiderata of this approach is achieved; operations conducted in connected sequences are represented by the composition of the corresponding linear operators. Moreover, it is worth noting that these linear operators form a convex set.

If $T:V \rightarrow V$ represents a filtering operation as above, then $e \circ T$ is a continuous linear functional on V belonging to $[0, e] = \{f \in V^* \mid 0 \le f \le e\}$. Conversely, every functional f in the dual order interval $[0, e]$ arises in this way. These functionals, called simple observables by C. M. Edwards [12], correspond to the effects of G. Ludwig [26]. The extreme points of $[0, e]$, the so-called decision effects of Ludwig, can be regarded as representing events. The observables are now introduced as effect valued measures $A: \mathscr{B} \rightarrow [0, e]$ defined on the real Borel algebra $\mathscr{B}$ (Edwards [12]).

(e) The Operational Approach

(Randall [33], Randall & Foulis [35], Dacey [8], Cook [6], Rüttimann [37]).

By a *physical operation* we mean instructions that describe a well defined, physically realizable, reproducible procedure and furthermore that specify what must be observed and what can be recorded as a consequence of an execution of this procedure. In particular, a physical operation must require that, as a consequence of each execution of the instructions, one and only one symbol from a specified set be recorded as the *outcome* of this realization of the physical operation.

In the empirical sciences, a well founded experimental program ultimately is concerned with some cohesive collection of physical operations – usually complete or exhaustive in some sense. In the operational approach, physical operations are taken as primitive and a cohesive collection of such operations is represented mathematically by a formal structure called a *manual*. (This terminology is meant to suggest a manual or catalogue of instructions.) Events, states, and observables appear as derived concepts. This apporach has the virtue of formally permitting all such derived concepts to be connected directly to the physical activities of an experimental science.

In both classical and quantum physics, an idealized model invariably entails an implicit manual of operations. As always, in such models, the operations in question may be realizable only 'in principle.' The problem of explaining a phenomenological manual in terms of an idealized model is clearly a matter of finding a suitable structure preserving map (morphism) between manuals. This matter is attended to in the sequel.

2. MANUALS OF OPERATIONS

It should be emphasized, to begin with, that the outcome of a realization of a physical operation is merely a *symbol*; it is not any real or imagined occurrence in the 'real world out there'. Also, note that if we delete or add details to the instructions for any physical operation, especially if we modify the outcome set in any way, we thereby define a new physical operation.

Evidently, the subjective judgment of the observer is implicit in every realization of a physical operation, not only in regard to the interpretation of the instructions, but also in connection with the decision as to which symbol to record as the outcome. In our view, if a competent observer believes that he has executed a particular physical operation and obtained a particular outcome, then, in fact, the operation has been realized and the outcome in question has indeed been secured. Each realization of a physical operation is to be understood here as a 'Ding an sich', isolated, with no 'before' and no 'after'. Physical history, as it were, begins and ends with each execution of a physical operation. To put the matter in more traditional terms, the various realizations of the admissible physical operations in a manual are always to be regarded as 'independent'.

If physical procedures are to be executed 'simultaneously', or carried out in 'connected sequence', or otherwise 'interwoven', then the instructions for such a *compound operation* must say so. When a compound physical operation is built up from more primitive physical operations, it is understood that each constituent operation may thereby lose its identity, since in such a synthesis it may be subject to interference. Notice, in regard to this, that the recording of the passage of time is often an important constituent operation of a compound operation.

Once we have assembled those physical operations of concern to us in a particular effort, we are almost inevitably moved, either by custom or by a particular intent, to identify certain outcomes of different operations. For instance, we often prefer to regard a number of outcomes of distinct physical operations as registering the same 'property', or, if you prefer, as representing the same 'measurement'. If a voltage is measured using different instruments – or even different methods – identical numerical results are ordinarily taken to be equivalent. Similarly, the instructions for carrying out the admissible procedures may all but dictate the indentification of certain outcomes on purely syntactic grounds. We surely wish to avoid the necessity of taking a stand on the 'acceptability' of any such identifications since we hope to keep our formal language as free as possible from ad hoc decisions. On the other hand, we wish our language to be able to handle this common practice of outcome identification and if we permit unrestricted identifications of outcomes, grammatical chaos could result. Accordingly, we shall subject the outcome identification process to certain mild constraints.

In order to make the nature of these constraints precise, let us suppose that we are considering a particular collection of admissible physical operations and that we have already made what we consider to be the appropriate outcome identifications. (Naturally, these identifications will often be motivated by some 'picture of the real world'.) As our first condition, which we shall refer to as the *determination condition*, we require that each outcome is so constituted that the set of all admissible physical operations capable of yielding this outcome can be discerned from the symbolic pattern of the outcome itself. Thus, when we assert that such an outcome has been obtained, we understand that what we are really asserting is that it was obtained as a consequence of an execution of one of these admissible physical operations – although it is not necessary to specify which one.

The second condition on such an identification process, which we shall call the *irredundancy condition*, requires that for each admissible physical operation $\mathscr{P}$, say with set of outcomes E, there is only one admissible physical operation – namely $\mathscr{P}$ itself – capable of yielding each and every outcome in E. An immediate consequence of this condition is that each physical operation not only determines, but is uniquely determined by the set of all of its outcomes. As a result, we can and shall identify each physical operation with its outcome set.

In view of the above, we propose to introduce as a formal mathematical representation for a collection, or manual, of admissible physical operations a non-empty set $\mathscr{A}$ of non-empty sets. Each set E in $\mathscr{A}$ is supposed to correspond uniquely to the set of all outcomes of one of the admissible physical operations and each admissible physical operation is supposed to be so represented. The condition of irredundancy now translates into the assertion that if $E, F \in \mathscr{A}$ with $E \subseteq F$, then $E = F$. The elements of $\mathscr{A}$

will henceforth be referred to simply as $\mathscr{A}$-*operations*, while the set theoretic union $X = \bigcup \mathscr{A}$ of all of the $\mathscr{A}$-operations will be referred to as the set of all $\mathscr{A}$-*outcomes*.

If x and y are two $\mathscr{A}$-outcomes with $x \neq y$, and if there exists an $\mathscr{A}$-operation E with $x, y \in E$, then we shall say that x is *orthogonal to* y and we shall write $x \perp y$. Notice that when $x \perp y$, then x and y reject each other operationally in the sense that an execution of E that yields the outcome x cannot yield the outcome y and vice versa. A set D of $\mathscr{A}$-outcomes will be called an *orthogonal* set provided that $x \perp y$ holds for all pairs x, y of distinct elements of D. Obviously, every $\mathscr{A}$-operation is an orthogonal set and every subset of an orthogonal set is an orthogonal set.

If E is an $\mathscr{A}$-operation, then a subset D of E will be called an *event* for E. As usual, if the physical operation corresponding to E is executed and an outcome $d \in D$ is obtained as a consequence, then we shall say that the event D has *occurred*. Naturally, an $\mathscr{A}$-event is defined to be a set of $\mathscr{A}$-outcomes that is an event for at least one $\mathscr{A}$-operation. When we assert that an $\mathscr{A}$-event D has occurred, we of course are asserting that it has occurred as a consequence of an execution of a physical operation corresponding to *some* $E \in \mathscr{A}$ for which $D \subseteq E$.

If A and B are two $\mathscr{A}$-events, then we shall say that A is *orthogonal* to B and write $A \perp B$ provided that $a \perp b$ holds for all $a \in A$ and all $b \in B$. Our third condition, named the *coherence condition*, in effect permits us to extend the interpretation for the orthogonality relation $\perp$ (as operational rejection) from $\mathscr{A}$-outcomes to $\mathscr{A}$-events. Specifically, the coherence condition requires that if A and B are $\mathscr{A}$-events with $A \perp B$, then there must exist an $\mathscr{A}$-operation E such that both A and B are events for E. In particular, the coherence condition requires that the union of two orthogonal $\mathscr{A}$-events is again an $\mathscr{A}$-event.

Let us summarize by making the following formal definitions: A *premanual* is defined to be a non-empty set $\mathscr{A}$ of non-empty sets. An element $E \in \mathscr{A}$ is called an $\mathscr{A}$-*operation*. The set theoretic union $X = \bigcup \mathscr{A}$ is called the set of $\mathscr{A}$-*outcomes*. We shall call $\mathscr{A}$ an *irredundant* premanual provided that $E, F \in \mathscr{A}$ and $E \subseteq F$ implies that $E = F$. Two $\mathscr{A}$-outcomes x, y in X are said to be *orthogonal*, in symbols $x \perp y$, provided that $x \neq y$ and there exists an $E \in \mathscr{A}$ with $x, y \in E$. A subset D of X is called an *orthogonal* set if $x \perp y$ holds for all $x, y \in D$ with $x \neq y$, while a subset D of X is called an $\mathscr{A}$-*event* if there exists an $E \in \mathscr{A}$ with $D \subseteq E$. If A and B are subsets of X, we say that A and B are *orthogonal* and we write $A \perp B$ provided that $a \perp b$ holds for all $a \in A$ and $b \in B$. We call the premanual

$\mathscr{A}$ *coherent* provided that the union of any two orthogonal $\mathscr{A}$-events is again an $\mathscr{A}$-event. A *manual* is defined to be an irredundant and coherent premanual.

The notion of coherence admits a number of natural variations. For instance, a manual is said to be *countably coherent* (respectively, *completely coherent*) if the union of every countable orthogonal family (respectively, every orthogonal family) of events is again an event.

A manual in which all events (hence, in particular, all operations) are finite is said to be *locally finite*. Similarly, if all events in a manual are countable, the manual is called *locally countable*. A manual consisting of a finite (respectively, countable) number of operations each of which is finite (respectively, countable) is said to be *totally finite* (respectively, *totally countable*).

A family of events is said to be *compatible* provided that there exists one operation that contains all of them. If A is a set of outcomes, we denote by $A^{\perp}$ the set of all outcomes that are orthogonal to every outcome in A. Thus, if D is an event, then $D^{\perp}$ is the set of all outcomes that operationally reject D. Two events A and B are called *equivalent* if $A^{\perp} = B^{\perp}$. Two events A and B are said to *commute* if there exist events A_0 and B_0 such that A is equivalent to A_0, B is equivalent to B_0, and the events A_0 and B_0 are compatible.

3. THE CLASSIFICATION OF MANUALS

In practice, physical operations of interest are normally synthesized from more primitive procedures and arrangement approvals. Although the decision to regard certain physical operations as being primitive must be largely subjective, there would appear to be no harm in supposing that the analysis of a given planned experimental program could begin with an identification of the various primitive physical operations from which the remaining physical operations are to be obtained.

It is customary to assume (often implicitly) that there exists – at least in principle – a 'grand physical operation' that can be regarded as a common refinement of all the physical operations under consideration. A well known example of such an operation in statistical physics is the grand canonical measurement of classical mechanics – the operation that was supposed to determine the location and momentum of every particle of a physical system. As a consequence, we shall refer to a manual consisting of a single operation as a *classical manual*.

The tenets of quantum physics rule out even the in principle existence of a single common refinement of the measurements of position and momentum. Thus, in general, we cannot make do with a *single* primitive physical operation. However, a collection of primitive physical operations can be represented mathematically by the manual consisting of their outcome sets, provided that we choose our notation so that outcome sets of different primitive physical operations are disjoint. The resulting manual $\mathscr{D}$ of primitive operations will then have the following property: E, F $\in \mathscr{D}$. $E \cap F \neq \emptyset$ implies $E = F$. We call such a manual *semi-classical*.

Suppose we have a single primitive physical operation $\mathscr{E}$ with outcome set E. Additional physical operations can now be obtained by an obvious process of *coarsening*. The outcome set for such a coarsening will be a partition of E and the corresponding coarsened operation is executed by first executing $\mathscr{E}$ to obtain (say) the outcome $x \in E$, but recording only the equivalence class that contains x. It is easy to see that the collection of all partitions of E, representing all possible coarsenings of $\mathscr{E}$ does in fact form a manual. Furthermore, in this manual, any two events commute.

In general, a manual in which any two events commute is called a *Boolean* manual (since its so-called 'logic' is a Boolean algebra [34]). The measurable spaces of analysis and probability theory provide an important class of Boolean manuals. Indeed, if $(X, \mathscr{F})$ is a measurable space, then the collection $\mathscr{B}$ of all countable partitions of X into $\mathscr{F}$-measurable sets is a locally countable Boolean manual (whose 'logic' is isomorphic to the σ-algebra $\mathscr{F}$). The operations in $\mathscr{B}$ are naturally regarded as coarsenings of a perhaps in principle operation with outcome set X; however, not all such coarsenings need appear in $\mathscr{B}$.

Now suppose that $\mathscr{A}$ is any manual and that D is an $\mathscr{A}$-event. If E is an $\mathscr{A}$-operation containing D, it is natural to regard the event $E \backslash D$ as a *complement* or *negation* of D relative to the operation E. Since D can be a subset of many $\mathscr{A}$-operations, then it can have many such relative complements. If $\mathscr{A}$ is a Boolean manual, then all such relative complements of D are equivalent.

In general, a manual in which all the relative complements of each event are equivalent is called a *Dacey* manual. (The 'logic' of such a manual is an orthomodular poset [8], [34].) Most manuals considered in practice are Dacey manuals. The Hilbert spaces of orthodox quantum mechanics provide an important class of Dacey manuals. Indeed, if $\mathscr{H}$ is a Hilbert space, then the collection $\mathscr{A}$ of all maximal orthogonal families of normalized vectors in $\mathscr{H}$ forms a Dacey (but not a Boolean) manual.

An operation in $\mathscr{A}$ can be regarded as a perhaps in princple measurement which determines the (pure) state of a physical system.

If L is a quantum logic, so that L is at least an orthomodular poset, then the collection $\mathscr{A}$ of all finite maximal orthogonal families of non-zero elements of L forms a locally finite Dacey manual (whose 'logic is isomorphic to L [34]). If L is σ-orthocomplete, one can also form a locally countable Dacey manual by using countable maximal orthogonal families of non-zero elements of L. In either case, the operations are interpreted as determinations of which of a complete set of mutually exclusive propositions is true.

Now assume that $\mathscr{A}$ is a given manual and that, for one reason or anoother, we wish to focus our attention not on all of the available operations in $\mathscr{A}$, but only on certain operations forming a non-empty subset $\mathscr{M}$ of $\mathscr{A}$. For instance, the operations in $\mathscr{M}$ many be operations that we actually intend to execute over a specific interval of time – in which case, $\mathscr{M}$ would have to be a finite set. Denote by M the set of all outcomes of all of the operations in $\mathscr{M}$, so that we are limiting our attention only to those $\mathscr{A}$-outcomes that belong to M. Unfortunately, although $\mathscr{M}$ inherits the irredundancy condition from $\mathscr{A}$, it does not necessarily inherit the coherence condition from $\mathscr{A}$. Naturally, if $\mathscr{M}$ is coherent, we refer to it as a *submanual* of $\mathscr{A}$. Even if $\mathscr{M}$ is a submanual of $\mathscr{A}$, it may happen that there are $\mathscr{M}$-outcomes that are orthogonal with respect to the parent manual $\mathscr{A}$, but not orthogonal with respect to $\mathscr{M}$. A submanual $\mathscr{M}$ for which this anomaly does not occur is referred to as an *induced* submanual.

If $\mathscr{M}$ is not a submanual of $\mathscr{A}$, it may be a consequence of the existence of $\mathscr{A}$-operations that are contained in M, but do not belong to $\mathscr{M}$. Thus, it appears reasonable to extend $\mathscr{M}$ to the set of operations $\bar{\mathscr{M}} = \{E \in \mathscr{A} \,|\, E \subseteq M\}$. Suitable examples show that even this device may fail and $\bar{\mathscr{M}}$ may still not enjoy coherence. Of course, we can view such a phenomenon as a 'defect' of the parent manual $\mathscr{A}$ and seek conditions on $\mathscr{A}$ to insure that $\bar{\mathscr{M}}$ has the coherence that we desire.

The above considerations motivate the following definition: A manual $\mathscr{A}$ is *hereditary* provided that, for each finite non-empty subset $\mathscr{M}$ of $\mathscr{A}$, $\bar{\mathscr{M}} = \{E \in \mathscr{A} \,|\, E \subseteq \bigcup \mathscr{M}\}$ is an induced submanual of $\mathscr{A}$. Of course, by an *hereditary Dacey* manual, we mean a manual that is both hereditary and Dacey. Notice that the manual

$$\mathscr{H} = \{\{a, b\}, \{b, c\}, \{c, d\}\}$$

is hereditary, but not Dacey; while the manual

$$\mathscr{D} = \{\{a, b, c\}, \{a, d\}, \{c, e\}\}$$

is Dacey, but not hereditary. (Actually, $\mathscr{D}$ is Boolean.) It can be shown that a manual $\mathscr{A}$ is hereditary Dacey if and only if it does not admit four distinct outcomes a, b, c, and d such that $a \perp b$, $b \perp c$, $c \perp d$, $a \notin \{c\}^\perp$, $a \notin \{d\}^\perp$, and $b \notin \{d\}^\perp$.

4. STATES

Let $\mathscr{A}$ be a manual and let $\mathscr{E} = \mathscr{E}(\mathscr{A})$ denote the set of all $\mathscr{A}$-events. Denote by $\mathbb{R}^\mathscr{E}$ the real vector space of all real valued functions defined on $\mathscr{E}$, with pointwise operations. If $\varDelta \subseteq \mathbb{R}^\mathscr{E}$, then $\mathrm{lin}(\varDelta)$ denotes the linear span of $\varDelta$ and $\mathrm{con}(\varDelta)$ denotes the convex hull of $\varDelta$. Let $\alpha \in \mathbb{R}^\mathscr{E}$. We say that α is *finitely additive* if, for any finite family $(D_i)_{i=1}^n$ of pairwise orthogonal $\mathscr{A}$-events, $\alpha(\bigcup_{i=1}^n D_i) = \sum_{i=1}^n \alpha(D_i)$. Similarly, we say that α is *countably additive* provided that, for any countable family $(D_i)_{i=1}^\infty$ of pairwise orthogonal $\mathscr{A}$-events such that $\bigcup_{i=1}^\infty D_i$ is an $\mathscr{A}$-event, we have $\alpha(\bigcup_{i=1}^\infty D_i) = \sum_{i=1}^\infty \alpha(D_i)$. Moreover, α is said to be *additive* if, for every family $(D_i)_{i \in I}$ of pairwise orthogonal $\mathscr{A}$-events such that $\bigcup_{i \in I} D_i$ is an $\mathscr{A}$-event, we have $\alpha(\bigcup_{i \in I} D_i) = \sum_{i \in I} \alpha(D_i)$, the latter sum being understood in the unordered sense. We say that α *respects equivalence* if $A^\perp = B^\perp$ implies that $\alpha(A) = \alpha(B)$, that it *respects operations* if E, $F \in \mathscr{A}$ implies that $\alpha(E) = \alpha(F)$, and that it is *operationally normalized* if $\alpha(E) = 1$ for all $E \in \mathscr{A}$. Naturally, α is said to be *positive valued* if $\alpha(D) \geq 0$ for all $D \in \mathscr{E}$.

A *state* is a positive valued, operationally normalized, finitely additive function $\alpha \in \mathbb{R}^\mathscr{E}$. The set of all such states is denoted by $\Sigma = \Sigma(\mathscr{A})$. It is easy to show that every state respects equivalence; if $\alpha \in \Sigma$ and A, $B \in \mathscr{E}$ with $A^\perp \subseteq B^\perp$, then $\alpha(B) \leq \alpha(A)$. An additive state will also be called a *regular* state and the set of all regular states is denoted by $\Omega = \Omega(\mathscr{A})$. (The restriction of a regular state to $\mathscr{A}$-outcomes is called a *weight function for* $\mathscr{A}$ [14].)

A regular state $\omega \in \Omega(\mathscr{A})$ is to be regarded as a logically possible complete stochastic model for the empirical situation described by the manual $\mathscr{A}$. This is to be understood in the sense that, for every $\mathscr{A}$-event D, $\omega(D)$ is interpreted as the 'long run relative frequency' with which the event D occurs as a consequence of executions of $\mathscr{A}$-operations E for which $D \subseteq E$. Notice that the implicit decision to ignore the actual operations that are executed is already present in the outcome identification process used to form the manual $\mathscr{A}$. The *dispersion free* states – that is, the states that can assume only the values 0 and 1 – are naturally regarded as deterministic models, in the sense that they predict the occurrence or non-occurrence of each event with statistical certainty.

It is surely desirable to have a lavish supply of potential stochastic models for a manual and the physical circumstances that it describes. Any ad hoc assumption assuring such a supply of models is a non-trivial constraint on the manual: nevertheless, in any realistic situation there always appears to be a generous supply of states in one of the following senses:

Let $\varDelta$ be a subset of $\sum(\mathscr{A})$.

(1) $\varDelta$ is *unital* if, for every non-empty $A \in \mathscr{E}$, there exists $\alpha \varDelta \in$ such that $\alpha(A) = 1$.

(2) $\varDelta$ is *separating* if, for $A, B \in \mathscr{E}$, $A^\perp \neq B^\perp$ implies that there exists $\alpha \in \varDelta$ such that $\alpha(A) \neq \alpha(B)$.

(3) $\varDelta$ is *full* if, for $A, B \in \mathscr{E}$, $A^\perp \nsubseteq B^\perp$ implies that there exists $\alpha \in \varDelta$ such that $\alpha(A) < \alpha(B)$.

(4) $\varDelta$ is *ultrafull* if it is full and, for $A, B \in \mathscr{E}$, $A^1 \subseteq B^1$ and $A^0 \subseteq B^0$ implies that $A^1 = B^1$ and $A^0 = B^0$. Here, if $D \in \mathscr{E}$, the notation D^t is defined by $D^t = \{\alpha \in \varDelta \mid \alpha(D) = t\}$.

(5) $\varDelta$ is *strong* if, for $A, B \in \mathscr{E}$, $B^1 \subseteq A^1$ implies that $A^\perp \subseteq B^\perp$.

It can be shown that strong implies ultrafull and unital, ultrafull implies full, and full implies separating. Moreover, if a manual admits a separating set of states, then it is necessarily a Dacey manual. H. Fischer and G. Rüttimann [13] have shown that $\varOmega(\mathscr{A})$ is unital if and only if, for every $E \in \mathscr{A}$, every completely additive normed measure on the power set of E can be extended to a regular state on $\mathscr{A}$.

A state $\alpha \in \sum(\mathscr{A})$ is said to be a *Jauch-Piron* state if it satisfies the following condition: if A, B, and C are $\mathscr{A}$-events with $\alpha(A) = \alpha(B) = 0$ and $A^\perp \cap B^\perp \subseteq C^\perp$, then $\alpha(C) = 0$. If $\mathscr{A}$ is a Boolean manual, then every state in $\sum(\mathscr{A})$ is a Jauch-Piron state. Similarly, in orthodox quantum mechanics, every state is Jauch-Piron. For these reasons, Jauch and Piron required the states for their proposition systems to satisfy the above condition. In this connection, the following theorem of G. Rüttimann [36] is of some interest:

Let $\mathscr{A}$ be a totally finite Dacey manual such that, for every pair A, B of $\mathscr{A}$-events, there exists an $\mathscr{A}$-event C such that $A^\perp \cap B^\perp = C^\perp$. (That is, the 'logic' of $\mathscr{A}$ is a finite orthomodular lattice [34].) Then $\mathscr{A}$ is a Boolean manual if and only if $\sum(\mathscr{A})$ is unital and consists entirely of Jauch-Piron states.

Recall that the quantum logic approach requires at least an orthomodular partially ordered set L and that a state is represented by a σ-orthoadditive map $\alpha: L \to [0, 1]$. If $\mathscr{A}$ is a Dacey manual, then the set L of equivalence classes of $\mathscr{A}$-events forms an orthomodular partially ordered set and every state $\alpha \in \sum(\mathscr{A})$ induces a map (also denoted by α) $\alpha: L \to [0, 1]$. Although α is finitely orthoadditive on L, it need not be σ-orthoadditive. Such additional features as σ-orthoadditivity may be imposed on the regular states as a consequence of the detailed structure of the manual $\mathscr{A}$. In particular, if $\mathscr{A}$ is σ-coherent, then the regular states are in fact σ-orthoadditive on L. A generalization of the Alexandroff theorem [11] obtained by T. Cook and O. Beaver provides another useful condition for σ-orthoadditivity [3]. Thus, our approach yields substantially the same mathematical system as does the quantum logic approach, but provides it with a rich operational infrastructure which is at best tacit in the usual quantum logic approach.

The base norm and order unit spaces of the ordered vector space approach also arise quite naturally in this operational formalism. In fact, let $\mathscr{A}$ be a manual and suppose that $\varDelta$ is a convex subset of $\sum(\mathscr{A})$. The set $\varDelta$ might be regarded as the physically admissible states [22] or perhaps as an appropriate statistical hypothesis [35]. Thus, it is reasonable to assume that, whenever D is a non-empty $\mathscr{A}$-event, then there exists $\alpha \in \varDelta$ such that $\alpha(D) > 0$. It follows that $(V, \varDelta)$, where $V = \operatorname{lin}(\varDelta)$, is a base norm space [30]. The *evaluation map* $e: V \to \mathbb{R}$, defined by $e(\alpha) = \alpha(E)$, where E is any operation in $\mathscr{A}$, is a continuous linear functional independent of the choice of E. Moreover, (V^*, e) is the order unit space dual to $(V, \varDelta)$.

Each $\mathscr{A}$-event D induces a continuous linear functional $f_D \in V^*$, defined by $f_D(\alpha) = \alpha(D)$. Notice that the functionals f_D, D running through $\mathscr{E}$, form a total set in that they separate the points of V. These functionals, restricted to the positive cone in V, or to its base $\varDelta$, can be regarded as intensity functionals or frequency functionals. Evidently, for $D \in \mathscr{E}$, f_D belongs to the dual order interval $[0, e]$. Note that, if A and B are equivalent $\mathscr{A}$-events, then $f_A = f_B$; while the converse holds if $\varDelta$ is separating. Also, if $B^\perp \subseteq A^\perp$, then $f_A \le f_B$; while the converse holds if $\varDelta$ is full.

In many important examples, the functionals f_D, $D \in \mathscr{E}$, are extreme

points, ext $[0, e]$, of the dual order interval. As a consequence, considerable effort has been expended to establish sufficient conditions to ensure
this feature. In this connection, T. Cook and G. Rüttimann have made
effective use of the notion of a Jordan-Hahn decomposition. A element
$v \in V$ is said to admit a *Jordan-Hahn decomposition* provided that there
exist $D \in \mathscr{E}$, states α_0 and α_1 in $\varDelta$, and non-negative real numbers t_0 and t_1
such that $f_D(\alpha_0) = 0$, $f_D(\alpha_1) = 1$, and $v = t_1\alpha_1 - t_0\alpha_0$. The set of states
$\varDelta$ is said to have the *Jordan-Hahn property* provided that every element
$v \in V$ (not already a scalar multiple of a state in $\varDelta$) admits a Jordan-Hahn
decomposition.

For the case in which the manual $\mathscr{A}$ is totally finite, Rüttimann [37]
has proved the following theorem: If $\varDelta$ is a full set of states which is topo-
logically closed in V, then $\varDelta$ is ultrafull and has the Jordan-Hahn property
if and only if $\{f_D | D \in \mathscr{E}\} = \text{ext}[0, e]$.

If the manual $\mathscr{A}$ is not totally finite, then, in general, V is infinite dimensional and delicate convergence problems arise. Thus, in order to obtain
results similar to the above when $\mathscr{A}$ is not totally finite, T. Cook [6] introduced a technical variation of the Jordan-Hahn property, called
monotonically $\mathscr{E}$-$\mathscr{P}$-Jordan generating, and obtained the following result.
If $\varDelta$ is strong and monotonically $\mathscr{E}$-$\mathscr{P}$-Jordan generating, then $\{f_D | D \in \mathscr{E}\}$
$\subseteq \text{ext}[0, e] \subseteq$ weak *-closure of $\{f_D | D \in \mathscr{E}\}$. In [7], T. Cook and G. Rüttimann proved the following related theorem. If $\varDelta$ is ultrafull, has the
Jordan-Hahn property, and all of the extreme points of $[0, e]$ are exposed
[1], then $\{f_D | D \in \mathscr{E}\} = \text{ext}[0, e]$. Related results can be found in the recent
work of E. Alfsen and F. Schultz [2].

The 'Marburg school,' under the direction of G. Lüdwig, initiated much
of the current interest in the structure of the dual order interval $[0, e]$
and the relationship between its extreme points (called *decision effects*)
and the propositions of quantum logic. The elements of $[0, e]$ (called
effects) have been interpreted as 'generalized events.' These matters were
discussed and reported at an important 1973 international conference in
Marburg [20].

5. MORPHISMS OF MANUALS

A scientist never walks into a laboratory without preconceived notions –
however self-evident or subtle. Thus when he must deal with a specific
manual (particularly a semiclassical manual of primitive operations or a
manual with none of the pleasant characteristics described in Section 3) he

is prone to make modifications suggested by his 'model of reality' and his current requirements.

We have already discussed the *phenomenological identification* of outcomes, wherein outcomes in distinct operations are identified. In addition, however, a scientist might wish to coarsen certain operations, and thus might choose to *collapse* a number of outcomes within these individual operations to single outcomes. For example, such a collapse might be indicated when a scientist feels that the details described by a particular operation are more than his experimental program calls for or could handle effectively. Conversely, he could be faced with a manual too crude for his purposes and as a consequence indulge in some *outcome refinement* – that is the replacing of outcomes by events containing more than one outcome.

Furthermore a 'model of reality' on the one hand might require that certain outcomes in a manual never occur and on the other hand might suggest outcomes that were overlooked in the original description of the manual. In the former case, a scientist would very likely *truncate* some of the operations in his manual by discarding the superfluous outcomes. In the latter case, he would be moved to *annex* the missing outcomes.

Of course a theory – or model of reality – generally fits relatively crude laboratory procedures into an idealized picture that is supposed to explain what is taking place. Thus it often appears advantageous to *immerse* such a crude manual into a more descriptive manual. Such an immersion should map outcomes to outcomes in a one-to-one fashion and map operations onto operations.

The term *morphism*, it would seem, should be reserved for those maps that can represent the execution of any combination of the above modifications. A study of the common properties of these procedures leads to the following formal definition: A *morphism* ϕ from a manual $\mathscr{A}$ to a manual $\mathscr{B}$, written $\mathscr{A} \xrightarrow{\phi} \mathscr{B}$, is a map from the $\mathscr{A}$-events into the $\mathscr{B}$-events such that

(i)   if $(A_i)_{i \in I}$ is a compatible family of $\mathscr{A}$-events, then $\phi(\bigcup_{i \in I} A_i) = \bigcup_{i \in I} \phi(A_i)$.

(ii)   if A and D are $\mathscr{A}$-events with $A^\perp \subseteq D^\perp$, then $\phi(A)^\perp \subseteq \phi(D)^\perp$.

Clearly, the composition of morphisms is again a morphism. Notice also that a morphism maps equivalent events to equivalent events and maps commuting events to commuting events.

Suppose that $\mathscr{A}$ and $\mathscr{B}$ are manuals with outcome sets X and Y, respectively. If $\mathscr{A} \xrightarrow{\phi} \mathscr{B}$, we define

(i) $\phi(x) = \phi(\{x\})$ for $x \in X$;

(ii)   zer $\phi = \{x \in X | \phi(x)$ is the empty event$\}$, called the *zero set* of ϕ;

(iii)   coz $\phi = X \backslash$zero ϕ, called the *cozero set* of ϕ;

(iv)   img $\phi = \{y \in Y | y \in \phi(x)$ for some $x \in X\}$, called the *image* of ϕ;

(v) $\mathscr{A}_\phi = \{E \cap$ coz $\phi | E \in \mathscr{A}\}$, called the *operational preimage* of ϕ; and

(vi) $\phi(\mathscr{A}) = \{\phi(E) | E \in \mathscr{A}\}$, called the *operational image* of ϕ.

Naturally, ϕ is said to have a *trivial* zero set if zer ϕ is empty. If, for each $x \in X$, $\phi(x)$ is a singleton, then we say that ϕ is *outcome preserving*. If, for each $E \in \mathscr{A}$, $\phi(E)$ is a $\mathscr{B}$-operation, then we say that ϕ is *operation preserving*.

The conditions that morphisms have been required to satisfy were made mild indeed in order to include most of the maps that are useful in the application and study of manuals. The more tractable morphisms that arise in practice are usually one of the following types: Let $\mathscr{A} \xrightarrow{\phi} \mathscr{B}$ be a morphism. We say that ϕ is

(i)   a *conditioning* morphism if, whenever A and D are orthogonal $\mathscr{A}$-events, then $\phi(A)$ and $\phi(D)$ are orthogonal $\mathscr{B}$-events;

(ii)   a *faithful* morphism if, whenever A and D are $\mathscr{A}$-events such that $\phi(A)$ and $\phi(D)$ are orthogonal, then A and D are orthogonal;

(iii)   an *interpretation* if it is an operation preserving conditioning morphism;

(iv)   an *injection* morphism if it is an outcome preserving morphism such that, for any two $\mathscr{A}$-outcomes x_1, x_2, $\phi(x_1) = \phi(x_2)$ implies that $x_1 = x_2$;

(v)   a *homomorphism* if $\phi(\mathscr{A})$ is a manual;

(vi)   a *homomorphism onto* $\mathscr{B}$ if $\phi(\mathscr{A}) = \mathscr{B}$;

(vii)   an *isomorphism* if it is an injection and a homomorphism onto $\mathscr{B}$; and

(viii) an *automorphism* if it is an isomorphism and $\mathscr{A} = \mathscr{B}$.

The most attractive feature of a conditioning morphism $\mathscr{A} \xrightarrow{\phi} \mathscr{B}$ is that it induces linear adjoint map ϕ^+: lin $\sum(\mathscr{B}) \rightarrow$ lin $\sum(\mathscr{A})$. Here $\phi^+(\alpha)$ is defined for $\alpha \in \sum(\mathscr{B})$ by $\phi^+(\alpha) = \alpha \circ \phi$; then ϕ^+ is extended to all of lin $\sum(\mathscr{B})$ by linearity. It is easy to see that ϕ^+ is a positive linear map such that $e(\phi^+(\alpha)) \leq e(\alpha)$, for α in the positive cone of lin $\sum(\mathscr{B})$. Thus, when $\mathscr{A} = \mathscr{B}$, ϕ^+ represents a 'filtering' operation in the sense of the ordered vector space approach.

If $\mathscr{A} \xrightarrow{\phi} \mathscr{B}$ is a conditioning morphism and α is a state for $\mathscr{B}$, such that $\phi^+(\alpha) \neq 0$, then $\alpha_\phi = [\phi^+(\alpha)]/[e(\phi^+(\alpha))]$ is a state for $\mathscr{A}$. Thus a conditioning morphism $\mathscr{A} \xrightarrow{\phi} \mathscr{B}$ allows us to 'pull back' stochastic models for $\mathscr{B}$ into stochastic models for $\mathscr{A}$.

An interpretation morphism $\mathscr{A} \xrightarrow{\phi} \mathscr{B}$ is so called because $\mathscr{A}$ can be thought of as a phenomenological manual of actual laboratory procedures, while $\mathscr{B}$ can be regarded as an idealized manual of in principle operations governed by some scientific theory. Each $\mathscr{A}$-operation E is mapped by ϕ into a (possibly) more refined $\mathscr{B}$-operation $\phi(E)$. Now the stochastic models assigned to $\mathscr{B}$ by the scientific theory can be pulled back linearly by ϕ^+ to $\mathscr{A}$, for ϕ^+ maps *all* of $\sum(\mathscr{B})$ into $\sum(\mathscr{A})$. It is in this sense that the morphism ϕ interprets, or explains, the phenomena associated with $\mathscr{A}$ in terms of the theory governing $\mathscr{B}$. If, in addition, such an interpretation is faithful, then the theory imposes no new orthogonalities (operational rejections) on the phenomenological manual $\mathscr{A}$. Such an interpretation of $\mathscr{A}$ should turn out to be a homomorphism whenever the governing theory implies that $\mathscr{A}$ is operationally self-contained.

We can now formally define the "modification procedures" mentioned earlier. Let $\mathscr{A} \xrightarrow{\phi} \mathscr{B}$. We say that ϕ is

(a)   a pure *phenomenological identification* if ϕ is an outcome preserving interpretation homomorphism onto $\mathscr{B}$;

(b)   a pure *collapse* if ϕ is a faithful outcome preserving homomorphism onto $\mathscr{B}$ such that, if $F \in \mathscr{B}$, then $\{x \in \bigcup \mathscr{A} \mid \phi(x) \subseteq F\} \in \mathscr{A}$;

(c)   a pure *outcome refinement* if ϕ is a homomorphism onto $\mathscr{B}$, ϕ has a trivial cozero set, and $\phi(x_1) \cap \phi(x_2) \neq \varnothing$ implies that $x_1 = x_2$, for all $\mathscr{A}$-outcomes x_1, x_2;

(d)   a pure *truncation* if ϕ is a homomorphism onto $\mathscr{B}$ such that, if $x_1, x_2 \in \text{coz } \phi$, then $\phi(x_1)$ and $\phi(x_2)$ are singleton $\mathscr{B}$-events and $\phi(x_1) = \phi(x_2)$ implies that $x_1 = x_2$;

(e)   a pure *annexation* if ϕ is a faithful injection morphism such that, if $F \in \mathscr{B}$, there exists an $E \in \mathscr{A}$ such that $\phi(E) \subseteq F$; and

(f)   a pure *immersion* if ϕ is an injection and an interpretation.

Notice that an injection morphism is automatically a conditioning morphism; hence, a pure annexation is automatically a conditioning morphism. Moreover, a pure outcome refinement is, in fact, a faithful interpretation. It is also easy to show that an isomorphism is the same thing as a pure immersion that is also a homomorphism onto. More generally, if $\mathscr{A} \xrightarrow{\phi} \mathscr{B}$ is an immersion (respectively, a faithful immersion), then $\phi(\mathscr{A})$ is a submanual of $\mathscr{B}$ (respectively, an induced submanual of $\mathscr{B}$) and $\mathscr{A} \xrightarrow{\phi} \phi(\mathscr{A})$ is an isomorphism.

It is interesting to observe that any manual $\mathscr{B}$ can be obtained as the operational image of a semi-classical manual $\mathscr{D}(\mathscr{B})$ under a pure phenomenological identification $\mathscr{D}(\mathscr{B}) \xrightarrow{\delta} \mathscr{B}$. Specifically, if $F \in \mathscr{B}$, let $E_F = \{(F, y) | y \in F\}$, and define $\mathscr{D}(\mathscr{B}) = \{E_F | F \in \mathscr{B}\}$. An event in $\mathscr{D}(\mathscr{B})$ has the form $F \times D$, where $D \subseteq F \in \mathscr{B}$. Define $\delta(F \times D) = D$.

Incidentally, the following technical fact is quite useful. Suppose that $\mathscr{B}$ is a Dacey manual and that $\mathscr{A}$ is any manual. Assume that ψ is a map from the set X of $\mathscr{A}$-outcomes to $\mathscr{B}$-events, such that $\psi(x_1) \perp \psi(x_2)$ whenever $x_1 \perp x_2$ and $\bigcup_{x \in E} \psi(x) \in \mathscr{B}$ whenever $E \in \mathscr{A}$. Then there exists a unique interpretation $\mathscr{A} \xrightarrow{\phi} \mathscr{B}$ such that $\phi(x) = \psi(x)$ for all $x \in X$. In particular, suppose that $\mathscr{B}$ is a Dacey manual and $\mathscr{A}$ is a submanual of $\mathscr{B}$. It follows that the inclusion map $i: \mathscr{E}(\mathscr{A}) \to \mathscr{E}(\mathscr{B})$ is an injection morphism. Furthermore, $\mathscr{A} \xrightarrow{i} \mathscr{B}$ is faithful if and only if $\mathscr{A}$ is an induced submanual of $\mathscr{B}$.

The ultimate aim of any experimental research program is surely to 'explain' laboratory observations in terms of some governing theory. In the language of morphisms, one seeks a faithful interpretation $\mathscr{A} \xrightarrow{\phi} \mathscr{B}$ from a manual $\mathscr{A}$ of laboratory operations into a manual $\mathscr{B}$ that serves as a mathematical model in the sense that it is adequately descriptive and mathematically tractable. Naturally, the mathematical structures inherent in classical mechanics, modern quantum physics, and contemporary statistics provide the most familiar models.

Many of the outstanding philosophical and pragmatic issues of modern physics are, in effect, matters concerning the existence of faithful interpretations of one kind or the other. The most obvious example is the controversial problem of 'hidden variables', which essentially asks whether conventional quantum mechanics admits a 'classical interpretation'. Much of the contentious character of the polemics surrounding such problems is a consequence of ill defined terms. Acceptable formal definitions, such as we have provided, reduce many of the philosophical problems to purely mathematical questions.

If we were to decide, for example, that a 'classical interpretation' means a faithful interpretation morphism into a classical manual, then the above question has a mathematically rigorous answer. Indeed, a manual $\mathscr{A}$ admits a faithful interpretation $\mathscr{A} \xrightarrow{\phi} \mathscr{B}$ into a classical manual $\mathscr{B}$ if and only if $\mathscr{A}$ admits a strong set of regular dispersion free states.

It is important to notice that, even if a manual $\mathscr{A}$ admits a faithful interpretation into a classical manual $\mathscr{B}$, there still may be states in $\Sigma(\mathscr{A})$ that have no interpretation in $\mathscr{B}$; that is, states that cannot be obtained by pulling back a state in $\Sigma(\mathscr{B})$. A simple example is the so-called *pentagon* manual

$$\mathscr{A} = \{\{a, b, c\}, \{c, d, e\}, \{e, f, g\}, \{g, h, i\}, \{i, j, a\}\}.$$

This manual admits a strong set of regular dispersion free states, but the state ω defined by

$$\omega(a) = \omega(c) = \omega(e) = \omega(g) = \omega(i) = \tfrac{1}{2},$$
$$\omega(b) = \omega(d) = \omega(f) = \omega(h) = \omega(j) = 0,$$

is an extreme point of the convex set $\Sigma(\mathscr{A})$ that cannot be pulled back from a state on any classical manual under the adjoint of a faithful interpretation morphism.

Although many authors, concerned with the hidden variable problem, would regard the existence of a strong set of dispersion free states as sufficient evidence to view a system as being 'essentially classical', the above pentagon manual should be a clear warning that, in some sense, this need not be the case.

Naturally any manual that might serve as a mathematical model would be expected to admit a lavish supply of states. It is therefore encouraging to note that a set of states which is strong (respectively, ultrafull, full, separating, or unital) pulls back under the adjoint of a faithful interpretation into a set of states that is again strong (respectively, ultrafull, full, separating, or unital).

As was mentioned earlier, many mathematical models are obtained from the Hilbert space formalism of orthodox quantum mechanics. Thus, if $\mathscr{H}$ is a Hilbert space, we define the *Hilbert space manual* for $\mathscr{H}$, $M(\mathscr{H})$, to be the set of all countable sets E of pairwise orthogonal non-zero projection operators on $\mathscr{H}$ for which $\sum_{P \in E} P$ converges strongly to 1. The question of just which manuals admit a faithful interpretation into a

Hilbert space manual is presently unanswered. The natural conjecture is that a strong set of Jauch-Piron states is a necessary condition for such an interpretation.

In this connection the following result of R. Wright is especially provocative. Suppose that $\mathcal{H}$ and $\mathcal{K}$ are Hilbert spaces, that $\mathcal{H}$ has dimension at least three, and that $\mathcal{M}(\mathcal{H}) \xrightarrow{\phi} \mathcal{M}(\mathcal{K})$ is an interpretation morphism. Let ψ be the map from projections on $\mathcal{H}$ into projections on $\mathcal{K}$, defined by the strongly convergent sum $\psi(P) = \sum_{Q \in \phi(P)} Q$. Then $\mathcal{K}$ admits an orthogonal decomposition $\mathcal{K} = \mathcal{K}_1 \oplus \mathcal{K}_2$ and there exist Hilbert spaces $\mathcal{H}_1$ and $\mathcal{H}_2$, a Hilbert space isomorphism $U_1 : \mathcal{H} \otimes \mathcal{H}_1 \to \mathcal{K}_1$, and a Hilbert space conjugate isomorphism $U_2 : \mathcal{H} \otimes \mathcal{H}_2 \to \mathcal{K}_2$ such that $\psi(P) = U_1(P \otimes 1)U_1^{-1} \oplus U_2(P \otimes 1)U_2^{-1}$ for every projection operator P on $\mathcal{H}$. It follows that ϕ induces an outcome preserving faithful interpretation $\mathcal{M}(\mathcal{H}) \xrightarrow{\phi} \mathcal{M}(\mathcal{K})$ and that the original morphism ϕ was, in fact, faithful [41].

If it is assumed that every quantum mechanical system can be represented by a Hilbert space and if we consider only interpretations that preserve the dynamics (determined by non-degenerate Hamiltonians), then R. Wright's result has the following important implication. If a first quantum mechanical system can be 'explained' in terms of a second such system, then the second system must consist of a replica of the first coupled with a third quantum mechanical system.

Some of the early investigations concerning special types of morphisms can be found in the work of J. Dacey [8] and W. Collins [5]. Current work on yet another special type of morphism appears in an article by H. Fischer and G. Rüttimann [13].

6. THE SYNTHESIS OF MANUALS

There are a host of ways in which operations (and hence, manuals of operations) are synthesized from more primitive procedures and arrangement approvals. Constructions of this sort and their mathematical representations are already implicit in the modification procedures represented by the pure morphisms in the previous section. Here we will be concerned with the mathematical representation of manuals constructed in various ways from combinations of more primitive manuals.

In [8], J. Dacey, generalizing a technique originally introduced in [33], obtained a construction for a manual in which two-stage compound operations can be represented. The construction is as follows: Let $\mathcal{I}$

denote a manual with outcome set I called the *initial manual*. For each $i \in I$, suppose that $\mathscr{A}_i$ is a manual with outcome set X_i called the ith *secondary manual*. Suppose that $E \in \mathscr{I}$ and that, for each outcome $i \in E$, we choose an operation $F_i \in \mathscr{A}_i$. We define a two-stage operation G as follows: To execute G, we first execute E to obtain (say) the outcome i, then, in connected sequence, we execute F_i to obtain (say) the outcome x. This two-stage procedure constitutes an execution of G and we shall record its outcome as the ordered pair (i, x). We propose to identify G with its set of outcomes, $G = \{(i, x) \mid i \in E$ and $x \in F_i\}$. The set of all such two-stage operations G is called the *Dacey sum* of the family $(\mathscr{A}_i \mid i \in I)$ over $\mathscr{I}$. If we denote this Dacey sum by $\mathscr{A}$, we quickly verify that $\mathscr{A}$ is indeed a manual and that the outcome set of $\mathscr{A}$ is the set $X = \{(i, x) \mid i \in I$ and $x \in X_i\}$. Furthermore, if $(i, x), (j, y) \in X$, then $(i, x) \perp (j, y)$ in $\mathscr{A}$ if and only if either $i \perp j$ in $\mathscr{I}$ or else $i = j$ and $x \perp y$ in $\mathscr{A}_i$. In other words, the orthogonality in $\mathscr{A}$ is *lexicographic*.

It is easy to check that the Dacey sum $\mathscr{A}$ will be a Dacey manual if and only if the initial manual $\mathscr{I}$ as well as each secondary manual $\mathscr{A}_i$ is a Dacey manual. There is a natural projection morphism $\mathscr{A} \xrightarrow{\pi} \mathscr{I}$ such that $\pi(i, x) = \{i\}$ for every $\mathscr{A}$-outcome (i, x). In fact π is an outcome preserving faithful homomorphism onto $\mathscr{I}$; it is a pure collapse if and only if every secondary manual $\mathscr{A}_i$ is a classical manual. On the other hand, we obtain a faithful interpretation homomorphism $\mathscr{I} \xrightarrow{\lambda} \mathscr{A}$ by choosing an operation $E_i \in \mathscr{A}_i$ for each $i \in I$ and defining $\lambda(D) = \bigcup_{i \in D} \{i\} \times E_i$ for each $\mathscr{I}$-event D. It is not difficult to see that λ is a composition of a pure outcome refinement followed by a pure faithful immersion.

Given a state $\alpha \in \sum(\mathscr{A})$ and an $\mathscr{I}$-outcome i for which $\alpha(\lambda(i)) \neq 0$, we can define the *operationally conditioned* state $\alpha_i \in \sum(\mathscr{A}_i)$ by $\alpha_i(D) = [\alpha(\{i\} \times D)]/[\alpha(\lambda(i))]$ for every $\mathscr{A}_i$-event D. Here, if Δ is a strong set of (regular) states for $\mathscr{A}$, then $\Delta_i = \{\alpha_i \mid \alpha \in \Delta$ and $\alpha(\lambda(i)) \neq 0\}$ is a strong set of (regular) states for $\mathscr{A}_i$. Indeed, so is the set $\lambda^+(\Delta)$ of states pulled back to $\mathscr{I}$.

Conversely, if the initial manual $\mathscr{I}$ and every secondary manual $\mathscr{A}_i$ carry strong sets of regular states, then so does the Dacey sum $\mathscr{A}$. To see this, let Δ be a strong set of regular states for $\mathscr{I}$ and, for every $i \in I$, let Δ_i be a strong set of regular states for $\mathscr{A}_i$. Denote by Λ the cartesian product $\Lambda = \Delta \times \prod_{i \in I} \Delta_i$. If $(\alpha, \beta) \in \Lambda$ and (i, x) is an $\mathscr{A}$-outcome, define $(\alpha, \beta)(i, x) = \alpha(i)\beta_i(x)$. Then (α, β) extends uniquely by additivity to a regular state on $\mathscr{A}$, which we continue to denote by (α, β). It is not difficult to verify that Λ is a strong set of regular states for $\mathscr{A}$,

An operation having just one outcome is called a *transformation*; the idea being that if such an operation is executed, there is just one possible outcome, namely, that the transformation in question has indeed been effected. Such a transformation might be an actual physical transformation in spacetime, or it could require the adjustment of apparatus. Of special interest are the purely temporal transformations whose instructions only require the elapse of a specified time interval. A manual $\mathscr{T}$ consisting just of transformations is called a *transformation* manual. Evidently, a Dacey sum is a transformation manual if and only if the initial manual as well as every secondary manual are transformation manuals. Naturally, the idea is that transformations carried out in connected sequence produce transformations. We shall consider these matters in some detail in the next section.

Once more, let $\mathscr{A}$ be the Dacey sum of the family $(\mathscr{A}_i \mid i \in I)$ over $\mathscr{I}$. If $\mathscr{I}$ is a transformation manual, then $\mathscr{A}$ is called the *disjoint sum* of the family $(\mathscr{A}_i \mid i \in I)$, and we write $\mathscr{A} = \sum_{i \in I} \mathscr{A}_i$. Notice that, if each $\mathscr{A}_i$ is classical, then the disjoint sum $\mathscr{A}$ is semiclassical. The disjoint sum is, in fact, the categorical direct sum in the subcategory obtained by using only interpretation morphisms.

On the other hand, if $\mathscr{I} = \{I\}$ is a classical manual, then the Dacey sum $\mathscr{A}$ is called the *orthogonal sum* of the family $(\mathscr{A}_i \mid i \in I)$, and we write $\mathscr{A} = \oplus_{i \in I} \mathscr{A}_i$. Notice that, if each $\mathscr{A}_i$ is classical (respectively, semi-classical), then the orthogonal sum $\mathscr{A}$ is classical (respectively, semi-classical). The orthogonal sum is a categorical product for morphisms.

A Dacey sum is hereditary Dacey if and only if the initial manual, as well as every secondary manual, are hereditary Dacey. Furthermore, a totally finite manual is hereditary Dacey if and only if it can be built up in a finite number of steps, starting with totally finite classical manuals, and iteratively forming disjoint and/or orthogonal sums.

Another special case of importance is obtained by requiring all of the secondary manuals to be identical. Thus, if, as above, $\mathscr{I}$ is the initial manual and $\mathscr{B} = \mathscr{A}_i$ for all $i \in I$ is the fixed secondary manual, then the Dacey sum $\mathscr{A}$ is called the *operational product* of $\mathscr{I}$ and $\mathscr{B}$, and we write $\mathscr{A} = \overrightarrow{\mathscr{I}\mathscr{B}}$, or simply $\mathscr{A} = \mathscr{I}\mathscr{B}$. In particular, we define the *operational square* of $\mathscr{B}$ by $\mathscr{B}^2 = \mathscr{B}\mathscr{B}$ and by induction, the *operational n^{th} power* of $\mathscr{B}$ by $\mathscr{B}^n = \mathscr{B}^{n-1}\mathscr{B}$. An operation in $\mathscr{B}^n$ is to be regarded as an n-stage operation requiring that certain of the operations of $\mathscr{B}$ be performed in a a connected sequence of n steps. In a natural way, one can even construct a limiting manual $\mathscr{B}^c = \lim_{n \to \infty} \mathscr{B}^n$, called the *compound* manual over $\mathscr{B}$. This is discussed in more detail in [35], [17].

Let $\mathscr{A}$ and $\mathscr{B}$ be arbitrary manuals. If A is an $\mathscr{A}$-event and B is a $\mathscr{B}$-event, we naturally write AB for the $\mathscr{A}\mathscr{B}$-event $AB = \{(a, b) \mid a \in A$ and $b \in B\}$. It can be shown that if $\alpha \in \sum(\mathscr{A}\mathscr{B})$, A is an $\mathscr{A}$-event, and B and C are equivalent $\mathscr{B}$-events, then $\alpha(AB) = \alpha(AC)$. However, if A and D are equivalent $\mathscr{A}$-events, then, in general, $\alpha(AB) \neq \alpha(DB)$. As an immediate consequence, the operational conditioning of states by equivalent events need not produce identical results.

Indeed, let A be an $\mathscr{A}$-event and let $\alpha \in \sum(\mathscr{A}\mathscr{B})$. Naturally, we define $\alpha(A) = \alpha(AF)$, where F is any $\mathscr{B}$-operation. Now, if $\alpha(A) \neq 0$, we define the *operationally conditioned* state $\alpha_A \in \sum(\mathscr{B})$ by $\alpha_A(B) = [\alpha(AB)]/[\alpha(A)]$ for every $\mathscr{B}$-event B. Thus, if, as above, A and D are equivalent $\mathscr{A}$-events and $\alpha(AB) \neq \alpha(DB)$, it follows that $\alpha(A) = \alpha(D)$ and consequently that $\alpha_A(B) \neq \alpha_D(B)$ as asserted above.

Evidently, if $\alpha \in \Omega(\mathscr{A}\mathscr{B})$, then $\alpha_A \in \Omega(\mathscr{B})$. If x is an $\mathscr{A}$-outcome and $\alpha \in \sum(\mathscr{A}\mathscr{B})$, we write $\alpha_{(x)}$ simply as α_x. We have shown [16] that if α is a *pure* regular state (that is, an extreme point of $\Omega(\mathscr{A}\mathscr{B})$) and $\alpha(\{x\}) \neq 0$, then α_x is again a pure regular state. However, operational conditioning by an event A consisting of more than one outcome almost always produces a *mixed* state (that is, not pure). This follows from the simple formula

$$\alpha_A = \sum_{a \in A} t_a \, \alpha_a,$$

where

$$t_a = \frac{\alpha(a)}{\alpha(A)},$$

so that

$$\sum_{a \in A} t_a = 1.$$

The above concepts find familiar realizations within the Hilbert space formalism of orthodox quantum mechanics. Thus let $\mathscr{H}$ be a Hilbert space and let $\mathscr{M}(\mathscr{H})$ be its corresponding Hilbert space manual. If the dimension of $\mathscr{H}$ is at least three, then, by Gleason's theorem [18], the regular states $\alpha \in \Omega(\mathscr{M}(\mathscr{H}))$ are in one-to-one correspondence $\alpha_W \leftrightarrow W$ with the positive trace class operators W with unit trace on $\mathscr{H}$ according to $\alpha_W(P) = \text{tr}(WP)$ for all projections P on $\mathscr{H}$. In order for us to represent concatenated measurements, we form the operational square $\mathscr{M}(\mathscr{H})^2$. Since $\mathscr{M}(\mathscr{H})$ has a strong set of regular states, then so does $\mathscr{M}(\mathscr{H})^2$; however, not all of these states are regarded as physically realizable within the conventional quantum mechanical formalism as, for example described in [19]). If we translate this conventional formalism into operational language, we find that the regular state $\alpha^W \in \Omega(\mathcal{M}(\mathcal{H})^2)$, determined by the density operator W, may be evaluated from the formula

$$\alpha^W((P, Q)) = \text{tr}(PWPQ),$$

for each $\mathcal{M}(\mathcal{H})^2$-outcome (P, Q). Recall here that P and Q are non-zero projections on $\mathcal{H}$. According to the conventional wisdom, these are the only admissible states.

If $\alpha^W((P, I)) = \alpha^W(P) = \text{tr}(PW) \neq 0$, then according to our definition, the operationally conditioned regular state α_P^W is given by

$$\alpha_P^W(Q) = \frac{\text{tr}(PWPQ)}{\text{tr}(PW)}.$$

In case $W = P_\phi$ is the projection onto the one dimensional subspace generated by the normalized wave function ϕ, we write α^ϕ for α^W and obtain

$$\alpha^\phi((P, Q)) = \| QP\phi \|^2,$$

$$\alpha^\phi(P) = \| P\phi \|^2, \quad \text{and, if } P\phi \neq 0,$$

$$\alpha_P^\phi(Q) = \frac{\| QP\phi \|^2}{\| P\phi \|^2}.$$

Now consider the case for which the normalized wave functions in $\mathcal{H}$ represent the physical states of deuterons (spin 1 particles). Let $\{P_+, P_0, P_-\} \in \mathcal{M}(\mathcal{H})$ represent a spin resolving Stern-Gerlach operation. Let $\phi \in \mathcal{H}$ be a given normalized wave function and let Q be an arbitrary non-zero projection on $\mathcal{H}$ (that is, an $\mathcal{M}(\mathcal{H})$-outcome). Denote by P the $\mathcal{M}(\mathcal{H})$-outcome $P = P_0 + P_-$ and denote by A the $\mathcal{M}(\mathcal{H})$-event $A = \{P_0, P_-\}$. Clearly, $\{P\}$ and A are equivalent events in $\mathcal{M}(\mathcal{H})$ – they both indicate the non-occurrence of $\{P_+\}$. In spite of this equivalence, the operationally conditioned states α_P^ϕ and α_A^ϕ are not necessarily the same; in fact, the former is pure and the latter is in general mixed. Explicitly,

$$\alpha_P^\phi(Q) = [\| QP_0\phi \|^2 + 2\mathcal{R}\langle QP_0\phi, QP_-\phi \rangle + \| QP_-\phi \|^2] \| P\phi \|^{-2},$$

while

$$\alpha_A^\phi(Q) = [\| QP_0\phi \|^2 + \| QP_-\phi \|^2] \| P\phi \|^{-2}.$$

The middle term in the first formula represents the well known 'interference' effect. The details of this and similar examples can be found in an article by R. Wright [42].

It is very curious indeed that the quantum mechanically admissible regular states on $\mathcal{M}(\mathcal{H})^2$ do not even form a strong set of states.

It is also interesting to observe that an operational product $\mathcal{A}\mathcal{B}$ is Boolean if and only if $\mathcal{A}$ is a classical manual and $\mathcal{B}$ is a Boolean manual. Therefore it is possible to have two manuals, within each of which all events commute, form an operational product in which some events do not commute. For example, let $\mathcal{A} = \{\{P_+, P_0, P_-\}, \{P_+, P\}\}$ and $\mathcal{B} = \{\{Q, I - Q\}\}$. Note that $\mathcal{A}$ and $\mathcal{B}$ are induced Boolean submanuals of the Hilbert space manual $\mathcal{M}(\mathcal{H})$. In particular, the singleton events $\{(P, Q)\}$ and $\{(P_0, I - Q)\}$ in $\mathcal{A}\mathcal{B}$ do not commute.

In addition to the operational product, there are a variety of other useful 'products' of manuals. For example, if one wishes to consider the 'simultaneous' execution of two operations – with possible bilateral interference – then one can use the so-called *cross product* $\mathcal{A} \times \mathcal{B} = \{E \times F|$ $E \in \mathcal{A}$ and $F \in \mathcal{B}\}$. Notice that the injection $\mathcal{A} \times \mathcal{B} \xrightarrow{i} \mathcal{A}\mathcal{B}$ is a pure immersion and that both of the projection maps $\mathcal{A} \times \mathcal{B} \xrightarrow{\pi_1} \mathcal{A}$ and $\mathcal{A} \times \mathcal{B} \xrightarrow{\pi_2} \mathcal{B}$ are homomorphisms onto when $\mathcal{A}$ and $\mathcal{B}$ are Dacey.

On the other hand, if one wishes to consider the 'independent' execution of two operations – with no lateral interference – then one can use the so-called *tensor product* $\mathcal{A} \otimes \mathcal{B}$. The outcome set for $\mathcal{A} \otimes \mathcal{B}$ is the same as the outcome set for $\mathcal{A}\mathcal{B}$ and again the natural injection $\mathcal{A}\mathcal{B} \xrightarrow{i} \mathcal{A} \otimes \mathcal{B}$ is a pure immersion. In $\mathcal{A} \otimes \mathcal{B}$, two outcomes (x, y) and (u, v) are orthogonal if and only if $x \perp u$ or $y \perp v$. These matters will be discussed in detail in other articles.

7. GROUPS OF TRANSFORMATIONS

An *action* of a manual $\mathcal{G}$ on a manual $\mathcal{A}$ is defined to be an outcome preserving morphism $\mathcal{G}\mathcal{A} \xrightarrow{*} \mathcal{A}$. If $*(g, x) = \{y\}$, we shall write $g*x = y$ and we always understand that the outcomes (g, x) and y are to be regarded as identical. In particular, an action $\mathcal{G}^2 \xrightarrow{\cdot} \mathcal{G}$ is called a *binary operation* on $\mathcal{G}$. If G is the set of $\mathcal{G}$-outcomes and $(G, \cdot)$ is a group (respectively, a semigroup), we call $\mathcal{G}^2 \xrightarrow{\cdot} \mathcal{G}$ a *group operation* (respectively, a *semigroup operation*). If, in addition, $\mathcal{G}$ is a transformation manual, then we refer to $\mathcal{G}$, equipped with the group (respectively, semigroup) operation as a *group of transformations* (respectively, a *semigroup of transformations*).

In this particular paper, $\mathscr{G}$ will always be understood to be a group of transformations. Furthermore, when we say that $\mathscr{G}$ *acts on* a manual $\mathscr{A}$, we will understand that the action $\mathscr{G}\mathscr{A} \xrightarrow{*} \mathscr{A}$ must have the following two properties:

(i)   For all $\mathscr{G}$-outcomes g, h and all $\mathscr{A}$-outcomes x, $(g \cdot h) * x = g * (h*x)$.

(ii)   If l is the identity outcome in $\mathscr{G}$ and x is any $\mathscr{A}$-outcome, then $l * x = x$.

Then, if g is any $\mathscr{G}$-outcome, the map ϕ_g from $\mathscr{A}$-outcomes to $\mathscr{A}$-outcomes defined by $\phi_g(x) = g * x$ induces a unique automorphism $\mathscr{A} \xrightarrow{\phi_g} \mathscr{A}$. Moreover, if g and h are $\mathscr{G}$-outcomes, $\phi_g \circ \phi_h = \phi_{g \cdot h}$. (Maps such as ϕ_g were called operational symmetries in [15].) It also follows that an action of a group of transformations on a manual is automatically a phenomenological identification.

Not let $\mathscr{A}$ be an arbitrary manual. Even if we are not given an action of $\mathscr{G}$ on $\mathscr{A}$, we can form the operational product $\mathscr{G}\mathscr{A}$, and then note that $\mathscr{G}$ acts naturally on $\mathscr{G}\mathscr{A}$, $\mathscr{G}(\mathscr{G}\mathscr{A}) \xrightarrow{\mathscr{N}} \mathscr{G}\mathscr{A}$ according to $\mathscr{N}(g, (h, x)) = (g \cdot h, x)$, where g and h are $\mathscr{G}$-outcomes and x is an $\mathscr{A}$-outcome. Henceforth, for simplicity, we write $\mathscr{N}(g, (h, x))$ as $g(h, x)$; so that $g(h, x) = (g \cdot h, x)$. Since $\mathscr{G}$ is a transformation manual, $\mathscr{G}\mathscr{A}$ is nothing more than a disjoint sum of copies of $\mathscr{A}$, indexed by $\mathscr{G}$-outcomes, and the natural action $\mathscr{N}$ simply permutes these copies.

As before, we may be in possession of a suitable theory, model of reality, or picture of the world that provides an adequate interpretation of the manual $\mathscr{G}\mathscr{A}$ in some mathematical model $\mathscr{B}$. We now turn our attention to a particularly important instance of such a situation. Thus, suppose that $\mathscr{H}$ is a subgroup of $\mathscr{G}$ (in the sense that $\mathscr{H}$ is a submanual of $\mathscr{G}$ and that the outcome set H of $\mathscr{H}$ is a subgroup of G) and assume that $\mathscr{H}$ acts in a known way on the manual $\mathscr{A}$, say $\mathscr{H}\mathscr{A} \xrightarrow{*} \mathscr{A}$. This action, presumably a consequence of the aforementioned world picture, enforces certain non-trivial outcome identifications (namely, that $h * x = y$ requires (h, x) to be identified with y).

In order to motivate the construction of the promised model $\mathscr{B}$, suppose that $g \in G$, $h \in H$, and $E \in \mathscr{A}$. Notice that the instructions for executing $\{g \cdot h\}E$ are word for word the same as the instructions for executing g $(\{h\} *E)$ except for the portions of these instructions pertaining to the form in which the outcomes are to be recorded. In the first case, we would record the outcome in the form $(g \cdot h, x)$ and in the second case, we would record it in the form $(g, h * x)$, where $x \in E$. Here, we shall clearly wish to regard the outcome $(g \cdot h, x)$ as being physically equivalent to the outcome $(g, h * x)$.

With the above in mind, we define the relation $\sim$ on the set $G \times X$ of all $\mathscr{G}\mathscr{A}$-outcomes as follows: For $g_1, g_2 \in G$ and $x_1, x_2 \in X$, the relation $(g_1, x_1) \sim (g_2, x_2)$ holds if and only if there exists $h \in H$ with $g_1 = g_2 \cdot h$ and $x_2 = h * x_1$. One verifies without difficulty that $\sim$ is a *bona fide* equivalence relation on $G \times X$ and that, if $(g_1, x_1) \perp (g_2, x_2)$ in $\mathscr{G}\mathscr{A}$, then $(g_1, x_1) \not\sim (g_2, x_2)$. We define $[g, x]$, for $(g, x) \in G \times X$, to be the equivalence class in $G \times X$ consisting of all (g', x') with $(g, x) \sim (g', x')$.

Let $Z = \{[g, x] | g \in G \text{ and } x \in X\}$ and, for $g \in G$ and $E \in \mathscr{A}$, define $[g, E] = \{[g, x] | x \in E\}$. Then $\mathscr{G}\mathscr{A}/\mathscr{H} = \{[g, E] | g \in G \text{ and } E \in \mathscr{A}\}$ is a manual; which, in fact, serves as our promised model. The interpretation morphism $\mathscr{G}\mathscr{A} \xrightarrow{\phi} \mathscr{G}\mathscr{A}/\mathscr{H}$ is, of course, defined by $\phi(g, x) = [g, x]$. As one might suspect, ϕ is a pure phenomenological identification. As with any pure phenomenological identification, the operation $\phi(\{g\}E)$ in $\mathscr{G}\mathscr{A}/\mathscr{H}$ is understood to be executed by executing $\{g\}E$ to obtain (say) the outcome (g, x), but recording the outcome as $\phi(g, x)$. There is a natural action of $\mathscr{G}$ on the manual $\mathscr{G}\mathscr{A}/\mathscr{H}$, $\mathscr{G}(\mathscr{G}\mathscr{A}/\mathscr{H}) \xrightarrow{\xi} \mathscr{G}\mathscr{A}/\mathscr{H}$, defined by $\xi (g', [g, x]) = [g' \cdot g, x]$, for $g', g \in G$ and $x \in X$. Henceforth, for simplicity, we write $\xi(g', [g, x])$ as $g'[g, x]$; so that $g'[g, x] = [g' \cdot g, x]$. We shall refer to the action ξ as the *expanded action* corresponding to the original action $\mathscr{H}\mathscr{A} \xrightarrow{*} \mathscr{A}$.

In general if one is given an interpreatation of a phenomenological manual in terms of a mathematical model, then there is always the elusive matter of unraveling the physically essential aspects of the interpretation and its purely syntactic component. The notion of an expanded action suggests how this might be accomplished for those interpretations that are also actions of groups.

Indeed, let $\mathscr{B} = \mathscr{G}\mathscr{A}/\mathscr{H}$, $Z = \bigcup \mathscr{B}$ and, for every $g \in G$, put $\mathscr{B}_g = \{[g, E] | E \in \mathscr{A}\}$. Notice that $\mathscr{B}_g$ is an induced submanual of $\mathscr{B}$ and that it is naturally isomorphic to the original manual $\mathscr{A}$. Since $\mathscr{B}_g = \mathscr{B}_{g'}$ if and only if $g \cdot H = g' \cdot H$, we can unambiguously define $\mathscr{B}_A = \mathscr{B}_a$ for any A in the space G/H of left cosets with any choice of a in A. Thus, Z is decomposed into mutually exclusive and exhaustive sets $Z_A = \bigcup \mathscr{B}_A$, as A runs through G/H; hence, $\mathscr{B}$ is a disjoint union of the induced submanuals $\mathscr{B}_A$, as A runs through G/H. Furthermore, for $A \in G/H$ and

$g \in G$, $g(Z_A) = Z_{g \cdot A}$ and $g(\mathscr{B}_A) = \mathscr{B}_{g \cdot A}$. Finally, the original action of $\mathscr{H}$ on $\mathscr{A}$ is evidently 'equivalent' to the action of $\mathscr{H}$ on $\mathscr{B}_H$. It is only this action of $\mathscr{H}$ on $\mathscr{B}_H$ that can involve any physically significant outcome identification – the 'rest' of the action of $\mathscr{G}$ on $\mathscr{B}$ is purely syntactic.

Suppose now that $\mathscr{B}$ is an arbitrary manual and that the group of transformations $\mathscr{G}$ acts on $\mathscr{B}$. Part of this action $\mathscr{G}\mathscr{B} \xrightarrow{*} \mathscr{B}$ may be 'purely syntactic' and part of it may be 'physically significant', and we now turn our attention to the problem of extracting the latter component. To this end, we invert the above argument and define an *operational system of imprimitivity* for the action $*$ to be a decomposition $Z = \bigcup_{i \in I} Z_i$ of the set $Z = \bigcup \mathscr{B}$ by the family $(Z_i \mid i \in I)$ of pairwise disjoint non-empty sets such that:

(i)   For each $i \in I$, $\mathscr{B}_i = \{F \in \mathscr{B} \mid F \subseteq Z_i\}$ is an induced. sub manual of $\mathscr{B}$ and $\mathscr{B} = \bigcup_{i \in I} \mathscr{B}_i$.

(ii)   The group G acts transitively on the set I in such a way that $g(Z_i) = Z_{gi}$ for all $g \in G$ and all $i \in I$.

Suppose we have such an operational system of imprimitivity. Choose and fix one of the indices $k \in I$. Let H be the subgroup of G consisting of all elements $h \in G$ such that $hk = k$ and let $\mathscr{H}$ be the corresponding submanual of $\mathscr{G}$. Define $\mathscr{A} = \mathscr{B}_k$, so that $\mathscr{A}$ is an induced submanual of $\mathscr{B}$. Evidently, we get an action of $\mathscr{H}$ on $\mathscr{A}$ by restriction of the action of $\mathscr{G}$ on $\mathscr{B}$. As before, this action of $\mathscr{H}$ on $\mathscr{A}$ gives rise to an expanded action of $\mathscr{G}$ on $\mathscr{G}\mathscr{A}/\mathscr{H}$ which is isomorphic to $\mathscr{B}$ in an obvious way. Furthermore, under this isomorphism, the original action of $\mathscr{G}$ on $\mathscr{B}$ is equivalent to the action of $\mathscr{G}$ on $\mathscr{G}\mathscr{A}/\mathscr{H}$.

The above discussion shows that the operational systems of imprimitivity for an action of $\mathscr{G}$ on $\mathscr{B}$ correspond in a one-to-one fashion to the ways in which this action can be regarded as an expansion of an action of some subgroup $\mathscr{H}$ of $\mathscr{G}$ on some induced submanual $\mathscr{A}$ of $\mathscr{B}$. Hence, we define an action of a group $\mathscr{G}$ on a manual $\mathscr{B}$ to be *operationally primitive* provided that there are no non-trivial operational systems of imprimitivity for this action. (An operational system of imprimitivity is trivial if the index set I contains only one element.) We regard an operationally primitive action of $\mathscr{G}$ on $\mathscr{B}$ to be an action involving no physically irrelevant syntactic outcome identifications.

Let Δ be a non-empty convex subset of a real vector space V. An *automorphism* of Δ is defined to be a bijective affine map from Δ onto Δ and the group of all such automorphisms is denoted by $\mathrm{Aut}(\Delta)$. Suppose that G is a group, that H is a subgroup of G, and that $U:H \to \mathrm{Aut}(\varDelta)$ is a group homomorphism; that is, U is a *representation* of H as automorphisms of $\varDelta$. Notice that $\varDelta^G$ is a convex subset of V^G and that

$$\varLambda = \{\lambda \in \varDelta^G \mid \lambda(g \cdot h) = U_h^{-1}(\lambda(g)) \text{ for all } g \in G, h \in H\}$$

is a convex subset of $\varDelta^G$, where $U_h = U(h)$. Motivated by G. Mackey [28], we define the *induced* group homomorphism (representation)

$$W:G \to \mathrm{Aut}(\varLambda) \text{ by } (W_g \lambda)(g') = \lambda(g^{-1} g') \text{ for all } g, g' \in G.$$

If $\mathscr{A}$ is any manual and $\Omega = \Omega(\mathscr{A})$, then an automorphism in Aut (Ω) is called a *stochastic symmetry* for $\mathscr{A}$. If $\mathscr{A} \xrightarrow{\phi} \mathscr{A}$ is an automorphism of the manual $\mathscr{A}$, then the restriction to Ω of the adjoint ϕ^{+} – which here we shall also denote by ϕ^{+} – is a stochastic symmetry. A group homomorphism $U:H \to \mathrm{Aut}(\Omega)$ is called a *stochastic representation* of the group H over the manual $\mathscr{A}$.

If $\mathscr{H}$ is any group of transformations acting on a manual $\mathscr{A}$ according to $\mathscr{H}\mathscr{A} \xrightarrow{*} \mathscr{A}$, then each $h \in H = \bigcup \mathscr{H}$ determines an automorphism ϕ_h of the manual $\mathscr{A}$ such that $\phi_h(x) = \{h * x\}$ for every $x \in X = \bigcup \mathscr{A}$. This automorphism, in turn, defines a stochastic symmetry $(\phi_h^{-1})^{+}$ in Aut $(\Omega (\mathscr{A}))$. The map $U:H \to \mathrm{Aut}(\Omega(\mathscr{A}))$, defined by $U(h) = U_h = (\phi_h^{-1})^{+}$ for $h \in H$, is then a stochastic representation of H over $\mathscr{A}$. Thus, any action of a group of transformations on a manual determines a stochastic representation of the group over the manual in a natural way.

Now if the preceding $\mathscr{H}$ is a subgroup of the group of transformations $\mathscr{G}$, so that H is a subgroup of $G = \bigcup \mathscr{G}$, then we can form the induced representation $W:G \to \mathrm{Aut}(\varLambda)$ as above. An operational version of this induced representation is now readily accessible as a consequence of our definition of the expanded action $\mathscr{G}(\mathscr{G}\mathscr{A}/\mathscr{H}) \xrightarrow{\xi} \mathscr{G}\mathscr{A}/\mathscr{H}$. Thus, let $T:G \to \mathrm{Aut}(\Omega(\mathscr{G}\mathscr{A}/\mathscr{H}))$ be the stochastic representation of G determined by the action ξ. In order to show that the representation W is equivalent to the stochastic representation T, we define the affine bijection $\Phi:\varLambda \to \Omega(\mathscr{G}\mathscr{A}/\mathscr{H})$ by

$$(\Phi(\lambda))([g, x]) = (\lambda(g))(x)$$

for $\lambda \in \varLambda$, $g \in G$, and $x \in X$. Using Φ, we define the group isomorphism $\Psi: \mathrm{Aut}(\varLambda) \to \mathrm{Aut}(\Omega(\mathscr{G}\mathscr{A}/\mathscr{H}))$ by

$$(\Psi(S))(\alpha) = \Phi(S(\Phi^{-1}(\alpha)))$$

for $S \in \mathrm{Aut}(\Lambda)$ and $\alpha \in \Omega(\mathscr{G}\mathscr{A}/\mathscr{H})$. Then $\Psi \circ W = T$ and, as promised, the two representations are equivalent.

In short, the stochastic representation determined by an expansion of an action is equivalent to the representation induced by the stochastic representation of the original action.

8. OBSERVABLES, RANDOM VARIABLES, AND PARAMETERS

Let $(K, \mathscr{F})$ be a measurable space and let $\mathscr{A}$ be a manual. The manual $\mathscr{M}(K, \mathscr{F})$ of all countable partitions of K into $\mathscr{F}$-measurable sets is a Boolean manual, and an interpretation morphism $\mathscr{M}(K, \mathscr{F}) \xrightarrow{\mathrm{A}} \mathscr{A}$ is called a $(K, \mathscr{F})$-*observable* for $\mathscr{A}$. Naturally, a real observable is understood to be an $(\mathbb{R}, \mathscr{B})$-observable, where $\mathscr{B}$ is the σ-algebra of real Borel sets.

If A is a $(K, \mathscr{F})$-observable for $\mathscr{A}$ and $C \in \mathscr{F}$, it is hereby agreed that whenever the event $\mathrm{A}(C)$ occurs, as a consequence of executing an operation $E \in \mathscr{A}$, A *has assumed some value in the set* C. Thus, to *measure* the observable A, with a prescribed precision, means to select a suitable $\mathscr{F}$-measurable partition $(C_i \mid i = 1, 2, 3, \cdots)$ and to execute the prescribed operation $E = \bigcup_{i=1}^{\infty} \mathrm{A}(C_i)$ in $\mathscr{A}$. An operation of this form is called a *measurement* of A. If $(K_i \; i = 1, 2, 3, \cdots)$ is any countable family of $\mathscr{F}$-measurable sets, then there exists a measurement E of A such that $\mathrm{A}(K_i)$ is an event for E for $i = 1, 2, 3, \cdots$; hence, $(\mathrm{A}(K_i) \mid i = 1, 2, 3, \cdots)$ is a compatible family of A-events.

If D is an $\mathscr{A}$-event, denote by $p(D)$ the equivalence class consisting of all $\mathscr{A}$-events that are equivalent to D and denote by $\Pi(\mathscr{A})$ the set of all such equivalence classes. If $p(A), p(B) \in \Pi(\mathscr{A})$, we define $p(A) \perp p(B)$ if $A \perp B$. Also, we define a partial order on $\Pi(\mathscr{A})$ by declaring $p(A) \leq p(B)$ if, for all $p(C) \in \Pi(\mathscr{A})$, $p(C) \perp p(B)$ implies that $p(C) \perp p(A)$. Furthermore, we put $0 = p(\emptyset)$ and $1 = p(E)$ for any $E \in \mathscr{A}$. The structure $(\Pi(\mathscr{A}), \leq, \perp)$ is called the *logic* for the manual $\mathscr{A}$ and is discussed in considerable detail in [34], [14], and [35]. As a matter of convenience, for the remainder of this section we shall suppose that $\mathscr{A}$ is a Dacey manual and consequently, that $\Pi(\mathscr{A})$ is an orthomodular poset.

Now, if A is a $(K, \mathscr{F})$-observable for $\mathscr{A}$, then the map $p\mathrm{A} : \mathscr{F} \to \Pi(\mathscr{A})$ defined by $p\mathrm{A}(C) = p(\mathrm{A}(C))$, for $C \in \mathscr{F}$, is a homomorphism The quantum logic approach, lacking an operational infrastructure, must represent its observables as homomorphisms such as $p\mathrm{A}$ [27].

For the remainder of this section we assume that A is a real observable on the Dacey manual $\mathscr{A}$. By a block B in the orthomodular poset $\Pi(\mathscr{A})$, we mean a maximal Boolean suborthomodular lattice in $\Pi(\mathscr{A})$. If B is such a block, we denote the Stone space of B by $S(B)$. It is useful to regard $S(B)$ as the outcome set for a 'virtual operation' that is a 'common refinement' for all of the operations $E \in \mathscr{A}$ affiliated with B in the sense that $p(D) \in B$ for every event $D \subseteq E$. This idea can be made more precise in terms of the so-called refinement ideals discussed in [14]. Evidently, any two operational propositions in the image of the mapping pA will commute in $\Pi(\mathscr{A})$; hence, there will exist at least one block B of $\Pi(\mathscr{A})$ such that $pA\ (C) \in B$ for every real Borel set C. We call such a B a pA-*block* in $\Pi(\mathscr{A})$.

Each operational proposition $p(D)$ in the pA-block B corresponds to a unique compact open subset $\phi(p(D))$ of the Stone space $S(B)$; ϕ is an isomorphism of the Boolean algebra B onto the algebra of all compact open subsets of $S(B)$. We define a B *random variable* to be a Baire measurable real-valued function on $S(B)$ and we say that such a B-random variable f *corresponds* to pA if, for each real Borel set C, the symmetric difference of $\phi(pA\ (C))$ and $f^{-1}(C)$ is a meager subset of $S(B)$. Notice that two B random variables f and g which correspond to the same pA are *equivalent* in the sense that $\{s \in S\ (B) \mid f(s) \neq g\ (s)\}$ is a meager subset of $S(B)$.

A random variable f corresponding to pA can be constructed as follows: First select a pA block B. For each real number t, let $M_t = \phi(pA(-\infty, t])$, so that M_t is a compact open subset of $S(B)$. Let N_0 be the intersection of all of the sets M_t and let N_1 be the intersection of all of the sets $S\ (B) \backslash M_t$, as t runs through the real numbers. Put $N = N_0 \cup N_1$ and $U = S\ (B) \backslash N$, so that N is a closed nowhere dense Baire subset of $S(B)$ and U is an open dense subset of $S(B)$. Define the real-valued function f on $S(B)$ by putting $f = 0$ on N and $f(s) = \inf\{t \mid s \in M_t\}$ for $s \in U$, noting that f is a Baire measurable function on $S(B)$ and that f is continuous on U. Obviously, the B-random variable f corresponds to the original pA. If g is and B-random variable, then g will correspond to pA if and only if g is equivalent to f. Furthermore, if pA is *bounded* in the sense that $pA(C) = 1$ for some bounded open interval C on the real line, then $U = S(B)$ and f is actually continuous on $S(B)$.

If B is a block in $\Pi(\mathscr{A})$, there will (in general) exist B-random variables that do not correspond to any pA; however, if B happens to be a σ-complete Boolean algebra, then any B-random variable g will correspond to some pA for which B is a pA-block. Clearly, if pA and $p\mathbb{B}$ arise from two observables A and B, and if B is both a pA-block and a pB-block, then $pA = pB$ if and only if pA and pB correspond to equivalent B-random variables.

Suppose now that $\alpha \in \Sigma(\mathscr{A})$, noting that α induces a finitely additive function α_B on the Boolean algebra B such that $\alpha_B(p(D)) = \alpha(D)$ for every $\mathscr{A}$-event D with $p(D) \in B$. Thus α_B induces a finitely additive probability measure α^* on the algebra of compact open subsets of $S(B)$. By [21], α^* admits a unique extension to a countably additive probability measure α_S defined on the Boolean σ-algebra of Baire subsets of $S(B)$. Thus, if f is a B-random variable, we can define the *expectation value* of f in the state α by $E(f, \alpha) = \int_S f d\alpha_S$, provided that this integral exists. If α_B is not only finitely additive, but countably additive on B, then α_S vanishes on all meager Baire subsets of $S(B)$; hence, in this case $E(f, \alpha) = E(g, \alpha)$ holds for equivalent B-random variables f and g.

If A is a real observable for $\mathscr{A}$, B is a pA-block in $\Pi(\mathscr{A})$, and $\alpha \in \Sigma(\mathscr{A})$, then we define the *expectation value* of A in the state α by $E(A, \alpha) = E(f, \alpha)$ where f is the particular B-random variable constructed as above to correspond to pA. Evidently, if α_B is countably additive on B, then $\alpha_B \circ pA$ is a Borel probability measure and $E(A, \alpha)$ coincides with the expected value of the corresponding probability distribution on $\mathbb{R}$.

The physical properties of objects – such as temperature, momentum, position, and energy – and the constants of statistical distributions – such as mode, mean, median, and variance – are naturally represented here as parameters. Indeed, if we define $\mathscr{I}$ to be the σ-algebra of subsets of $\Sigma(\mathscr{A})$ generated by sets of the form $\alpha^{-1}(C)$, where $\alpha \in \Sigma(\mathscr{A})$ and C is a Borel subset of $[0, 1]$, then a real *parameter* is simply an $\mathscr{I}$-measurable function $q: \Delta \to \mathbb{R}$ that assigns real numbers to the states of $\mathscr{A}$ that belong to a set $\Delta \in \mathscr{I}$. Since the sets in $\mathscr{I}$ might be regarded as inductively supportable statistical hypotheses or laws, $\mathscr{I}$ has been referred to as the *inductive logic* for the manual $\mathscr{A}$. The matter of how one provides such support by means of statistical inference has been discussed elsewhere [35].

Naturally, there is considerable interplay among observables, random variables, and parameters. We have already indicated a connection between observables and random variables. An analogous connection between observables and parameters might be arranged as follows: A real observable Q *corresponds* to a real parameter $q: \Delta \to \mathbb{R}$ if and only if, for every $\alpha \in \Delta$ and every real Borel set C, $\alpha(Q(C)) = 1$ if and only if $q(\alpha) \in C$.

9. CONCLUDING REMARKS

It is our hope that the reader, in spite of the somewhat technical nature of much of the preceding material, shares with us the excitement of watching an operational language of remarkable precision and versatility unfold and develop. Ideas that are difficult to apprehend in the vernacular of conventional science, can be expressed with clarity and precision in the operational idiom advocated here.

REFERENCES

[1] Alfsen, E. M., *Compact Convex Sets and Boundary Integrals*, Springer-Verlag, New York, 1971.

[2] Alfsen, E. M. and Shultz, F. W., *Non-Commutative Spectral Theory for Affine Function Spaces on Convex Sets*, Part I, Preprint Series, Inst. of Math., Univ. of Oslo, Norway 1974

[3] Beaver, O. R. and Cook, T. A., 'States on Quantum Logics and their Connection with a Theorem of Alexandroff', *Proc. A.M.S.* **67** (1977), 133–134.

[4] Birkhoff, G. and von Neumann, J., 'The Logic of Quantum Mechanics', *Ann. of Math.*, **37** (1936), 823–843.

[5] Collins, W. R. 'A Category of Sample Spaces', Unpublished Ph.D. dissertation, University of Massachusetts, Amherst, 1971.

[6] Cook, T. A., 'The Geometry of Generalized Quantum Logics', *Int. J. Theor. Phys.*, To appear (1977).

[7] Cook T. A. and Rüttimann, G. T., 'Symmetries on Quantum Logics', Submitted to *Proc. A.M.S.*

[8] Dacey, J. C., Jr., 'Orthomodular Spaces', Unpublished Ph.D. dissertation, University of Massachusetts, Amherst, 1968.

[9] Davies, E. B. and Lewis, 'An Operational Approach to Quantum Probability', *Commun. Math. Phys.* **17** (1970), 239–260.

[10] Dirac, P. A. M., *The Principles of Quantum Mechanics*, Clarendon, Oxford, 1930.

[11] Dunford, N. and Schwarz, J., *Linear Operators*, Part I, Interscience Pub. Inc., New York, 1957.

[12] Edwards, C. M., 'The Operational Approach to Algebraic Quantum Theory I', *Commun. Math. Phys.* **16** (1970), 207–230.

[13] Fischer, H. and Rüttimann, G. T., 'Fields of Manuals', To appear.

[14] Foulis. D. J. and Randall, C. H., 'Operational Statistics I, Basic Concepts', *J. Math. Phys.* **13** (1972), 1667–1675.

[15] Foulis, D. J. and Randall, C. H., 'Empirical Logic and Quantum Mechanics', *Synthese* **29** (1974), 81–111.

[16] Foulis, D. J. and Randall, C. H. 'The Stability of Pure Weights Under Conditioning', *Glasgow Math. J.* **15** (1974), 5–12.

[17] Foulis, D. J. and Randall, C. H., 'The Empirical Logic Approach to the physical Sciences', in A. Hartkämper and H. Neumann (eds.), *Foundations of Quantum Mechanicsand Ordered Linear Spaces*, Springer-Verlag, New York, 1974, 230–249.

[18] Gleason, A. M., 'Measures on the Closed Subspaces of a Hilbert Space', *J. Math. Mechanics* 6 (1953), 885–893.

[19] Haag, R. and Kastler, D., 'An Algebraic Approach to Quantum Field Theory,' *J. Math. Phys.* 5 (1964), 848–861.

[20] Hartkämper, A. and Neumann, H. (eds.), *Foundations of Quantum Mechanics and Ordered Linear Spaces*, Springer-Verlag, New York, 1974.

[21] Heider, L. J., 'A Representation Theory for Measures on Boolean Algebras', *Michigan Math. J.* 5 (1958), 213–221.

[22] Hein, C., 'The Poincare sample space', *Foundations of Phys.* 7 (1977), 597–608

[23] Jauch, J. M., *Foundations of Quantum Mechanics*, Addison Wesley Publ. Co., Reading, Mass., 1968.

[24] Jordan, P., von Neumann, J. and Wigner, E., 'On an Algebraic Generalization of the Quantum Mechanics Formalism', *Ann. Math.* 35 (1934), 29–64.

[25] Kolmogorov, A. N., *Foundations of the Theory of Probability*, 2nd Edition, Chelsea Publ. Co., New York, 1956 (German ed. 1933.)

[26] Ludwig, G., 'Versuch einer axiomatischen Grundlegung der Quanten Mechanik und allgemeinerer physikalischer Theorien', *Z. Physik* 181 (1964), 233–260.

[27] Mackey, G. W., *Mathematical Foundations of Quantum Mechanics*, W. A. Benjamin, Inc., New York, 1963.

[28] Mackey, G. W., *Induced Representations of Groups and Quantum Mechanics*, W. A. Benjamin, Inc., New York, 1968.

[29] Mielnik, B., 'Geometry of Quantum States', *Commun. Math. Phys.* 9 (1968), 55–80.

[30] Nagel, R. J., 'Order Unit and Base Norm Spaces', in A. Hartkämper and H. Neumann (eds.), *Foundations of Quantum Mechanics and Ordered Linear Spaces*, Springer-Verlag, New York, 1974, 23–29.

[31] Pool, J.C.T., 'Baer *-Semigroups and the Logic of Quantum Mechanics', *Commun. Math. Phys.* 9 (1968), 118–141.

[32] Piron, C., 'Axiomatique quantique', *Helv. Phys. Acta* 37 (1964), 439–468.

[33] Randall, C. H., 'A Mathematical Foundation for Empirical Science', Ph.D. dissertation, Rensselaer Polytechnic Institute, Troy, N.Y., 1966.

[34] Randall, C. H. and Foulis D. J., 'Operational Statistics II, Manuals of Operations and their Logics', *J. Math. Phys.* 14 (1973), 1472–1480.

[35] Randall, C. H. and Foulis, D. J., 'A mathematical setting for inductive reasoning', in C. Hooker (ed.), *Foundations of Probability Theory, Statistical Inference, and Statistical Theories of Science*, Vol. III, D. Reidel Publ. Co., Dordrecht: Holland, 1976, 169–205.

[36] Rüttimann, G. T., 'Jauch-Piron States', *J. Math. Phys.* 18 (1977), 189–193.

[37] Rüttimann, G. T. 'Jordan-Hahn Decomposition of Signed Weights on Finite Orthogonality Spaces', *Comment. Math. Helvetica* 52 (1977), 129–144.

[38] Segal, I. E., 'Postulates for General Quantum Mechanics', *Ann. Math.* 48 (1947), 930–948.

[39] Stone, M. H., 'Postulates for the Barycentric Calculus', *Ann. Math. Pure Appl.* **29** (1949), 25–30.
[40] von Neumann, J., *Mathematical Foundations of Quantum Mechanics*, Princeton University Press, Princeton, N.J., 1955.
[41] Wright, R., 'The Structure of Projection-Valued States; A Generalization of Wigner's Theorem', *Int. J. Theor. Phys.* **16** (1977), 567–573.
[42] Wright, R., 'Spin manuals': Empirical Logic Talks Quantum Mechanics,' in A. R. Marlow (ed.) Mathematical Foundations of Quantum Theory, Academic Press, 1978, 177–254.

E.-W. STACHOW*

COMPLETENESS OF QUANTUM LOGIC

ABSTRACT: This paper is based on a semantic foundation of quantum logic which makes use of dialog-games. In the first part of the paper the dialogic method is introduced and under the conditions of quantum mechanical measurements the rules of a dialog-game about quantum mechanical propositions are established. In the second part of the paper the quantum mechanical dialog-game is replaced by a calculus of quantum logic. As the main part of the paper we show that the calculus of quantum logic is complete and consistent with respect to the dialogic semantics. Since the dialog-game does not involve the 'excluded middle' the calculus represents a calculus of effective (intuitionistic) quantum logic. In a forthcoming paper it is shown that this calculus is equivalent to a calculus of sequents and more interestingly to a calculus of propositions. With the addition of the 'excluded middle' the latter calculus is a model for the lattice of subspaces of a Hilbert space.

A. INTRODUCTION

1. *General Remarks*

Since quantum mechanics came into being its epistemological interpretation had led to serious difficulties because it differs from some concepts of classical physics which had been supposed to be generally true. In particular Bohr restricted the applicability of classical notions and replaced the concept of objectivity of all observables by the concept of complementarity. This interpretation is known as the Copenhagen interpretation and is basic to all orthodox interpretations of quantum mechanics. Starting from the Hilbert space formalism of quantum mechanics, Birkhoff and von Neumann [1] and later Jauch and Piron [2] investigated its algebraic structure which turned out to be a model of an orthomodular lattice. Since the subspaces of Hilbert space can be interpreted as quantum mechanical propositions, this lattice represents a propositional calculus L_q, usually called quantum logical calculus, since it differs from the calculus of classical logic in some respects. The calculus L_q has been the

---

*On leave of absence from the Institut für Theoretische Physik der Universität zu Köln, Köln, W.-Germany.

Hooker (ed.), Physical Theory as Logico-Operational Structure, 203–243.
All Rights Reserved.

starting point of an axiomatic reconstruction of quantum mechanics [3]. It is indeed not obvious that the propostional calculus L_q is also a logical calculus. Therefore a semantic foundation is required by which the truth of quantum mechanical propositions is defined. In ordinary logics there are several well-known semantic foundations. The Boolean calculus of classical logic can, for instance, be interpreted by means of a bivalent function which associates each proposition with one of the two values 'true' and 'false'. As Kamber, Gleason, Kochen and Specker [4] have shown there is no such function for L_q but only for the Boolean sublattices of L_q. These Boolean sublattices consist of all those propositions which are objective with repect to a certain state $|\phi\rangle$ of the quantum mechanical system. The logic of those restricted classes of propositions is then the classical logic. Of special interest, however, is a semantic foundation of the whole calculus L_q concerning general quantum mechanical propositions. By means of such a foundation there is no distinction between objective and non-objective propositions. In this sense all propositions can be considered to be objective and the orthodox concept of complementarity becomes unnecessary. Several attempts were aimed at defining generalized value-functions on the calculus L_q. Reichenbach [5] introduced a three-valued logic and von Weizsäcker [6] even proposed a complex-valued logic to remove the difficulties which arise from a valuation of all propositions of L_q. But these many-value interpretations can always be applied and do not necessarily result from quantum mechanical conditions. This paper addresses itself to a semantic approach to quantum logic which makes use of the dialogic method [7]. The dialogic method has been applied with success on the implicative lattice L_i of effective logic which can also not be interpreted by a bivaluation of the propositions. If the 'excluded middle' which represents the value-definiteness of propositions is included, the dialogic method concerning classical propositions can be regarded as an interpretation of the Boolean lattice L_B of classical logic [8]. In this paper a generalisation of the dialogic method is developed which under certain conditions coming from quantum mechanical measurements admits applications to quantum mechanical propositions. Preliminary results are given in [9].

2. Outline of the Paper

To demonstrate the completeness (and consistency) of quantum logic, given by means of a quantum logical calculus, we show that all formulas which are deducible in the calculus can be proven within the dialogic procedure and vice versa. The dialogic procedure has to be understood as a foundation of quantum logic. This means that the object-language and, as far as necessary for our aim in this paper, the meta-language of quantum mechanical propositions are not only interpreted but rather based and developed by means of the dialogic approach. To emphasize this important point I decided to introduce the language in a genetic conception which elucidates the various stages in the development. In doing this we have to distinguish there main stages:

   a. the operational dialogic definition of the logical notions,
   b. the formal representation of the notions within a dialog-game about propostions,
   c. the incorporation of the rules of the dialog-game into the formal scheme which then, as a calculus, allows a deduction of all true quantum mechanical propositions.

In the following a preliminary description of the three stages will be given.

To a: The operational method starts with a certain concept about propositions which is laid out in part B. 1.a. We consider only those propositions which are proof-definite, i.e. for which a proof-procedure is established. (In general this does not mean that the proof-procedure decides between the truth and falsity of a proposition.) If a proposition is asserted there should be two basic attitudes with respect to the assertion: The truth of the propostion might be called in question; upon this request there should be the obligation to prove the proposition. This suggests that a dialog be performed about the asserted proposition. The person who asserts the proposition is called the proponent. Some other person, the opponent, may try to refute the asserted proposition and participate in a dialog about it. By an argument I shall mean any assertion which occurs during the execution of a dialog. A dialog then consists of a series of arguments which are stated in turn and which are either attacks or defences. If one of the participants has no argument to continue the dialog we say that he has lost the dialog and the other one has won the dialog. Such a position of a dialog is called a final position. Dialogs are teachable and learnable action schemes which make propositions reasonable. The truth of a proposition should not depend on an individual dialog which is won by the proponent. The proponent should rather have a strategy of success against all potential opponents or (equivalently) against all possible arguments of one opponent. The next step in the development of dialogs is to make precise which arguments may be used in a dialog. At first the logical connectives are dialogically defined by means of the different possibilities to attack and to defend compound propositions where only the formal structure of compound propositions is taken into account. This is executed in part B. 1.b. Since dialogs will be used to establish the truth of propositions we have to postulate their finiteness. This means, on the one hand that after a finite number of steps the dialogs must arrive at certain propositions which are not composed of other propositions and which therefore are called elementary propositions. According to the general concept of propositions the elementary propositions, though not dialog-definite, are assumed to be proof-definite. Whenever a participant asserts an elementary proposition he assumes the obligation to give a material proof of it. An example from mathematics of such an elementary proposition would be 'a formula f is deducible in the calculus K', and from physics 'the magnitude A of the system γ has the value a'. In the latter case we prove the elementary proposition by means of a measurement. On the other hand, to make the dialogs finite we have to restrict the repetitions of attacks and defences (until now arbitrary) to a finite number of possibilities. This is easily done under the condition of unrestricted availability of propositions which is introduced in B. 1.b. If a proposition is unrestrictedly available like the above mathematical proposition repeated attacks and defences are redundant because the participants can always refer to the first proofs in the dialog. The property of unrestricted availability is an interior property of elementary propositions because it depends on the general conditions to demonstrate elementary propositions in a dialog.

If γ is a quantum mechanical system the above physical proposition is generally not unrestrictedly available, as pointed out in part B. 2.a. Quantum mechanical propositions satisfy the weaker condition of restricted availability in a dialog. This means that a previous proposition is still available in a later position of a dialog if and only if it is commensurable with all propositions which have been proven in the meantime. The consequences of this poperty of quantum mechanical propositions for the dialogs is described in B. 2.a. A particular material proposition is defined which asserts the commensurability of two propositions, and the proof-definiteness of this commensurability-proposition is described. In this way it is made clear how the dialogic procedure establishes a pragmatic foundation of quantum logic. The material rules of the dialog result from the general conditions of quantum mechanical measurements, namely the restricted availability, and guarantee that whenever a propositions is under consideration it's material proof is available.

To prove the truth of a proposition the proponent has to possess a strategy of success against all possible arguments of the opponent. For demonstrating strategies of success the dialogic procedure can be simplified, and the material rules of the dialog can be replaced by formal rules. In the resulting formal dialogs material proofs of elementary propositions and commensurabilities are no longer performed. Since the proponent has to win the dialog against all possible arguments of the oppponent it is sufficient to consider the most advantageous case for the opponent in which he is able to prove all elementary propositions he states. That is why he is allowed to state elementary propositions in any position of a dialog. On the other hand the proponent is forced to state elementary propositions only if they have been stated by the opponent previously and are still available. If we reintroduce material proofs into the formal dialog the proponent will never be in a position where he might have to prove a new elementary proposition and fail in this proof. Furthermore, the formal rules of the quantum dialog guarantee that the proponent is able to refer to all previous propositions which are true (in the opponent's most advantageous case) and still available. In this way formal dialogs serve to establish strategies of success for the proponent and therefore the truth of propositions.

To b: After the logical notions have been defined operationally a formal game-theoretical representation is given. By means of this representation (see α to δ and D1 to D4) the language of the quantum dialog-game is precisely defined. The reader should not be confused by the two representations: the first operational one which serves as a foundation of the logical notions, and the second formal one which allows a systematic and precise presentation. The positions of the dialog are given by formal figures z, whereas the transitions between two positions z and z', designated by the relation $z \, R \, z'$ and determined by the rules R of the dialog-game, are not yet incorporated into the formal scheme. This last step of the formalization is done in stage c where the calculus of quantum logic is established. As a preparation for the calculus we reduce the positions z to the content of symbols which uniquely indicate the positions with respect to the possible continuations of the dialog. Only propositions which are still available and obligations to defend which still have to be performed are left. Symbols, redundant for the further dialog, are eliminated even though they had to be introduced at first to make the whole dialogic procedure understandable. For obligations to defend, new symbols, namely potential arguments, are introduced. Finally, positions of success, ie. positions for which the proponent has a strategy of success, are defined within the formal language by means of the recursive scheme I, which starts at the final positions of success (see the end of B. 2.b).

To c: In the last stage the rules of the quantum mechanical dialog-game are incorporated into the recursive scheme I. This total formalisation is represented by the calculus T_n in part C. 1. which still depends on the integer n. This restricts the proponent's possibilities to attack to at most n times. Since we generally postulate this number to be arbitrarily chosen by the proponent, but to be finite, we have to consider the union of all calculi T_n with respect to $n(n < \infty)$. As we show in part C. 3. the union can be replaced by the calculus T_{eff} of effective quantum logic.

It should already be noted in this introduction that the rules of the calculus, as well as of the scheme I, do not mean logical inferences. They mean, rather, transitions from already established figures to new ones. It is an important point of the operational dialogic procedure that the meta-logic has not to be presupposed, i.e. for the deductions of the figures in the calculus no logical inferences are necessary. The transitions, by means of which the deductions are established, correspond to schematic actions as they are performed in the dialogic procedure. Therefore they belong to the syntax as well as to the semantics of propositions. Indeed the syntax-semantics distinction collapses in this scheme on account of the underlying pragmatic theory of truth which is involved in the dialog-game.

B. THE QUANTUM DIALOGIC METHOD

Before investigating the special kind of dialogs about quantum mechanical propositions it is useful to introduce the dialogic method in a more general sense. In this way it can be seen how we are justified in regarding the dialogic method as a semantic foundation of logic.

1. *The Dialogic Foundation of Logic*

The dialogic method represents a formalized manner of argumentation which serves as a proof-procedure for a proposition under consideration.

The essential point is that a proposition is not primarily characterized by its truth-values (value-definiteness) but by certain proof attempts, the correctness of which is established by means of the dialog-rules. In this sense a proposition is generally called *proof-definite*, in the case of dialogic proofs a proposition is called *dialog-definite*.

The first step in the formulation of the dialogic method is to introduce the concept of a dialog, which explains how a dialog should be carried out by the two participants without specifying which arguments can be stated and how arguments are to be handled in the course of a dialog.

a. *The Concept of a Dialog*

The concept of a dialog is formulated in an *operational* manner by means of the *frame-rules*:

$F1$: At the beginning of a dialog one of the two participants, the proponent (P), asserts the initial argument. In this way the initial position of the dialog is established.

$F2$: The second participant, the opponent (O), attempts to refute this assertion. The dialog then consists of a series of arguments which are stated in turn by the two participants and which obey certain rules.

$F3$: Arguments are either attacks on or defences of previous arguments, but not both.

$F4$: a: The participants have the right to invoke an attack at any position of a dialog.

b: On the other hand, having been attacked, the participants are obliged to defend in the reverse succession of the respective attacks, at the latest when there is no opportunity of attack left. I.e. the latest obligation of defence has to be performed first.

$F5$: If one of the participants has no argument to continue he loses the dialog. In this case the other one wins. In this way the final position of the dialog is established.

These frame-rules are close to those suggested and discussed by Lorenz [8] and will not be debated here. In a *formal* game-theoretical formulation the concept of a *dialog-game* is given by means of:

α. a class Z of game positions $z_1, z_2, \ldots \in Z$,

β. a two-place relation R defined on Z which is established by the rules of the dialog. (In $z_i R z_{i+1}$, z_i is called the R-predecessor of z_{i+1} and z_{i+1} is called the R-successor of z_i. Positions without an R-predecessor are initial positions, positions without an R-successor are final positions.)

γ. a division of Z into disjoint move-classes M_o and M_p,

δ. a division of the final positions E into disjoint success-classes E_o and E_p.

A dialog consists of a series of positions $z_i (0 \leq i)$ where z_0 is the initial position and $z_n R z_{n+1}$ is always satisfied. If z_n belongs to the move-class M_o this means that O has the right to continue the dialog by an R-successor z_{n+1}. In the same way M_p is defined. The initial position z_0 always belongs to M_o. If a final position belongs to E_o O wins the dialog and P loses the dialog. On the other hand P wins and O loses the dialog if a final position belongs to E_p.

By means of this concept of a dialog the following logical notions are defined: A *proposition* is an argument about which a dialog can be carried out, i.e. which can be asserted as initial argument in a dialog.

A proposition A is *true* if and only if P has a strategy of success within a dialog about A, i.e. P wins the dialog irrespective of the arguments of O. A proposition A is *false* if and only if P has a strategy of success against A, i.e. if O asserts A in a dialog P wins the further dialog about O's A irrespective of the arguments of O.

Since by means of the dialog-game it is established how the truth of a proposition can be determined propositions are dialog-definite. For the meta-proposition 'the proposition A is true' we write $\vdash_D A$. Moreover we demand that a practically useful dialog be finite. To obtain useful dialogs further rules must be added. These rules would determine, in particular, which arguments may be stated and how often they may be stated in a dialog.

b. *The Handling of Arguments within a Dialog*

Whereas the frame-rules as formal rules are independent of the special choice of arguments, the *argument-rules* take into account interior properties of propositions. There are some argument-rules, however, which refer to the interior properties only in a restricted sense and which therefore are added to the formal rules. Such rules concern the *logical connectives*. Since arguments are divided into attacks and defences they can be distinguished by particular possibilities of attack and defence. If only the formal structure of a connective is taken into account we have the following requirement: attacks on a certain connective should consist only of the subpropositions of the respective connective and furthermore of challenges to defend; defences should consist only of subpropositions. One can show by means of dialog-equivalence that the number of dialogically distinct connectives is reduced to the following three two-place connectives and one one-place connective:

$A_f 1$:

connective defined as	possibilities of attack	of defence
conjunction $A \wedge B$	1?	A
	2?	B
disjunction $A \vee B$	?	A
		B
material implication $A \rightarrow B$	A	B
negation $\neg A$	A	

On an attack of a negation no defence is possible.

In the formal representation of the dialog-game the operational formulation is replaced by:

$D1$: The positions z of a dialog-game are defined by a class of column-pairs $\overline{z_o|z_p}$ which satisfy the following conditions:

a. The left column consists of all arguments of O and the right column consists of all arguments of P. The succession of arguments is enumerated by the rows which begin at zero.

b. The argument of P in row zero is always a proposition; the argument of O in row zero is always the empty argument [ ].

c. The arguments are assigned the following numbers:
A number in parentheses (i) on the right side of an O-argument or on the left side of a P-argument indicates that the respective argument attacks an argument of the second participant in row i.
A number in carets $\langle i \rangle$ indicates that the respective argument is a defence against an attack of an argument in row i.
If a participant does not defend, the empty argument [ ] is placed in the respective row. He then has to continue in the next row.

$D2$: The move-classes M_o and M_p and the success-classes are defined as follows:

a. If the last empty argument of a position z is placed in the right column then $z \in M_p$, if it is placed in the left column then $z \in M_o$.

b. Final positions $z \in M_p$ belong to E_o, final positions $z \in M_o$ belong to E_p.

To illustrate the above definitions we assume that P has stated the proposition $A \rightarrow B$ in row i of a dialog. Now it is O's turn and he may attack by asserting A. If O attacks with A, P has to defend by stating B. To this position the dialog reads:

0	[ ]			$\cdot$
$\vdots$	$\vdots$			$\vdots$
i	$\cdot$			$A \rightarrow B$
$i+1$	A	(i)	$\langle i \rangle$	B

By the definition of the connectives there is no restriction on the possibilities to repeat the attack A. If, on the other hand, A or respectively B are attacked, the further dialog depends on the connective structure of A and B. As we see the dialog will only come to an end if the procedure of decomposing a proposition by attacks and defences cannot be continued arbitrarily. Therefore certain propositions must be specified which are themselves not composed of subpropositions and which we call *elementary propositions*. These elementary propositions, though not dialog-definite, are assumed to be proof-definite, i.e. it is assumed that for every elementary proposition a procedure is known which provides evidence for the respective proposition and which is not within the frame of the dialog. The proofs are then called material proofs and the propositions material propositions. In the case of dialogs about quantum mechanical propositions elementary propositions are proved by quantum mechanical measurements. In the operational procedure elementary propositions are a consequence of the postulate of finite dialogs. In the formal representation of the dialog-game they are introduced as primary propositions:

$D3$: We assume that a class of elementary propositions S_e and the connective-signs $\wedge, \vee, \rightarrow, \neg$ are given. Then propositions (denoted by $A, B, \ldots$) are defined recursively by:
  a. elementary propositions are propositions,
  b. if A is a proposition then $\neg A$ is a proposition; if A and B are propositions then $A \wedge B$, $A \vee B$, and $A \rightarrow B$ are propositions.

Now we can complete the dialogical treatment of propositions by the demand that every participant in a dialog who asserts an elementary proposition assume the obligation to show its evidence. We call such dialogs in which elementary propositions are to be proved *material dialogs*. If a? denotes the challenge to prove a and a! a successful proof of a we have the following material argument-rule:

A_m 2:

elementary proposition	possibility of attack	of defence
a	a?	a!

Since a? and a! are not proper arguments in the strict dialogic sense the frame-rule $F4b$ concerning elementary propositions has to be weakened:

A_m 3: If one of the participants cannot defend against an attack a? he may assume his last proper obligation of defence before a?.

The implication of this rule may be seen in the following dialog about a disjunction $a \lor b$, where a and b are elementary propositions:

0	[ ]			$a \lor b$
1	?	(0)	$\langle 0 \rangle$	a
2	a?	(1)		[ ]
3	[ ]		$\langle 0 \rangle$	b
4	b?	(3)		[ ]
5	[ ]		$\langle 0 \rangle$	a
⋮				⋮

Here the dialog should not already have been lost for P in row 2 where P could not prove a. There should still be the other possibility to defend the proposition $a \lor b$ by asserting b. Even if P cannot succeed in demonstrating b there is no argument so far to forbid subsequent attempts to defend $a \lor b$ by either of the two propositions a and b.

However, to establish finite dialogs the opportunities of attack and defence must be restricted. It is demanded that these restrictions do not change the possibilities of success for O and P. Otherwise they would be arbitrary and impose *ad hoc* limitations on the above dialog-rules. Therefore certain special conditions of demonstrating elementary propositions must be taken into account to cut off arbitrary repetitions in a dialog.

Consider again the above dialog about $a \lor b$. We could cut off this dialog in the easiest way if a and b satisfied the condition of *unrestricted availability*. This condition means that whenever a demonstration of a proposition has been performed the same demonstration always yields the same result in arbitrary repetitions. If in the above dialog the proofs of a and b have failed they would never succeed in subsequent attempts. In this case the above dialog would never be won. Therefore the dialog about $a \lor b$ can be cut off after one attempt at proving each of a and b. In a similar way dialogs about the other connectives can be treated under the condition of unrestricted availability. By adding appropriate argument-rules finite material dialogs are established. The condition of unrestricted

availability is fulfilled with respect to many examples of elementary pro-
position-classes S_e. For instance, all formal deductions in mathematics
remain conclusive once they have been established. Also, all propositions
about classical physical systems satisfy this property. It is assumed that at
least in principle the outcome of an experiment is not affected by other
experiments. Since the time between a series of experiments can be imagined to be arbitrarily small one is justified in speaking of simultaneous
measurements and of simultaneous propositions in a dialog.

In general specific conditions for demonstrating elementary propositions
result from the experience which at first can be derived from unrestricted
dialogs. They then determine further argument-rules which guarantee the
finiteness of the dialog-game.

Suppose that we are now in possession of material dialogs with respect
to a certain class S_e of elementary propositions. We direct our interest
towards those propositions for which P has a strategy of success independently of the evidence for the elementary propositions. A simple example may be the proposition $a \to a$. The dialog is until row 2:

$$
\begin{array}{cc|c}
0 & [\,] & a \to a \\
1 & a \quad (0) & [\,] \\
2 & & (1) \quad a?
\end{array}
$$

In this position of the dialog O is obliged to defend a. In case he fails he
loses the dialog, i.e. P wins. In case he succeeds and can state $a!$, P has to
continue the dialog by defending against the attack in row 1. If the proof
of a is still conclusive, i.e. a is still available, P can refer to O's proof of a
and wins the dialog again. Thus the proposition $a \to a$ is true independently of the proof-result of a. Therefore $a \to a$ is called a *formally* true
proposition.

We could now examine the formal demonstrability of a proposition by
considering all possibilities of a dialog with respect to the proof-results of
the elementary propositions. However, in general it would be difficult to
investigate all such possibilities and to formalize this procedure in an
appropriate manner. Therefore, a different way is chosen which replaces
the argument-rules of the material dialog by formal argument-rules which
generate so-called *formal dialogs*. The essential point is that in formal
dialogs elementary propositions are no longer challenged to be proved.
This is obtained by forbidding that elementary propositions be attacked
and by restricting P in his right to assert elementary propositions. The
proponent should be allowed to state an elementary proposition only if it

has been asserted by the opponent previously. Moreover, the respective proposition must be *available*. This means, as already has been pointed out, that the previous proof must still be conclusive in the respective position of the dialog. If in this position material proofs of elementary propositions are reintroduced P would be sure to win the dialog in any case, as can easily be seen in the above dialog about $a \rightarrow a$. The formal argument-rule concerning elementary propositions is:

A_f 2: a: Elementary propositions are not attackable.

b: O is allowed to state elementary propositions in every position of a dialog. P is allowed to state elementary propositions only if they have been asserted by O previously and if they are still available in the respective position of the dialog.

Since P's dialogic possibilities depend on O's previous arguments P must have far-reaching opportunities to attack previous propositions of O and to evoke elementary propositions from O. After having stated a proposition a participant should be obliged to defend it against an attack only as long as it is still available. Thus we have the following argument-rule:

A_f 3: P is allowed to attack propositions of O only as long as they are available in the respective position of the dialog.

According to the material dialog-rules further formal argument-rules must be added to specify how the availability of propositions is determined and to cut off infinite strategies. On the whole, the formal argument-rules must guarantee that in the case of reintroduction of material proofs of elementary propositions, propositions are determined to be true in formal dialogs if and only if they are true in all respective material dialogs.

Suppose that a formal dialog-game about a certain class S_e of elementary propositions has been established. By means of such a dialog-game we are then able to enumerate all possible strategies of success by a calculus. This calculus consists of special positions of a dialog, namely those for which P has a strategy of success. This procedure will now be executed in detail in the case of quantum mechanical propositions.

2. *Dialogs about Quantum Mechanical Propositions*

By a quantum mechanical proposition we mean any proposition with respect to a given quantum mechanical system which can be stated as initial argument in a dialog, the rules of which are now to be investigated. The initial position in a quantum dialog is the following: A quantum mechanical system γ is given; an initial proposition with respect to γ is stated. (This does not mean that the system is given in a certain state since this notion is not yet defined. The information about the state, usually obtained by means of a preparation-procedure, would depend on a foregoing dialog but is irrelevant with respect to any initial position of a dialog. We only assume that the system under consideration exists.) The quantum dialog has to satisfy the frame-rules $F1$ to $F5$ and the argument-rule A_f1, which are formal rules in any dialog.

a. *Material Quantum Dialogs*

With A_m2 the question arises: How are elementary quantum mechanical propositions to be proven? We require that elementary propositions with respect to a certain system be demonstrated by measurements on the respective system. For every elementary proposition a there exists a class of measurements which are accepted as admissible proof-attempts for a. In this way elementary quantum mechanical propositions are proof-definite. In general it is not assumed that the elementary propositions in addition are value-definite, i.e. that some admissible measurements with respect to a are yes-no experiments which decide between the truth and falsity of a proposition.

As an example of a quantum mechanical system we consider an electron which is assumed to move in a certain direction (x^+-direction). An example of an elementary proposition is the proposition "the electron has 'spin-up' in the direction a in the $y - z$ plane". An admissible proof-attempt for this proposition would be to perform a Stern-Gerlach experiment which is oriented in the direction a. If the outcome of this experiment is positive, i.e. if we find the electron having the right deviation in the magnetic field, the above elementary proposition is proven. If the outcome of the experiment is negative, i.e. if we do not find the electron having the right deviation we say that the proof failed.

In addition, we postulate that for every elementary proposition there exists, at least in principle, a measurement with the following property: Assume that the outcome of the measurement was positive. Then a subsequent proof by means of the same or an identical measurement with respect to the same system always yields a positive outcome again if no other action on the system is allowed in the meantime. A Stern-Gerlach experiment can be assumed to be such a measurement, often called a measurement of the first kind. At least it can be imagined how to refine this experiment such that it satisfies the above property. In our dialogs we shall not repeat proofs if they have already been established but *refer* to the previous proofs. Therefore, measurements of the first kind are not necessary for proofs in the dialogs but they should exist 'at least in principle' to make it reasonable to refer at a later time to a succesful proof.

Whereas the above property can be accepted with respect to classical systems as well as for quantum mechanical systems the classical property of unrestricted availability cannot be maintained in the quantum mechanical case. There are elementary propositions a and b which are incommensurable in the following sense: There exists no measurement for a and no measurement for b which satisfies the condition that in an arbitrary series of alternating measurements of a and b with respect to the same system under consideration the outcomes are always the same.

As an example we consider again the above proposition "the electron has 'spin-up' in the direction a in the y-z plane" and, in addition, the proposition "the electron has 'spin-up' in the direction b in the y-z plane" where the angle between a and b is assumed to be 30°. For the respective proofs we use two Stern-Gerlach experiments oriented in the a and b directions. It is an emperical fact that if a series of these Stern-Gerlach experiments are performed on the electron in turn, the outcomes of each experiment are not always the same. This means, according to our above definition, that both propositions are incommensurable.

What are the consequences of the incommensurability of two propositions a and b for quantum dialogs about connectives? Consider a dialog about the proposition $a \lor b$ where the system under consideration for instance is again an electron and where a and b are the above spin-propositions. On O's attack against the disjuction, P defends by stating the proposition a. Upon request he has to perform a Stern-Gerlach experiment to prove a. If the outcome of this experiment is negative P uses the other possibility to defend $a \lor b$ by asserting b. The proof-attempt then consists in performing another Stern-Gerlach experiment on the same electron. Even if the proof of b also fails P has the possibility to win the dialog because of the incommensurability of a and b. Since in a sufficiently large series of measurements of a and b a successful proof of one of the two propositions is obtained, P has a strategy of success in the dialog about $a \lor b$:

0	[ ]			$a \vee b$
1	?	(0)	⟨0⟩	a
2	a?	(1)		[ ]
3	[ ]		⟨0⟩	b
4	b?	(3)		[ ]
5	[ ]		⟨0⟩	a
⋮	⋮			⋮
n	a?	$(n-1)$	$\langle n-1 \rangle$	a!.

On the other hand, consider a dialog about the proposition $a \wedge b$:

0	[ ]			$a \wedge b$
1	1?	(0)	⟨0⟩	a
2	a?	(1)	⟨1⟩	a!
3	2?	(0)	⟨0⟩	b
4	b?	(3)	⟨3⟩	b!
⋮	⋮			⋮
$n-1$	1?	(0)	⟨0⟩	a
n	a?	$(n-1)$		[ ].

P is certain to lose the dialog in case a and b are incommensurable because, after a sufficiently large number of attacks, he will fail in a measurement of a or b.

This is also the case in a dialog about the proposition $a \rightarrow b$:

0	[ ]			$a \rightarrow b$
1	a	(0)		[ ]
2	a!	⟨1⟩	(1)	a?
3	[ ]		⟨0⟩	b
4	b?	(3)	⟨3⟩	b!
⋮	⋮			⋮
$n-3$	a	(0)		[ ]
$n-2$	a!	$\langle n-3 \rangle$	$(n-3)$	a?
$n-1$	[ ]		⟨0⟩	b
n	b?	$(n-1)$		[ ].

In case a and b are commensurable, i.e. their proofs always yield the same results, the dialog about $a \vee b$ would be lost for P after row 4, where he failed in both proof-attempts. On the other hand, P would win the dialogs about $a \wedge b$ and $a \rightarrow b$ if he succeeded in all proofs until row 4. Since P is able, in the case of the commensurability of a and b, to refer in the further dialog to his previous proofs he possesses a strategy of success in the last two dialogs. We see that in a quantum dialog propositions which have been proven are not always available but only under certain conditions which are determined by means of the commensurability of propositions. The commensurability and incommensurability of propositions can be used to obtain finite quantum dialogs. In this way they represent a general property of quantum mechanical propositions which we call the *restricted availability*. The precise meaning of the property of restricted availability is given in the following.

We assume that for every pair of propositions A and B there exists the proposition 'A and B are commensurable'. This *commensurability proposition* is dialogically defined as a material proposition, denoted by $k(A, B)$. The proof-definiteness is established in the following way: Consider at first elementary propositions a and b. Whenever $k(a, b)$ is asserted in a dialog with respect to a system γ, $k(a, b)$ may be attacked by $k(a, b)$?. The respective defence consists in a series of measurements on the same system under consideration which are performed in turn. $k(a, b)$ is proven to be true if and only if in an arbitrary succession of measurements of a and b the outcomes are always the same. In this case we write for the successful proof $k(a, b)$!. In an analogous way the commensurability of compound propositions is establised. The proof of $k(A, B)$ is performed by means of a series of dialogs about A and B with respect to the same system under consideration. $k(A, B)$ is proved to be true if and only if in an arbitrary succession of dialogs about A and B the final positions always belong to the same success-classes. In this case we write $k(A, B)$!. The respective argument-rule in the dialog is:

$A_m\ 4:$

commensurability	attack	defence
$k(A, B)$	$k(A, B)$?	$k(A, B)$!

The incommensurability of A and B is asserted by $\neg k(A, B)$.

In general, a proof of $k(A, B)$ consists of an infinite number of dialogs about A and B. In this case we assume that, after a sufficient number of dialogs in which the results did not change, both participants accept $k(A, B)$. By using commensurability-attacks in appropriate positions of a dialog finite material dialogs are obtained. This would be established by additional argument-rules to $A3_m$ which are formulated in [10] and which will not be given here. We are now prepared to investigate formal dialogs about quantum mechanical propositions.

b. *Formal Quantum Dialogs*

In the material dialogs as described above we have two kinds of material propositions: the elementary propositions and the commensurability-propositions which base themselves on elementary propositions again. The proofs of these material propositions have to be eliminated in formal dialogs. To get more easily through this problem, we eliminate them one after the other. The first step then is to replace the material argument-rules $A_m 2$ and $A_m 3$ by the formal argument-rules $A_f 2$ and $A_f 3$ and to specify how the availability of quantum mechanical propositions has to be determined.

(i) *Formal dialogs with material commensurabilities.* At first we have to exclude that by means of the rules $A_f 2$ and $A_f 3$ certain propositions are infinitely available. Since P has to possess a strategy of success within a finite dialog we require that he place a bound on himself limiting the number of times he may take an elementary proposition over and attack a proposition of O. He may do this by choosing a number n after having asserted the initial argument. Since the class of formally provable propositions depends on this number n, we at first consider a dialog-game, determined by a fixed n (n-dialog-game), and finally take the union of all n-dialog-games with respect to n ($n < \infty$).

Since P has to defend his strategy of success against all arguments of O it would be enough for O to assert his most advantageous argument. Moreover, it can be shown that he would not improve his strategy of success if he had more than one possibility to attack. Therefore we restrict the possibilities of P and O by the following argument-rule:

$A_f 4_n$: *a.* P is allowed to take over the same elementary proposition at most n times. P is allowed to attack the same proposition of O at most n times.

   b. O is allowed to attack propositions of P at most 1 time. O has to decide if he is going to defend against an attack by P or if he is going to attack by himself. In case he defends he no longer has the right to attack; in case he attacks the respective defence is no longer possible.

Within the formal representation of the dialog-game we characterize the availability of propositions in the follwoing way:

$D1:d$. To each proposition of O an index i is assigned which indicates the number of times the respective proposition is available. According to rule A_f4_n each proposition obtains the availability n after it has been stated by O. If a proposition is taken over or attacked by P the availability reduces by one.

Another essential restriction on the availability results from the general incommensurability of quantum mechanical propositions. Here we have to take into account that a previously proved proposition is sure to be true in a later position of a dialog only if it is commensurable with all succeeding propositions. However, only O's propositions are relevant because P has to take over all elementary propositions from O. Thus, in the case that O has stated a new proposition A, the availabilities of all previous propositions of O which are not commensurable with A have to be reduced to zero, whereas the availability of commensurable previous propositions remains unchanged. To establish an appropriate argument-rule those dialogic positions have to be investigated in which commensurabilities could be called in question. It turns out, as demonstrated elsewhere [10], that not in all positions of a dialog, where O states a new proposition, are commensurabilities of previous propositions uncertain. Only in the following three positions p_j are the availabilities of previous propositions in question:

p_j1: O attacks a material implication $A \to B$ by A:

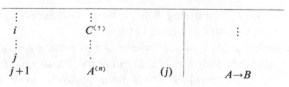

p_j2: O attacks a negation $\neg A$ by A:

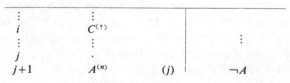

p_j3: O defends a disjunction $A \lor B$ against an attack of P:

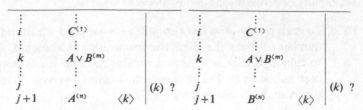

Now we can demand that the onus is on P to change the availability of C in the right way. P is challenged to do so in row j of the above positions p_j before O sets the new proposition. If P does not change the availability of C to zero O may question the commensurability in row $j + 1$ instead of asserting the new proposition. Since a reduction of the availability does not improve P's strategy of success P is urged to change the availability to zero as it would be the case in material dialogs. The respective argument-rules are formulated in $A5$: a and $A5$: b which followbelow. There may be propositions which P has proved by means of propositions of O, no longer available in a certain position of a dialog, and which are themselves still available in the respective position of the dialog. Then after having reduced the availabilities to zero, P would no longer have the possibility to refer to these propositions and his strategy of success would be shortened with respect to his possibilities in material dialogs. Therefore the following argument-rule $A5$:c. must be established which in addition to $A5$: a and b. enables P to refer to all propositions which are proved by O or by P previously and which are still available:

$A5$: a. In a position p_j of a dialog P is allowed to reduce the availability of all previous propositions of O to zero.

   b. In a position p_j of a dialog, where O can state a new proposition A, O is allowed to ask by $k(C, A)$? instead of asserting A. Here C is any previous proposition of O with non-zero availability. The defence against $k(C, A)$? consists in a demonstration of $k(C, A)$.

   c. In a position p_j of a dialog P is allowed to challenge O to assert a proposition B as hypothesis. Thereupon O may either assert the proposition B and continue the dialog in row $j + 1$ or he may ask by $\widetilde{\mathfrak{A}} \to B$?. $\mathfrak{A}$ represents the conjunction of all previous propositions of O with non-zero availabilities.

   On this attack P has to defend by asserting the new initial argument $\widetilde{\mathfrak{A}} \to B$.

To exclude infinite strategies of P by means of $A5$: c. we have to add:

A_f6_n: P is allowed to make use of $A5$: c. at most n times.

In the formal description of the dialog-game hypotheses are treated in the following way:

$D1$: e. Propositions which are challenged by P as hypotheses are placed in the rows -1 to $-n$. In the case that a new initial proposition is stated the previous hypotheses vanish in the new position of the dialog. If the previous hypotheses are stated in the rows -1 to $-m (m \leq n)$, the hypotheses of the new dialog have to be placed in the rows from $-m - 1$ to $-n$. If n hypotheses are stated no further hypothesis can be challenged by P.

An example for a dialog in which P has to use the rule $A5$ for establishing a strategy of success is the dialog about the proposition $a \rightarrow (((a \vee b) \rightarrow c) \rightarrow (a \vee b))$. This dialog reads to row 1 :

0	[ ]			$a \rightarrow (((a \vee b) \rightarrow c) \rightarrow (a \vee b))$
1	$a^{(n)}$	(0)	$\langle 0 \rangle$	$((a \vee b) \rightarrow c) \rightarrow (a \vee b)$

In this position of the dialog P has to reduce the availability of a to zero because a and $(a \vee b) \rightarrow c$, which is the next possible argument for O, cannot be proved to be commensurable. The reader should accept the incommensurability here. For a proof I refer to the later scheme K to demonstrate formal commensurabilities (see part (ii)). P has a strategy of success in the continuation of the above dialog if he succeeds in demonstrating $a \vee b$ after O's attack by $(a \vee b) \xrightarrow{\cdot} c$. Since $a \rightarrow (a \vee b)$ is formally true he applies, before reducing the availability of a to zero, rule $A5$: c. by challenging O to state the hypothesis $a \vee b$. After this step he reduces the availability of a to zero whereas he preserves the availability of the hypothesis $a \vee b$ because $a \vee b$ and $(a \vee b) \rightarrow c$ are formally commensurable. We then have the following position of the dialog:

-1	$a \vee b^{(n)}$			
0	[ ]			$a \rightarrow (((a \vee b) \rightarrow c) \rightarrow (a \vee b))$
1	$a^{(0)}$	(0)	$\langle 0 \rangle$	$((a \vee b) \rightarrow c) \rightarrow (a \vee b)$
2	$(a \vee b) \rightarrow c^{(n)}$	(1)	$\langle 1 \rangle$	$a \vee b$

Now it can easily be seen that P has a strategy of success in the further dialog. In row 3, O attacks P's proposition $a \vee b$. Upon this attack, P cannot defend, but he attacks O's hypothesis $a \vee b$:

-1	$a \vee b^{(n-1)}$			
0	[ ]			$a \to (((a \vee b) \to c) \to (a \vee b))$
1	$a^{(0)}$	(0)	$\langle 0 \rangle$	$((a \vee b) \to c) \to (a \vee b)$
2	$(a \vee b) \to c^{(n)}$	(1)	$\langle 1 \rangle$	$a \vee b$
3	?	(2)		[ ]
4			(-1)	?

This is again a position where the rule $A5$ applies. Since $(a \vee b) \to c$ is not commensurable with O's potential defences a and b, P has to reduce the availability of $(a \vee b) \to c$ to zero. The availability of the hypothesis $a \vee b$ has not to be changed because a and b are formally commensurable with $a \vee b$. After P's attack on the hypothesis, O has to defend by a or b. In both cases P can take over the elementary proposition to defend on O's attack in row 3. Since elementary propositions are unattackable O has no further argument. Therefore P wins the dialog in any case:

-1	$a \vee b^{(n-1)}$			
0	[ ]			$a \to (((a \vee b) \to c) \to (a \vee b))$
1	$a^{(0)}$	(0)	$\langle 0 \rangle$	$((a \vee b) \to c) \to (a \vee b)$
2	$(a \vee b) \to c^{(0)}$	(1)	$\langle 1 \rangle$	$a \vee b$
3	?	(2)		[ ]
4	$a^{(n-1)} \mid b^{(n-1)}$ $\langle -1 \rangle$		(-1)	?
5	[ ]		$\langle 2 \rangle$	$a \mid b$

This proves that P has a strategy of success for the proposition $a \to (((a \vee b) \to c) \to (a \vee b))$.

By means of the argument-rule $A5$ formal quantum dialogs are established in which the proponent can refer to all and only to previously proven propositions. Therefore, if material proofs are reintroduced he will never be challenged to prove a new elementary proposition, the proof of which is not available and hence uncertain. The final step to complete the formal quantum dialog-game is to take into account only formal commensurabilities and to give a proof-procedure for them.

(ii) *Formal dialogs with formal commensurabilities.* The demonstration that certain commensurabilities $k(A, B)$ are always true independently of the elementary propositions, which A and B consist of, is according to the previous definition: material dialogs about A and B are carried out in turn. If it is certain that all possible material dialogs about A and B have always the same results in arbitrary repetitions of the dialogs $k(A, B)$ is proved. By means of this definition it is obvious that the commensurabilities $k(A, A)$, $k(A, k(A, B))$ and $k(A, k(B, A))$ can be proved to be true. Furthermore it can be shown that $k(A, A \wedge B)$ $k(A, A \vee B)$, $k(A, A \rightarrow B)$ and $k(A, \neg A)$ are true [10]. Besides these formally true commensurabilities there are certain rules between commensurabilities which can be shown to be true irrespective of the proof-results of the elementary propositions. In this way we arrive at the following *scheme K for proving commensurabilities* (according to the usual representation, this scheme consists of beginnings and rules, denoted by the double-arrow '$\Rightarrow$' which means a transition in this scheme from already established figures to new ones.)

$K1.$ $a) \Rightarrow k(A, A)$ $b) \Rightarrow k(A, k(A, B))$ $c) \Rightarrow k(A, k(B, A))$
      $d) \Rightarrow k(A, A \vee B)$ $e) \Rightarrow k(A, A \vee B)$ $f) \Rightarrow k(A, A \rightarrow B)$
      $g) \Rightarrow k(A, \neg A)$

$K2.$ $k(A, B) \Rightarrow k(B, A)$

$K3.$ $\vdash_D A \rightarrow B \Rightarrow k(A, B)$

$K4.$ $k(A_1, B) \,\&\, k(A_2, B) \Rightarrow k(A_1 * A_2, B)$
        with $* \in \{k(\,,), \wedge, \vee, \rightarrow\}$

$K5.$ $k(A, B) \Rightarrow k(\neg A, B)$

Finally we show that the material commensurability-proposition is dialog-equivalent to a certain connected proposition.

At first we assume that we have a position p_j (as defined above) in a dialog in which C is the only proposition with non-zero availability. O may continue the dialog by asserting A. With respect to this position we prove the following equivalence:

$E1$: If $k(C, A)$ can be proved, then there exists a strategy of success for the dialog about the connected proposition $C \rightarrow (A \rightarrow C)$ and vice versa, i.e.: $\vdash_K k(C, A)$ iff $\vdash_D C \rightarrow (A \rightarrow C)$.

To demonstrate this equivalence we show in one direction that on the premise of $\vdash_K k(C, A)$ P wins the formal dialog about $C \rightarrow (A \rightarrow C)$. In this dialog, O having attacked by C, P defends by asserting $A \rightarrow C$. In this position of the dialog, P has to take into account the availability of C, because O has the right to attack in the next row by $k(C, A)$?. Since $k(C,A)$ is presumed to be true then P leaves the availability of C unchanged and O continues by asserting A. Thereupon P defends by C:

0	[ ]			$C \rightarrow (A \rightarrow C)$
1	$C^{(n)}$	(0)	$\langle 0 \rangle$	$A \rightarrow C$
2	$A^{(n)}$	(1)	$\langle 1 \rangle$	C

In the further dialog P possesses a strategy of success because if O attacks C then P attacks O's C in row 1 and from row 3 on he can take over every argument from O.

To demonstrate the other direction one shows that on the premise of a strategy of success for $C \rightarrow (A \rightarrow C)$ an obligation to prove $k(C, A)$ in a dialog can be prevented. A proof of $k(C, A)$ may be challenged in a position p_j. In this position now P uses $A5: c.$ to demand the proposition $A \rightarrow C$ as hypothesis from O. Thereupon O has the two possibilities either to attack by $C \rightarrow (A \rightarrow C)$? or to continue the dialog by asserting A. Since it is assumed that P has a strategy of success for $C \rightarrow (A \rightarrow C)$, O chooses the second possibility:

-1	$A \rightarrow C^{(n)}$	
$\vdots$	$\vdots$	
i	$C^{(0)}$	$\cdot$
$\vdots$	$\vdots$	$\cdot$
j	$\cdot$	$\cdot$
$j+1$	$A^{(n)}$	

P then continues the dialog by attacking the hypothesis by A. If O then attacks A, P attacks O's A and possesses a strategy of success by taking over the respective arguments from O. Therefore O defends by asserting C:

-1	$A \rightarrow C^{(n-1)}$		
$\vdots$	$\vdots$	$\cdot$	
i	$C^{(0)}$	$\cdot$	
$\vdots$	$\vdots$	$\cdot$	
j	$\cdot$		
$j+1$	$A^{(m)}$		
$j+2$	$C^{(n)} \langle -1 \rangle$	$(-1)\,A$	

In this position the proposition C is available after O has asserted A.

In the general case a position p_j consists of several previous propositions $C_1, ..., C_r$ with non-zero availabilities. In this case we can prove the following equivalence [10]:

$E2$: If the material commensurabilities $k(C_1, A)$, ..., $k(C_r, A)$ can be proved, then there exists a strategy of success for the connected proposition $(C_1 \wedge ... \wedge C_r) \rightarrow (A \rightarrow (C_1 \wedge ... \wedge C_r))$ and vice versa, i.e.: $\vdash_K k(C_1, A)$ & ... & $\vdash_K k(C_r, A)$ iff $\vdash_D (C_1 \wedge ... \wedge C_r) \rightarrow (A \rightarrow (C_1 \wedge ... \wedge C_r))$

The above formal commensurability-scheme K can now be incorporated into the formal dialog-game by adding the following argument-rules to the previous ones:

$A_f 7$: a. O is not allowed to attack the initial argument $A \to ((A \to B) \to A)$.

b. P is allowed to replace the initial argument $A \to (B \to A)$ by the new initial argument $B \to (A \to B)$.

c. P is allowed to replace the initial argument $(A_1 * A_2) \to (B \to (A_1 * A_2))$ by the new initial arguments $A_1 \to (B \to A_1)$ or by $A_2 \to (B \to A_2)$ ($* \in \{\wedge, \vee, \to\}$) where O has to choose between the two possibilities.

d. P is allowed to replace the initial argument $\neg A \to (B \to \neg A)$ by the new initial argument $A \to (B \to A)$.

It is due to the special form of the commensurabiltiy-connective that the commensurabilities $K1$, except f), and the commensurability-rule $K3$ can be proved in a formal dialog and therefore are not established by $A_f 7$. To exclude infinite strategies by means of this rule we add:

$A_f 7$: e. P is allowed to state the same initial argument at most one time.

According to the introduction of the commensurability-connective the argument-rule $A5$ has to be reformulated by replacing $A5$:b. by the formal rule:

$A_f 5$: b. In a position p_j of a dialog, where O can state a new proposition A, O is allowed to ask by $k(\mathfrak{A}, A)$? instead of asserting A. Here $\mathfrak{A}$ is the conjunction of all previous propositions of O with non-zero availabilities. The defence against $k(\mathfrak{A}, A)$? consists in asserting the new initial argument $\mathfrak{A} \to (A \to \mathfrak{A})$.

Because only formal commensurabilities are taken into account the argument-rules $A_f 1$ to $A_f 7$ allow to demonstrate all formally true propositions by means of the n-dialog-game. The formal representation of this dialog-game is given by $D1$ to $D3$ and finally by:

$D4$: In a formal dialog arguments are given by:
   a. propositions,
   b. the following doubts : ?, 1?, 2?, $k(A, B)$?, $A \to B$?,
   c. the empty argument [ ].

To obtain utmost simple game-positions we now reduce the content of a dialog to its most essential part. By starting at the previous positions (defined by $D1$ to $D4$) we obtain the *reduced* positions in the following way:

$S1$: In the right column of the dialog-scheme all attacks against which O has defended and all arguments which O has attacked are eliminated. Furthermore, all empty arguments are eliminated, except the last one in case P has not defended the respective argument.

$S2$: In the left column all propositions with non-zero availabilities are eliminated.

Furthermore, all empty arguments are eliminated, except the last one in case it stands in the last row and the last row is not row zero.

$S3$: In both columns all remaining doubts and the attack- and defence-numbers are eliminated.

$S4$: The remaining empty arguments are replaced by the so-called *potential* arguments:

   a. $[A]$, if A is the only possible defence in the respective position,

   b. $[A, B]$ if A and B are the possible defences,

   c. $[\ ]$, i.e. the empty argument is left unchanged, if against an attack on a negation no defence is possible.

$S5$: After execution of the steps $S1$ to $S4$ the move-classes M_o and M_p are determined as follows:

   a. if there is no potential argument in both columns then $z \in M_o$,

   b. if there is one potential argument in both columns then $z \in M_o$,

   c. if there is only one potential argument (which has to stand in the right column according to $D1$. then $z \in M_p$.

$S6$: Positions which differ only by the succession of the arguments are identified.

In the following considerations only those positions of a dialog are interesting which permit a strategy of success for P in the subsequent continuation of the dialog. Starting at the final positions of success $z \in E_p$ the whole class G of positions of success can be defined by the following *recursive* scheme:

$I1$: Final positions of success in the formal dialog-game are positions of success, i.e.

$$z \in E_p \Rightarrow z \in G.$$

$I2$: If G is a class of positions of success then all predecessors $z \in M_p$ of positions $z' \in G$ are positions of success, i.e.:

$$z \in M_p \,\&\, (\exists_{z'} \, zRz' \,\&\, z' \in G) \Rightarrow z \in G.$$

$I3$: If G is a class of positions of success then all those predecessors $z \in M_o$ of positions $z' \in G$ for which all successors are members of G are positions of success, i.e.:

$$z \in M_o \,\&\, (\forall_{z' s. t. zRz'} \, z' \in G) \Rightarrow z \in G.$$

3. Further Remarks on the Quantum Dialog-Game

In part B. 1.a. we defined truth and falsity of a proposition. In addition we can define two other semantic notions 'satisfiability' and 'refutability' in the following way:

A proposition A is *satisfiable* if and only if O has a successful argument for A, i.e. if O asserts A in a dialog he has an argument such that he wins the dialog about A.

A proposition A is *refutable* if and only if O has a sucessful argument against A, i.e. if P asserts A in a dialog O has an argument such that O wins the dialog about A.

By means of these definitions it is easy to demonstrate the follwing rules in which '$\Rightarrow$' and '$\daleth$' are meta-linguistic abbreviations for 'then' and 'not'.

1.	a.	A true	$\Rightarrow$	A satisfiable
	b.	A false	$\Rightarrow$	A refutable
2.	a.	A true	$\Leftrightarrow$	$(\neg A)$ false
	b.	A false	$\Leftrightarrow$	$(\neg A)$ true
	c.	A satisfiable	$\Leftrightarrow$	$(\neg A)$ refutable
	d.	A refutable	$\Leftrightarrow$	$(\neg A)$ satisfiable
3.	a.	A true	$\Rightarrow$	$\daleth A$ refutable
	b.	A false	$\Rightarrow$	$\daleth A$ satisfiable
	c.	A satisfiable	$\Rightarrow$	$\daleth A$ false
	d.	A refutable	$\Rightarrow$	$\daleth A$ true

Finally let us compare the quantum dialog-game with the dialog-game about unrestrictedly available propositions which can be regarded as a foundation of the usual logical calculi and which is given in [8]. It is a consequence of the general condition of unrestricted availability that in the latter dialog-game the availabilities are only restricted by means of the argument-rule $A_f 4_n$. But this implies that, with respect to the quantum dialog, P can only improve his strategy of success. If we denote the semantic notions concerning the dialog with *unrestricted availabilities* by '*' the following rules can be demonstrated:

1. $\forall_A$ (A true $\Rightarrow$ A true*)

2. $\exists_A$ (A satisfiable & A false*)

An easy example to show the invalidity of the reverse direction of 1. is the proposition $A \rightarrow (B \rightarrow A)$. This proposition is true* because P has a strategy of success in the unrestricted dialog:

0	[ ]			$A \rightarrow (B \rightarrow A)$
1	$A^{(n)}$	(0)	$\langle 0 \rangle$	$B \rightarrow A$
2	$B^{(n)}$	(1)	$\langle 1 \rangle$	A
⋮				⋮

In the quantum dialog O has the additional possibility to ask $k(A, B)$? in row 2. Since this argument is successful in case A and B are not commensurable, $A \rightarrow (B \rightarrow A)$ is refutable. On the other hand $\neg (A \rightarrow (B \rightarrow A))$ is false* as the following unrestricted dialog shows:

i	$\neg (A \rightarrow (B \rightarrow A))$		[ ]
$i+1$		(i)	$A \rightarrow (B \rightarrow A)$
⋮			⋮

In the quantum dialog, however, $\neg (A \rightarrow (B \rightarrow A))$ is satisfiable because O can state $k(A, B)$? in the further dialog. This gives an example for rule 2.

C. THE CALCULUS OF QUANTUM LOGIC

The main point of this paper is to show that the above n-dialog-game about quantum mechanical porpositions can be replaced by a calculus of quantum logic. This calculus allows us to deduce all formally true propositions which are initial positions in an n-dialog.

1. The Tableaux-Calculus T_n

We call the figures of this calculus tableaux. The tableaux represent certain reduced positions of a dialog by means of the following one-one mapping:

$$|A \leftrightarrow \quad \| A$$

$$\mathfrak{A}^{(\alpha)}|A \leftrightarrow \mathfrak{A}^{(\alpha)}\| A$$

$$\mathfrak{A}^{(\alpha)}|\Gamma \leftrightarrow \mathfrak{A}^{(\alpha)}\| \Gamma.$$

In these figures $\mathfrak{A}$ denotes a system of propositions of O in which hypotheses are numbered according to $S3$. (α) designates a respective system of availabilities; furthermore Γ designates a potential argument. By means of this mapping only those positions of a dialog are associated with tableaux in which O's column consists of propositions or is empty and in which P's column consists of either one proposition or one potential argument. Therefore the argument in P's column establishes to which move-class the respective position belongs. In a position $\mathfrak{A}^{(\alpha)}| A \ P$ has moved and it is then O's turn. A tableau $\mathfrak{A}^{(\alpha)}\|A$ is called an O-tableau. On the other hand, in a position $\mathfrak{A}^{(\alpha)} | \Gamma \ O$ has moved and the next turn is on P. Therefore $\mathfrak{A}^{(\alpha)} \| \Gamma$ is called a P-tableau.

In the following *tableaux-calculus* T_n, the double-arrow '$\Rightarrow$' designates transitions between tableaux. If a tableau is a beginning, i.e. has no predecessor, we write the double-arrow without a premise.

(1.1) $\qquad \Rightarrow \begin{array}{c} \mathfrak{A}^{(\alpha)} \\ a^{(\mu)} \end{array} \Big\| a$

(1.2) $\qquad \Rightarrow \| A \to ((A \to B) \to A)$

(2.1) $\qquad \mathfrak{A}^{(\alpha)} \| [A] \ \& \ \mathfrak{A}^{(\alpha)} \| [B] \qquad\qquad \Rightarrow \mathfrak{A}^{(\alpha)} \| A \wedge B$

(2.2) $\qquad \mathfrak{A}^{(\alpha)} \| [A, B] \qquad\qquad\qquad \Rightarrow \mathfrak{A}^{(\alpha)} \| A \vee B$

(2.3.1) $\qquad \| \hat{\mathfrak{A}} \to (A \to \hat{\mathfrak{A}}) \ \& \ \begin{array}{c} \mathfrak{A}^{(\alpha)} \\ A^{(n)} \end{array} \Big\| [B] \qquad \Rightarrow \mathfrak{A}^{(\alpha)} \| A \to B$

(2.3.2) $\qquad \mathfrak{A}^{(\alpha)} \| A \to B \qquad\qquad\qquad \Rightarrow \begin{array}{c} \mathfrak{A}^{(\alpha)} \\ C^{(\mu)} \end{array} \Big\| A \to B$

(2.3.3) $\qquad \| \hat{\mathfrak{A}} \to C \ \& \ \begin{array}{c} \mathfrak{A}^{(\alpha)} \\ C^{(n)} \end{array} \Big\| A \to B \qquad \Rightarrow \mathfrak{A}^{(\alpha)} \| A \to B$

(2.4.1) $\qquad \| \hat{\mathfrak{A}} \to (A \to \hat{\mathfrak{A}}) \ \& \ \begin{array}{c} \mathfrak{A}^{(\alpha)} \\ A^{(n)} \end{array} \Big\| [\ \ ] \qquad \Rightarrow \mathfrak{A}^{(\alpha)} \| \neg A$

(2.4.2) $\qquad \mathfrak{A}^{(\alpha)} \| \neg A \qquad\qquad\qquad\qquad \Rightarrow \begin{array}{c} \mathfrak{A}^{(\alpha)} \\ C^{(\mu)} \end{array} \Big\| \neg A$

$(2.4.3)$ $\left\| \hat{\mathfrak{A}} \to C \ \& \ \begin{matrix} \mathfrak{A}^{(\alpha)} \\ C^{(n)} \end{matrix} \right\|_{\neg A}$ $\Rightarrow \mathfrak{A}^{(\alpha)} \|_{\neg A}$

$(3.1.1)$ $\begin{matrix} \mathfrak{A}^{(\alpha)} \\ A \wedge B^{(\mu-1)} \\ A^{(n)} \end{matrix} \Big\|_{\Gamma}$ $\Rightarrow \begin{matrix} \mathfrak{A}^{(\alpha)} \\ A \wedge B^{(\mu)} \end{matrix} \Big\|_{\Gamma}$

$(3.1.2)$ $\begin{matrix} \mathfrak{A}^{(\alpha)} \\ A \wedge B^{(\mu-1)} \\ B^{(n)} \end{matrix} \Big\|_{\Gamma}$ $\Rightarrow \begin{matrix} \mathfrak{A}^{(\alpha)} \\ A \wedge B^{(\mu)} \end{matrix} \Big\|_{\Gamma}$

(3.2) $\left. \begin{matrix} \|(\hat{\mathfrak{A}} \wedge (A \vee B)) \to \\ (A \to (\hat{\mathfrak{A}} \wedge (A \vee B))) \ \& \\ \|(\hat{\mathfrak{A}} \wedge (A \vee B)) \to \\ (B \to (\hat{\mathfrak{A}} \wedge (A \vee B))) \ \& \\ \begin{matrix} \mathfrak{A}^{(\alpha)} \\ A \vee B^{(\mu-1)} \\ A^{(n)} \end{matrix} \Big\|_{\Gamma} \ \& \ \begin{matrix} \mathfrak{A}^{(\alpha)} \\ A \vee B^{(\mu-1)} \\ B^{(n)} \end{matrix} \Big\|_{\Gamma} \end{matrix} \right\}$ $\Rightarrow \begin{matrix} \mathfrak{A}^{(\alpha)} \\ A \vee B^{(\mu)} \end{matrix} \Big\|_{\Gamma}$

(3.3) $\begin{matrix} \mathfrak{A}^{(\alpha)} \\ A \to B^{(\mu-1)} \end{matrix} \Big\|_{A} \ \& \ \begin{matrix} \mathfrak{A}^{(\alpha)} \\ A \to B^{(\mu-1)} \\ B^{(n)} \end{matrix} \Big\|_{\Gamma}$ $\Rightarrow \begin{matrix} \mathfrak{A}^{(\alpha)} \\ A \to B^{(\mu)} \end{matrix} \Big\|_{\Gamma}$

(3.4) $\begin{matrix} \mathfrak{A}^{(\alpha)} \\ \neg A^{(\mu-1)} \end{matrix} \Big\|_{A}$ $\Rightarrow \begin{matrix} \mathfrak{A}^{(\alpha)} \\ \neg A^{(\mu)} \end{matrix} \Big\|_{\Gamma}$

(3.5) $\mathfrak{A}^{(\alpha)} \|_{\Gamma}$ $\Rightarrow \begin{matrix} \mathfrak{A}^{(\alpha)} \\ C^{(\mu)} \end{matrix} \Big\|_{\Gamma}$

(3.6) $\left\| \hat{\mathfrak{A}} \to C \ \& \ \begin{matrix} \mathfrak{A}^{(\alpha)} \\ C^{(n)} \end{matrix} \right\|_{\Gamma}$ $\Rightarrow \mathfrak{A}^{(\alpha)} \|_{\Gamma}$

(4.1) $\| A \to (B \to A)$ $\Rightarrow \| B \to (A \to B)$

(4.2) $\| A_1 \to (B \to A_1) \ \& \ \| A_2 \to (B \to A_2)$ $\Rightarrow \|(A_1 * A_2) \to$ $(B \to (A_1 * A_2))$

(4.3) $\| A \to (B \to A)$ $\Rightarrow \| \neg A \to$ $(B \to \neg A)$

(5.1) $\mathfrak{A}^{(\alpha)} \| A$ $\Rightarrow \mathfrak{A}^{(\alpha)} \| [A]$

$(5.2.1)$ $\mathfrak{A}^{(\alpha)} \| A$ $\Rightarrow \mathfrak{A}^{(\alpha)} \| [A, B]$

$(5.2.2)$ $\mathfrak{A}^{(\alpha)} \| B$ $\Rightarrow \mathfrak{A}^{(\alpha)} \| [A, B]$

with $* \in \{\wedge, \vee, \to\}$

2. *The Completeness and Consistency-Proof of T_n*

THEOREM 1: *A quantum mechanical proposition A is true in the n-dialog-game D_n if and only if the P-tableau $\| A$ can be deduced in the tableaux-calculus T_n, i.e.:*

$$\vdash_{D_n} A \quad \text{iff} \quad \vdash_{T_n} \| A.$$

Proof: A. *Completeness*: The proof is performed by means of induction with respect to the recursive scheme I (which establishes the positions of success in the dialog-game). Here we have the following possibilities:

1. A position of success is a final position of success $z \in E_p$. As can easily be seen there are exactly the two positions $\genfrac{}{}{0pt}{}{\mathfrak{A}^{(\alpha)}}{a^{(\mu)}}\Big|_a$ and $|A \rightarrowtail ((A \rightarrow B) \rightarrowtail A)$ for which O has no subsequent move. The respective tableaux $\genfrac{}{}{0pt}{}{\mathfrak{A}^{(\alpha)}}{a^{(\mu)}}\Big\|_a$ and $\| A \rightarrowtail ((A \rightarrow B) \rightarrowtail A)$ are however deducible as beginnings in T_n.

2. A position $z \in G$ is a member of M_p (in this case a position $\mathfrak{A}^{(\alpha)} | \Gamma$). According to $I2$ then at least one of the subsequent positions must be a position of success. Here we have to distinguish the following successors:

2.1. In case P defends, and $z \in M_p$ is:

a) $\mathfrak{A}^{(\alpha)} | [A]$. The successor then is $\mathfrak{A}^{(\alpha)} | A$ and the respective tableau is $\mathfrak{A}^{(\alpha)} \| A$. By assumption $\mathfrak{A}^{(\alpha)} \| A$ is deducible. This tableau, however, is the premise in rule 5.1. of the calculus T_n. Therefore the tableau $\mathfrak{A}^{(\alpha)} \| [A]$ also is deducible.

b) $\mathfrak{A}^{(\alpha)} | [A, B]$. The successors are $\mathfrak{A}^{(\alpha)} | A$ or $\mathfrak{A}^{(\alpha)} | B$ and the corresponding tableaux $\mathfrak{A}^{(\alpha)} \| A$ or $\mathfrak{A}^{(\alpha)} \| B$. These tableaux are assumed to be deducible. Then by means of 5.2. $\mathfrak{A}^{(\alpha)} \| [A, B]$ is deducible.

2.2. In case P attacks, and the attacked proposition is:

a) $A \wedge B$. The successors are

$$\genfrac{}{}{0pt}{}{\mathfrak{A}^{(\alpha)}}{\genfrac{}{}{0pt}{}{A \wedge B^{(\mu-1)}}{[A]}}\Big|\Gamma \quad \text{or} \quad \genfrac{}{}{0pt}{}{\mathfrak{A}^{(\alpha)}}{\genfrac{}{}{0pt}{}{A \wedge B^{(\mu-1)}}{[B]}}\Big|\Gamma \;.$$

If these positions are positions of success then the respective successors

$$\genfrac{}{}{0pt}{}{\mathfrak{A}^{(\alpha)}}{\genfrac{}{}{0pt}{}{A \wedge B^{(\mu-1)}}{A^{(n)}}}\Big|\Gamma \quad \text{or} \quad \genfrac{}{}{0pt}{}{\mathfrak{A}^{(\alpha)}}{\genfrac{}{}{0pt}{}{A \wedge B^{(\mu-1)}}{B^{(n)}}}\Big|\Gamma \;.$$

are positions of success. By assumption, the corresponding tableaux are deducible. By means of (3.1) $\mathop{A \wedge B^{(\mu)}}\limits^{\mathfrak{A}^{(\alpha)}} \Big\|_\Gamma$ also is deducible

b) $A \vee B$. Here the successor is

$$\mathop{\substack{\mathfrak{A}^{(\alpha)} \\ A \vee B^{(\mu-1)} \\ [A, B]}}\Bigg|_\Gamma .$$

In this position P may use the argument-rules A_f 5: $a.$ and $c.$ In case he reduces the availability of a proposition C to zero the subsequent position is

$$\mathop{\substack{\mathfrak{A}'^{(\beta)} \\ A \vee B^{(\mu-1)} \\ [A, B]}}\Bigg|_\Gamma \quad \text{with } \mathfrak{A}^{(\alpha)} \equiv \left.\begin{cases} \mathfrak{A}'^{(\beta)} \\ C^{(\nu)} \end{cases}\right. .$$

If this position is a position of success, $\mathop{A \vee B^{(\mu)}}\limits^{\mathfrak{A}'^{(\beta)}} \Big|_\Gamma$ is a position of success. By assumption the corresponding tableau is deducible, therefore by means of (3.5) $\mathop{A \vee B^{(\mu)}}\limits^{\mathfrak{A}^{(\alpha)}} \Big\|_\Gamma$ also is deducible.

In case P challenges O to state the hypothesis C, the two successors are

$$|(\widehat{\mathfrak{A}} \wedge (A \vee B)) \to C \quad \& \quad \mathop{\substack{\mathfrak{A}^{(\alpha)} \\ A \vee B^{(\mu-1)} \\ [A, B] \\ C^{(n)}}}\Bigg|_\Gamma .$$

If they are positions of success then

$$\mathop{\substack{\mathfrak{A}^{(\alpha)} \\ A \vee B^{(\mu)} \\ C^{(n)}}}\Bigg|_\Gamma$$

is a position of success. Since by assumption the tableaux

$$\|(\widehat{\mathfrak{A}} \wedge (A \vee B)) \to C \text{ and } \mathop{\substack{\mathfrak{A}^{(\alpha)} \\ A \vee B^{(\mu)} \\ C^{(n)}}}\Bigg\|_\Gamma$$

are deducible, with (3.6)

$$\mathop{A \vee B^{(\mu)}}\limits^{\mathfrak{A}^{(\alpha)}} \Big\|_\Gamma \text{ also is deducible.}$$

After P has potentially made use of A_f 5, O has to continue the dialog. The successors

are $|(\hat{\mathfrak{A}} \wedge (A \vee B)) \to (A \to (\hat{\mathfrak{A}} \wedge (A \vee B)))$ &
$|(\hat{\mathfrak{A}} \wedge (A \vee B)) \to (B \to (\hat{\mathfrak{A}} \wedge (A \vee B)))$ &

$$\& \begin{array}{c} \mathfrak{A}^{(\alpha)} \\ A \vee B^{(\mu-1)} \\ A^{(n)} \end{array} \Bigg| \Gamma \quad \& \begin{array}{c} \mathfrak{A}^{(\alpha)} \\ A \vee B^{(\mu-1)} \\ B^{(n)} \end{array} \Bigg| \Gamma .$$

They are positions of success and, by assumption, the corresponding tableaux are deducible. With (3.2) it then follows that $\begin{array}{c} \mathfrak{A}^{(\alpha)} \\ A \vee B^{(\mu)} \end{array} \Big\| \Gamma$ is deducible.

c) $A \to B$. The subsequent position is

$$\begin{array}{c} \mathfrak{A}^{(\alpha)} \\ A \to B^{(\mu-1)} \\ [B] \end{array} \Bigg| \Gamma .$$

If this position is a position of success then all possible subsequent positions are positions of success. In case O states B, the successor is

$$\begin{array}{c} \mathfrak{A}^{(\alpha)} \\ A \to B^{(\mu-1)} \\ B^{(n)} \end{array} \Bigg| \Gamma ;$$

in case O attacks P's proposition A, and the sucessor is a member of G, the we have

$$\begin{array}{c} \mathfrak{A}^{(\alpha)} \\ A \to B^{(\mu-1)} \end{array} \Big| A \in G. \text{ By assumption the tableaux } \begin{array}{c} \mathfrak{A}^{(\alpha)} \\ A \to B^{(\mu-1)} \\ B^{(n)} \end{array} \Bigg\| \Gamma$$

and $\begin{array}{c} \mathfrak{A}^{(\alpha)} \\ A \to B^{(\mu-1)} \end{array} \Big\| A$ are deducible, then by means of (3.3) $\begin{array}{c} \mathfrak{A}^{(\alpha)} \\ A \to B^{(\mu)} \end{array} \Big\| \Gamma$ also is deducible.

d) $\neg A$. The successor is $\begin{array}{c} \mathfrak{A}^{(\alpha)} \\ \neg A^{(\mu-1)} \\ [\quad] \end{array} \Bigg| \begin{array}{c} \Gamma \\ A \end{array}$.

If this position is a member of G, then we have $\begin{array}{c} \mathfrak{A}^{(\alpha)} \\ \neg A^{(\mu-1)} \end{array} \Big| A \in G$ because O has no other possibility than to attack A. By assumption the tableau $\begin{array}{c} \mathfrak{A}^{(\alpha)} \\ \neg A^{(\mu-1)} \end{array} \Big\| A$ is deducible; therefore, by means of (3.4) $\begin{array}{c} \mathfrak{A}^{(\alpha)} \\ \neg A^{(\mu)} \end{array} \Big\| \Gamma$ also is deducible.

3. A position $z \in G$ is a member of M_o (in this case a position $\mathfrak{A}^{(\alpha)} | A$ which is not a final position). According to $I3$ then all subsequent positions

must be positions of success. We have to distinguish the following possibilities: z is a position

a) $\mathfrak{A}^{(\alpha)} | A \wedge B$. In this case the successors are $\mathfrak{A}^{(\alpha)} | [A]$ and $\mathfrak{A}^{(\alpha)} | [B]$. The corresponding tableaux $\mathfrak{A}^{(\alpha)} \| [A]$ and $\mathfrak{A}^{(\alpha)} \| [B]$ are deducible by assumption; therefore, by means of (2.1) $\mathfrak{A}^{(\alpha)} \| A \wedge B$ also is deducible.

b) $\mathfrak{A}^{(\alpha)} | A \vee B$. Here the successor is $\mathfrak{A}^{(\alpha)} | [A, B]$. By assumption, the corresponding tableau $\mathfrak{A}^{(\alpha)} \| [A, B]$ is deducible; by means of (2.2) $\mathfrak{A}^{(\alpha)} \| A \vee B$ also is deducible

c) $\mathfrak{A}^{(\alpha)} | A \to B$. In this position P can use the argument-rule A_f 5: a. and c. In case he reduces the availability of a proposition C to zero, the subsequent position is $\mathfrak{A}'^{(\beta)} | A \to B$ with $\mathfrak{A}^{(\alpha)} \equiv \begin{cases} \mathfrak{A}'^{(\beta)} \\ C^{(\mu)} \end{cases}$. By assumption the corresponding tableau is deducible. Then it follows by means of (2.3.2) that $\mathfrak{A}^{(\alpha)} \| A \to B$ also is deducible. In case P challenges O to state the hypothesis C, the two successors are $\begin{matrix} \mathfrak{A}^{(\alpha)} \\ C^{(n)} \end{matrix} \Big| A \to B$ and $| \mathfrak{\hat{A}} \to C$. By assumption the corresponding tableaux $\begin{matrix} \mathfrak{A}^{(\alpha)} \\ C^{(n)} \end{matrix} \Big\| A \to B$ and $\| \mathfrak{\hat{A}} \to C$ are deducible; therefore, by means of (2.3.3) $\mathfrak{A}^{(\alpha)} \| A \to B$ also is deducible. After P has potentially made use of $A_f 5$, O has to continue the dialog. Then the successors are $| \mathfrak{\hat{A}} \to (A \to \mathfrak{\hat{A}})$ and $\begin{matrix} \mathfrak{A}^{(\alpha)} \\ A^{(n)} \end{matrix} \Big| [B]$. By means of the assumption that the corresponding tableaux are deducible and by means of (2.3.1) we conclude that $\mathfrak{A}^{(\alpha)} \| A \to B$ also is deducible.

d) $\mathfrak{A}^{(\alpha)} | \neg A$. In this position P may again use $A_f 5$. In case he applies this rule we conclude as above that by means of the assumption and (here) of (2.4.2) and (2.4.3) the tableau $\mathfrak{A}^{(\alpha)} \| \neg A$ is deducible. In case O continues the dialog the successors are $| \mathfrak{\hat{A}} \to (A \to \mathfrak{\hat{A}})$ and $\begin{matrix} \mathfrak{A}^{(\alpha)} \\ A^{(n)} \end{matrix} \Big| [\ ]$. By assumption the corresponding tableaux $\| \mathfrak{\hat{A}} \to (A \to \mathfrak{\hat{A}})$ and $\begin{matrix} \mathfrak{A}^{(\alpha)} \\ A^{(n)} \end{matrix} \Big\| [\ ]$ are deducible; therefore, by means of (2.4.1) $\mathfrak{A}^{(\alpha)} \| \neg A$ is also.

e) $| B \to (A \to B)$. With respect to this position we have to consider the case that P makes use of the argument-rule $A_f 7$. If P replaces the initial argument by the new initial argument $A \to (B \to A)$, the corresponding tableau is assumed to be deducible, and by means of (4.1) $\| B \to (A \to B)$ also is proved to be deducible. If P replaces the initial argument $(A_1 * A_2) \to (A \to (A_1 * A_2))$ by the new initial argument $A_1 \to (A \to A_1)$ or

$A_2 \to (A \to A_2)$, the assumption and (4.2) imply that $\|(A_1*A_2) \to (A \to (A_1*A_2))$ is deducible.

If P replaces the initial argument $\neg A_1 \to (A \to \neg A_1)$ by the new initial argument $A_1 \to (A \to A_1)$ we prove the deducibility of $\| \neg A_1 \to (A \to \neg A_1)$ by means of the assumption and (4.3).

B. *Consistency*: The proof of this direction is performed by means of induction with respect to the calculus T_n. Here we have to distinguish the following cases:

1. The beginnings of T_n are positions of success, namely the reduced final

   position $\left.\dfrac{\mathfrak{A}^{(\alpha)}}{a^{(\mu)}}\right|_a$ and $|A \to ((A \to B) \to A$.

2. The positions which correspond to the premises of the rules (2.), (3.1) to (3.4), (4.) and (5.) are positions of success by assumption. Such a position can always be obtained starting at a position which corresponds to a conclusion of the above rules. Therefore these positions are also positions of success.

3. The position which corresponds to the premise of (3.5) is assumed to be a position of success. Now we prove the following admissibility with respect to I:

$$\mathfrak{A}^{(\alpha)} \mid \theta \in G \text{ then } \left.\frac{\mathfrak{A}^{(\alpha)}}{C^{(\mu)}}\right|_\theta \in G \text{ with } \theta \in \{\Gamma, A\}.$$

The proof is performed by means of induction on the premises of $\mathfrak{A}^{(\alpha)}|\theta$. Therefore we have to distinguish the following cases: $\mathfrak{A}^{(\alpha)}|\theta$ is

a) a final position. Then we have $\mathfrak{A}^{(\alpha)}\,|a$ or$|\,A \to ((A \to B) \to A)$. If the latter position is a position of success, $C^{(\mu)}|\,A \to ((A \to B) \to A)$ also is a position of success, because from this position the premise can be obtained by means of the argument-rule $A_f5: a$. If $\mathfrak{A}^{(\alpha)}|a$ is a member

of G, then we have $a \in \mathfrak{A}$ and therefore $\left.\dfrac{\mathfrak{A}^{(\alpha)}}{C^{(\mu)}}\right|_a \in G$.

b) a position of M_p(i.e. a position $\mathfrak{A}^{(\alpha)}|\Gamma$). Here we have the possibilities:

(i) P defends with A. We have $\left.\dfrac{\mathfrak{A}^{(\alpha)}}{C^{(\mu)}}\right|_A \in G$ by assumption. Then it follows

   that also $\left.\dfrac{\mathfrak{A}^{(\alpha)}}{C^{(\mu)}}\right|_\Gamma \in G$.

(ii) P attacks, namely the proposition

   $\alpha)$ $A_1 \wedge A_2$, where $\mathfrak{A}^{(\alpha)} \equiv \begin{cases} \mathfrak{A}'^{(\beta)} \\ A_1 \wedge A_2^{(\nu)} \end{cases}$. In this case the successors are

$$\left. \begin{matrix} \mathfrak{A}'^{(\beta)} \\ A_1 \wedge A_2^{(\nu-1)} \\ A^{(n)} \end{matrix} \right|_{\Gamma} \quad \text{or} \quad \left. \begin{matrix} \mathfrak{A}'^{(\beta)} \\ A_1 \wedge A_2^{(\nu-1)} \\ A_2^{(n)} \end{matrix} \right|_{\Gamma} \quad \text{respectively.}$$

By assumption $\left. \begin{matrix} \mathfrak{A}^{(\alpha)} \\ A_1 \wedge A_2^{(\nu-1)} \\ A_1^{(n)} \\ C^{(\mu)} \end{matrix} \right|_{\Gamma}$ or $\left. \begin{matrix} \mathfrak{A}^{(\alpha)} \\ A_1 \wedge A_2^{(\nu-1)} \\ A_2^{(n)} \\ C^{(\mu)} \end{matrix} \right|_{\Gamma}$ respectively

are members of G. But then $\left. \begin{matrix} \mathfrak{A}^{(\alpha)} \\ C^{(\mu)} \end{matrix} \right|_{\Gamma} \in G$.

β) $A_1 \vee A_2$, where $\mathfrak{A}^{(\alpha)} \equiv \begin{cases} \mathfrak{A}'^{(\beta)} \\ A_1 \vee A_2^{(\nu)} \end{cases}$. Here, the conclusion $\left. \begin{matrix} \mathfrak{A}^{(\alpha)} \\ C^{(\mu)} \end{matrix} \right|_{\Gamma}$

$\in G$ follows immediately, because from the position

$$\left. \begin{matrix} \mathfrak{A}'^{(\beta)} \\ A_1 \vee A_2^{(\nu)} \\ C^{(\mu)} \end{matrix} \right|_{\Gamma} P \text{ can gain the position } \left. \begin{matrix} \mathfrak{A}'^{(\beta)} \\ A_1 \vee A_2^{(\nu)} \end{matrix} \right|_{\Gamma}$$

by means of the argument-rule $A_f 5: a$.

γ) $A_1 \rightarrow A_2$, where $\mathfrak{A}^{(\alpha)} \equiv \begin{cases} \mathfrak{A}'^{(\beta)} \\ A_1 \rightarrow A_2^{(\nu)} \end{cases}$. If the assumption is applied

on the subsequent positions we obtain $\left. \begin{matrix} \mathfrak{A}^{(\alpha)} \\ C^{(\mu)} \end{matrix} \right|_{\Gamma} \in G$.

δ) $\neg A_1$, where $\mathfrak{A}^{(\alpha)} \equiv \begin{cases} \mathfrak{A}'^{(\beta)} \\ \neg A_1^{(\nu)} \end{cases}$. By means of the assumption,

applied on the subsequent position, we obtain $\left. \begin{matrix} \mathfrak{A}^{(\alpha)} \\ C^{(\mu)} \end{matrix} \right|_{\Gamma} \in G$.

c) a position of M_o (i.e. a position $\mathfrak{A}^{(\alpha)} | A$ with $A \not\equiv a$). We have the following cases: $\mathfrak{A}^{(\alpha)} | A$ is

(i) $\mathfrak{A}^{(\alpha)} | A_1 \wedge A_2$. Here the successors are $\mathfrak{A}^{(\alpha)} | [A]$ and $\mathfrak{A}^{(\alpha)} [B]$. Then we have by assumption $\left. \begin{matrix} \mathfrak{A}^{(\alpha)} \\ C^{(\mu)} \end{matrix} \right| [A_1] \in G$ and $\left. \begin{matrix} \mathfrak{A}^{(\alpha)} \\ C^{(\mu)} \end{matrix} \right| [A_2] \in G$. From this it

follows that $\left. \begin{matrix} \mathfrak{A}^{(\alpha)} \\ C^{(\mu)} \end{matrix} \right| A_1 \wedge A_2 \in G$ is also.

(ii) $\mathfrak{A}^{(\alpha)} | A_1 \vee A_2$. In this case the successor is $\mathfrak{A}^{(\alpha)} | [A_1, A_2]$. From

the assumption $\left. \begin{matrix} \mathfrak{A}^{(\alpha)} \\ C^{(\mu)} \end{matrix} \right| [A_1, A_2] \in G$ it follows that $\left. \begin{matrix} \mathfrak{A}^{(\alpha)} \\ C^{(\mu)} \end{matrix} \right| A_1 \vee A_2 \in G$ is

also.

(iii) $\mathfrak{A}^{(\alpha)} | A_1 \rightarrow A_2$. It follws immediately that $\left. \begin{matrix} \mathfrak{A}^{(\alpha)} \\ C^{(\mu)} \end{matrix} \right| A_1 \rightarrow A_2 \in G$ because, starting at this position, P can gain $\left. \begin{matrix} \mathfrak{A}^{(\alpha)} \\ C^{(\mu)} \end{matrix} \right| A_1 \rightarrow A_2$ by means

of $A_f 5: a$.

(iv) $\mathfrak{A}^{(\alpha)}|\neg A_1$. In this case it follows by means of $A_f 5$: a. that
$\dfrac{\mathfrak{A}^{(\alpha)}|}{C^{(\mu)}|\neg A_1} \in G.$

By means of this admissibility and the assumption $\mathfrak{A}^{(\alpha)}|\Gamma \in G$ it follows that $\dfrac{\mathfrak{A}^{(\alpha)}|}{C^{(\mu)}|\Gamma} \in G.$

4. The positions which correspond to the premise of (3.6) are assumed to be positions of success. To prove that the conclusion too is a position of success we prove the following admissibility:

$$|\hat{\mathfrak{A}} \to C \in G \ \& \ \dfrac{\mathfrak{A}^{(\alpha)}|}{C^{(n)}|\theta} \in G \ \text{then} \ \mathfrak{A}^{(\alpha)}|\theta \in G \ \text{with} \ \theta \in \{\Gamma, A \not\equiv a\}.$$

The proof is performed by means of induction with respect to the subpropositions of C.

a) If C is an elementary proposition c, we have to prove:

$$|\hat{\mathfrak{A}} \to c \in G \ \& \ \dfrac{\mathfrak{A}^{(\alpha)}|}{c^{(n)}|\theta} \in G \ \text{then} \ \mathfrak{A}^{(\alpha)}|\theta \in G.$$

To prove this admissibility we perform an induction on the premises of $\dfrac{\mathfrak{A}^{(\alpha)}|}{c^{(n)}|\theta}$. Then we have the following cases: $\dfrac{\mathfrak{A}^{(\alpha)}|}{c^{(n)}|\theta}$ is

(i) a final position. The only relevant position is $\dfrac{\mathfrak{A}^{(\alpha)}|}{c^{(n)}|a}$, where a is either contained in $\mathfrak{A}$ or is c. In the first case it follows immediately that $\dfrac{\mathfrak{A}^{(\alpha)}|}{c^{(n)}|a} \in G$. In the second case it can easily be shown that $\mathfrak{A}^{(\alpha)}|c \in G$ follows from $|\hat{\mathfrak{A}} \to c \in G$.

(ii) a member of M_p (i.e. $\dfrac{\mathfrak{A}^{(\alpha)}|}{c^{(n)}|\Gamma}$). In case P defends, we conclude as in 3.b) (i), In case P attacks, we conclude as in 3.b) (ii), applying $A_f 5 : c$ in case β).

(iii) a member of M_o (i.e. $\dfrac{\mathfrak{A}^{(\alpha)}|}{c^{(n)}|A}$ with $A \not\equiv a$). Here we conclude as in 3.c), applying $A_f 5: c$. in the cases (iii) and (iv).
This proves the above admissibility for $C \equiv c$.

b) Now we assume that the above admissibility has been proved for all subpropositions of C. Then we have to prove it for C. Again we use induction on the premises of $\dfrac{\mathfrak{A}^{(\alpha)}|}{C^{(n)}|\theta}$. Concerning the above proof in

4.a), we have to regard also the cases that P attacks the proposition C. Here we have to distinguish:

α) $C \equiv C_1 \wedge C_2$. The successors are

$$\begin{array}{c|c} \mathfrak{A}^{(\alpha)} & \\ C_1 \wedge C_2^{(\mu-1)} & \\ [C_1] & \Gamma \end{array} \quad \text{respectively} \quad \begin{array}{c|c} \mathfrak{A}^{(\alpha)} & \\ C_1 \wedge C_2^{(\mu-1)} & \\ [C_2] & \Gamma \end{array}.$$

If these positions are positions of success then

$$\begin{array}{c|c} \mathfrak{A}^{(\alpha)} & \\ C_1 \wedge C_2^{(\mu-1)} & \\ C_1^{(n)} & \Gamma \end{array} \quad \text{and} \quad \begin{array}{c|c} \mathfrak{A}^{(\alpha)} & \\ C_1 \wedge C_2^{(\mu-1)} & \\ C_2^{(n)} & \Gamma \end{array}$$

are positions of success. By means of the assumption we have $\begin{array}{c|c} \mathfrak{A}^{(\alpha)} & \\ C_1^{(n)} & \Gamma \end{array}$ $\in G$ and $\begin{array}{c|c} \mathfrak{A}^{(\alpha)} & \\ C_2^{(n)} & \Gamma \end{array} \in G$. Furthermore it can be shown that if $|\widehat{\mathfrak{A}} \rightarrow (C_1 \wedge C_2) \in G$, then $|\widehat{\mathfrak{A}} \rightarrow C_1 \in G$ and $|\widehat{\mathfrak{A}} \rightarrow C_2 \in G$. From this it follows that $\mathfrak{A}^{(\alpha)}|\Gamma \in G$.

β) $C \equiv C_1 \vee C_2$. In this case $\mathfrak{A}^{(\alpha)}|\Gamma \in G$ follows by means of the rule $A_f 5 : c$.

γ) $C \equiv C_1 \rightarrow C_2$. If the successors of this position are positions of success then we have

$$\begin{array}{c|c} \mathfrak{A}^{(\alpha)} & \\ C_1 \rightarrow C_2^{(\mu-1)} & \\ C_2^{(n)} & \Gamma \end{array} \in G \quad \text{and} \quad \begin{array}{c|c} \mathfrak{A}^{(\alpha)} & \\ C_1 \rightarrow C_2^{(\mu-1)} & \\ & C_1 \end{array} \in G.$$

By means of the assumption it follows that $\begin{array}{c|c} \mathfrak{A}^{(\alpha)} & \\ C_2^{(n)} & \Gamma \end{array} \in G$ and $\mathfrak{A}^{(\alpha)}|C_1 \in G$. Then we have $|\widehat{\mathfrak{A}} \rightarrow C_1 \in G$ also. In this case it can be shown that if $|\widehat{\mathfrak{A}} \rightarrow (C_1 \rightarrow C_2) \in G$ then $\begin{array}{c|c} \mathfrak{A}^{(\alpha)} & \\ C_1^{(n)} & [C_2] \end{array} \in G$. With $|\widehat{\mathfrak{A}} \rightarrow C_1 \in G$ it follows that $\mathfrak{A}^{(\alpha)}|[C_2] \in G$. But this implies $|\widehat{\mathfrak{A}} \rightarrow C_2 \in G$. With $\begin{array}{c|c} \mathfrak{A}^{(\alpha)} & \\ C_2^{(n)} & \Gamma \end{array} \in G$ we finally have $\mathfrak{A}^{(\alpha)}|\Gamma \in G$.

δ) $C \equiv \neg\, C_1$. If the successor of this position is a position of success then $\begin{array}{c|c} \mathfrak{A}^{(\alpha)} & \\ \neg C_1^{(\mu-1)} & C_1 \end{array} \in G$. By means of the assumption it follows that $\mathfrak{A}^{(\alpha)}|C_1 \in G$, and from this it follows that $|\widehat{\mathfrak{A}} \rightarrow C_1 \in G$. In this case we can show that if $|\widehat{\mathfrak{A}} \rightarrow \neg\, C_1 \in G$ then $\begin{array}{c|c} \mathfrak{A}^{(\alpha)} & \\ C_1^{(n)} & \Gamma \end{array} \in G$. But with $|\widehat{\mathfrak{A}} \rightarrow C_1 \in G$ we have $\mathfrak{A}^{(\alpha)}|\Gamma \in G$ also. Since B. is now proved, the proof of theorem I is finished.

3. *The Tableaux-Calculus* T_{eff} *of Effective Quantum Logic*

In order to establish the whole class of formally true propositions we have as already announced in 2.b. (i), to take the union of all dialog-games D_n with respect to n, and respectively the union of all tableaux-calculi T_n. By means of the following theorem II the union of all calculi T_n can be replaced by the *tableaux-calculus* T_{eff} of *effective* quantum logic which reads:

(1.1) $\qquad \Rightarrow \dfrac{\mathfrak{A}}{a} \Big\| a$

(1.2) $\qquad \Rightarrow \| A \to ((A \to B) \to A)$

(2.1) $\qquad \mathfrak{A} \, \|[A] \,\&\, \mathfrak{A} \, \|[B] \hspace{4cm} \Rightarrow \mathfrak{A} \| A \wedge B$

(2.2) $\qquad\qquad \mathfrak{A} \|[A, B] \hspace{4cm} \Rightarrow \mathfrak{A} \| A \vee B$

(2.3.1) $\qquad \| \hat{\mathfrak{A}} \to (A \to \tilde{\mathfrak{A}}) \,\&\, \dfrac{\mathfrak{A}}{A}\Big\|[B] \hspace{2.5cm} \Rightarrow \mathfrak{A} \| A \to B$

(2.3.2) $\qquad\qquad \mathfrak{A} \| A \to B \hspace{4cm} \Rightarrow \dfrac{\mathfrak{A}}{C}\Big\| A \to B$

(2.3.3) $\qquad \| \hat{\mathfrak{A}} \to C \,\&\, \dfrac{\mathfrak{A}}{C}\Big\| A \to B \hspace{3cm} \Rightarrow \mathfrak{A} \| A \to B$

(2.4.1) $\qquad \| \hat{\mathfrak{A}} \to (A \to \tilde{\mathfrak{A}}) \,\&\, \dfrac{\mathfrak{A}}{A}\Big\|[\,] \hspace{2.5cm} \Rightarrow \mathfrak{A} \| \neg A$

(2.4.2) $\qquad\qquad \mathfrak{A} \| \neg A \hspace{4cm} \Rightarrow \dfrac{\mathfrak{A}}{C}\Big\| \neg A$

(2.4,3) $\qquad \| \hat{\mathfrak{A}} \to C \,\&\, \dfrac{\mathfrak{A}}{C}\Big\| \neg A \hspace{3cm} \Rightarrow \mathfrak{A} \| \neg A$

(3.1.1) $\qquad \begin{array}{c} \mathfrak{A} \\ A \wedge B \\ A \end{array}\Big\| \Gamma \hspace{3.5cm} \Rightarrow \begin{array}{c} \mathfrak{A} \\ A \wedge B \end{array}\Big\| \Gamma$

(3.1.2) $\qquad \begin{array}{c} \mathfrak{A} \\ A \wedge B \\ B \end{array}\Big\| \Gamma \hspace{3.5cm} \Rightarrow \begin{array}{c} \mathfrak{A} \\ A \wedge B \end{array}\Big\| \Gamma$

(3.2) $\qquad \left. \begin{array}{l} \|(\hat{\mathfrak{A}} \wedge (A \vee B)) \to (A \to (\tilde{\mathfrak{A}} \wedge (A \vee B))) \,\& \\ \|(\hat{\mathfrak{A}} \wedge (A \vee B)) \to (B \to (\tilde{\mathfrak{A}} \wedge (A \vee B))) \,\& \\ \begin{array}{cc} \mathfrak{A} & \mathfrak{A} \\ A \vee B \Big\| \;\&\; A \vee B \Big\| \\ A \quad \Gamma \qquad B \quad \Gamma \end{array} \end{array} \right\} \Rightarrow \begin{array}{c} \mathfrak{A} \\ A \vee B \end{array}\Big\| \Gamma$

(3.3) $\quad \dfrac{\mathfrak{A}}{A \to B}\Big\|A \ \& \ A \to \dfrac{\mathfrak{A}}{B} \Big\|_\Gamma \qquad\qquad\qquad \Rightarrow \dfrac{\mathfrak{A}}{A \to B}\Big\|_\Gamma$

(3.4) $\qquad\qquad \dfrac{\mathfrak{A}}{\neg A}\Big\|A \qquad\qquad\qquad\qquad \Rightarrow \dfrac{\mathfrak{A}}{\neg A}\Big\|_\Gamma$

(3.5) $\qquad\qquad\quad \mathfrak{A}\|\Gamma \qquad\qquad\qquad\qquad\qquad \Rightarrow \dfrac{\mathfrak{A}}{C}\Big\|_\Gamma$

(3.6) $\qquad \Big\|\widehat{\mathfrak{A}} \to C \ \& \ \dfrac{\mathfrak{A}}{C}\Big\|_\Gamma \qquad\qquad\qquad \Rightarrow \mathfrak{A}\|\Gamma$

(4.1) $\qquad \|A \to (B \to A) \qquad\qquad\qquad\qquad \Rightarrow \|B \to$
$\qquad\qquad\qquad\qquad\qquad\qquad\qquad\qquad\qquad\quad (A \to B)$

(4.2) $\quad \|A_1 \to (B \to A_1) \ \& \ \|A_2 \to (B \to A_2) \qquad \Rightarrow \|(A_1 * A_2) \to$
$\qquad\qquad\qquad\qquad\qquad\qquad\qquad\qquad\qquad\quad (B \to (A_1 * A_2))$

(4.3) $\qquad \|A \to (B \to A) \qquad\qquad\qquad\qquad \Rightarrow \|\neg A \to$
$\qquad\qquad\qquad\qquad\qquad\qquad\qquad\qquad\qquad\quad (B \to \neg A)$

(5.1) $\qquad\qquad \mathfrak{A}\|A \qquad\qquad\qquad\qquad\qquad \Rightarrow \mathfrak{A}\|[A]$

(5.2.1) $\qquad\qquad \mathfrak{A}\|A \qquad\qquad\qquad\qquad\qquad \Rightarrow \mathfrak{A}\|[A, B]$

(5.2.2) $\qquad\qquad \mathfrak{A}\|B \qquad\qquad\qquad\qquad\qquad \Rightarrow \mathfrak{A}\|[A, B]$

$$\text{with } * \in \{\wedge, \vee, \to\}$$

THEOREM II: *A tableau* $\|A$ *can be deduced in the calculus* T_{eff} *if and only if there exists a number n, such that the tableau* $\|A$ *can be deduced in* T_n, *i.e.:*

$$\vdash_{T_{\text{eff}}} \|A \text{ iff } \underset{n}{\exists} \vdash_{T_n} \|A.$$

Proof: A) We assume that there exists a deduction of $\|A$ in T_{eff}. Then the number of the applications of (3.1) to (3.5) with respect to the same proposition is limited. Also the applications of (2.3.3), (2.4.3) and (3.6) are limited. If a tableau $\|B$ occurs again in the deduction of $\|A$, the repetition can be eliminated so that a tableau $\|B$ occurs only once in a deduction. By means of an appropriate choice of availabilities a deduction of $\|A$ in T_{eff} can be replaced by a deduction in T_n with a suitable n.

B) We assume that there exits a deduction of $\|A$ in T_n. Then all rules of T_n change into rules of T_{eff} if the availabilities are eliminated, except in the case $\mu = 1$. In this case, however, the corresponding rules of T_{eff} are deducible by means of (3.5).

Department of Philosophy
University of Western Ontario

ACKNOWLEDGEMENT

I wish to thank Prof. P. Mittelstaedt and, among other members of the Institut für Theoretische Physik der Universität zu Köln, especially Mr. H.-M. Denecke for many valuable discussions concerning the dialogic approach to quantum logic.

REFERENCES

[1] Birkhoff, G. and von Neumann J., *Ann. of Math.* **37**, 823 (1936)
[2] Jauch, J. M. and Piron, C. *Helv. Phys. Acta* **36**, 827 (1963) Piron, C., *Helv. Phys. Acta* **37**, 439 (1964)
[3] Jauch, J. M., *Foundations of Quantum Mechanics*, Addison-Welsey Publ. Co. Reading, Mass. (1968)
[4] Kamber, F., *Math. Ann.* **158**, 158 (1965)
    Gleason, A. M., *J. of Math. and Mech.* **6**, 885 (1957)
    Kochen, S. and E. P. Specker, *J. of Math. and Mech.* **17**, 59 (1967)
[5] Reichenbach, H., *Philosophical Foundations of Quantum Mechanics*, Benjamin (1963)
[6] von Weizsäcker, C. F., *Naturwiss.*, **42**, 547 (1955)
[7] Lorenzen, P., *Metamathematik*, Bibliographisches Institut, Mannheim (1962)
    Kamlah, W. und Lorenzen, P. *Logische Propädeutik*, Bibliographisches Institut, Mannheim (1973)
[8] Lorenz, K., *Arch. f. Math. Logik und Grundlagenforschung*, **11**, (1968)
[9] Mittelstaedt, P., *Philosophical Problems of Modern Physics*, Reidel, Dordrecht, (1975)
    Mittelstaedt, P., and E. -W. Stachow, *Found. of Physics* **4**, 355 (1974)
[10] Stachow, E. -W., Dissertation, Universität zu Köln (1975)

E. -W. STACHOW*

QUANTUM LOGICAL CALCULI
AND LATTICE STRUCTURES

Abstract. In a preceding paper [1] it was shown that quantum logic, given by the tableaux-calculus T_{eff}, is complete and consistent with respect to the dialogic foundation of logics. Since in formal dialogs the special property of the 'value-definiteness' of propositions is not postulated, the calculus T_{eff} represents a calculus of effective (intuitionistic) quantum logic.

Beginning with the tableaux-calculus the equivalence of T_{eff} to calculi which use more familiar figures such as sequents and implications can be investigated. In this paper we present a sequents-calculus of Gentzen-type and a propositional calculus of Brouwer-type which are shown to be equivalent to T_{eff}. The effective propositional calculus provides an interpretation for a lattice structure, called quasi-implicative lattice. If, in addition, the value-definiteness of quantum mechanical propositions is postulated, a propositional calculus is obtained which provides an interpretation for a quasi-modular orthocomplemented lattice which, as is well-known, has as a model the lattice of subspaces of a Hilbert space.

CONTENTS

I. *The tableaux-calculus* T_{eff} represents a full formalisation of the dialog-game D_{eff} about quantum mechanical propositions, cf. [1] . Since T_{eff} replaces the dialog-game, its language is still rich but, as is shown in this paper, can be reduced to the more familiar language of propositional logic. We show the equivalence of a 'Gentzen-like' calculus S_{eff} and a 'Brouwer-like' calculus Q_{eff} to T_{eff}.

II. *Some meta-logical notions* are introduced in order to give a precise language for establishing the equivalence of T_{eff}, S_{eff} and Q_{eff} by means of the deducibility and admissibility of rules in the calculi. The meta-logical notions are naturally defined by means of meta-dialogs. In particular, the meta-logical material implication establishes what is called an admissible rule in a calculus. The class of admissible rules includes the class of deducible rules, also called theorems in the literature, and the class of deducible rules includes the constitutive rules of the calculus.

III. *The calculus of sequents* S_{eff} can be considered as the quantum logical equivalent to the Gentzen-like calculus G3 in Kleene [2, p. 481] . The main

245

Hooker (ed.), Physical Theory as Logico-Operational Structure, 245–284.
All Rights Reserved. Reprinted from Journal of Philosophical Logic, 7 (1978).
Copyright © 1978 by D. Reidel Publishing Company, Dordrecht, Holland.

differences between S_{eff} and the intuitionistic Gentzen-calculus are discussed in Section III.1. In Section III.2 we shall give an outline of the equivalence proof (Theorem I) of S_{eff} to T_{eff}.

IV. *The propositional calculus* Q_{eff} can be considered as the quantum logical equivalent to the Brower-like calculus *Br* in Lorenzen [3, p. 80]. The difference between Q_{eff} and *Br* and some properties of Q_{eff} are discussed in Sections IV. 1 and 2. An outline of the equivalence proof (Theorem II) of Q_{eff} to S_{eff} is given in Section IV.3. Q_{eff} has the particular property that all admissible rules of Q_{eff} are also deducible rules of Q_{eff} (Theorem III of Section IV.4). This means that the admissible rules which are meta-logical inferences between the figures (implications) of Q_{eff} can entirely be replaced by the language of the calculus. In this respect the language of the calculus is syntactically complete. In Section IV.5 we consider a lattice structure, called quasi-implicative lattice L_{qi}. The lattice ordering relation and the operations can easily be interpreted by means of the language of Q_{eff}. The axioms of L_{qi} are formulated in a language which is also the meta-language of Q_{eff}. But, since Q_{eff} is syntactically complete, the lattice L_{qi} can entirely be described by means of the language of Q_{eff}. In this way Q_{eff} provides an interpretation of the lattice structure L_{qi}.

V. *The calculi of value-definite quantum logic* can be considered as the quantum logical equivalent to classical logic. In order to establish the extension of T_{eff}, S_{eff} and Q_{eff} to the corresponding full quantum logical calculi, the value-definiteness is introduced as a particular material property, corresponding to yes–no measurements of elementary quantum mechanical propositions, into the material dialog-game [1, p. 251]. The extension of the formal dialog-game [1, p. 255], consists of an additional argument rule, as shown in Section V.1, and leads to the calculi T, S, and Q of the full quantum logic (Section V.2). The propositional calculus Q is shown in Section V.3 to provide an interpretation of a lattice structure which is isomorphic to the orthocomplemented quasi-modular lattice L_q which has the lattice of the subspaces of a Hilbert space as a model.

I. THE TABLEAUX-CALCULUS T_{eff}

We briefly repeat the tableaux-calculus T_{eff} and some basic notions which are precisely defined in [1].

In the following, a, A, B, C, A_1, A_2 are used as variables for *propositions* which are either *elementary* propositions, denoted by a, or *compound* propositions. The connectives, the conjunction, disjunction, material implication and negation, are denoted by $\wedge, \vee, \rightarrow$ and $\neg$ respectively. $\mathfrak{C}$ denotes a finite column system of zero or more propositions which is invariant with respect to the succession of the propositions. $\hat{\mathfrak{C}}$ denotes the conjunction of all propositions of $\mathfrak{C}$. Parentheses are used within compound propositions. Γ is a symbol for a *potential argument* which is either $[A]$, $[B]$, $[A, B]$ or the empty argument $[\ ]$, cf. [1, p. 264]. By means of the above formulae and symbols *tableaux*-figures are established which consist of two columns, separated by the tableaux-sign $\|$ and containing a finite system of zero or more propositions in the left side column and one proposition or a potential argument in the right side column. For the denotation of the *beginnings* (axioms) and the *constitutive rules* of the *tableaux-calculus* T_{eff} we use the sign $\Rightarrow$, and in case of several premises in the rules we use the sign &. Note that the beginnings and constitutive rules have an *operational* meaning within the formal dialog-game.

The beginnings of T_{eff} are

$$T_{\mathrm{eff}}\ (1.1)\quad \Rightarrow\quad \begin{array}{c}\mathfrak{C}\\ a\end{array}\Big\|\,a$$

$$T_{\mathrm{eff}}\ (1.2)\quad \Rightarrow\quad \|A \rightarrow ((A \rightarrow B) \rightarrow A).$$

The constitutive rules of T_{eff} are

$$T_{\mathrm{eff}}\ (2.1)\qquad\qquad \mathfrak{C}\|[A] \ \& \ \mathfrak{C}\|[B] \ \Rightarrow\ \mathfrak{C}\|A \wedge B$$

$$T_{\mathrm{eff}}\ (2.2)\qquad\qquad \mathfrak{C}\|[A, B] \ \Rightarrow\ \mathfrak{C}\|A \vee B$$

$$T_{\mathrm{eff}}\ (2.3.1)\quad \|\hat{\mathfrak{C}} \rightarrow (A \rightarrow \hat{\mathfrak{C}}) \ \& \ \begin{array}{c}\mathfrak{C}\\ A\end{array}\Big\|_{[B]} \ \Rightarrow\ \mathfrak{C}\|A \rightarrow B$$

$$T_{\mathrm{eff}}\ (2.3.2)\qquad\qquad \mathfrak{C}\|A \rightarrow B \Rightarrow\ \begin{array}{c}\mathfrak{C}\\ C\end{array}\Big\|\,A \rightarrow B$$

$$T_{\mathrm{eff}}\ (2.3.3)\quad \|\hat{\mathfrak{C}} \rightarrow C \ \& \ \begin{array}{c}\mathfrak{C}\\ C\end{array}\Big\|_{A \rightarrow B} \ \Rightarrow\ \mathfrak{C}\|A \rightarrow B$$

$$T_{\mathrm{eff}}\ (2.4.1)\quad \|\hat{\mathfrak{C}} \rightarrow (A \rightarrow \hat{\mathfrak{C}}) \ \& \ \begin{array}{c}\mathfrak{C}\\ A\end{array}\Big\|_{[\ ]} \ \Rightarrow\ \mathfrak{C}\|\neg A$$

248 E.-W. STACHOW

T_{eff} (2.4.2) $\mathcal{Q} \,\|\, \neg A \;\Rightarrow\; {\textstyle{\mathcal{Q} \atop C}} \,\Big\|\, \neg A$

T_{eff} (2.4.3) $\| \,\hat{\mathcal{Q}} \to C \,\&\, {\textstyle{\mathcal{Q} \atop C}} \,\Big\|\, \neg A \;\Rightarrow\; \mathcal{Q} \,\|\, \neg A$

T_{eff} (3.1.1) ${\textstyle{\mathcal{Q} \atop {A \wedge B \atop A}}} \,\Big\|\, \Gamma \;\Rightarrow\; {\textstyle{\mathcal{Q} \atop A \wedge B}} \,\Big\|\, \Gamma$

T_{eff} (3.1.2) ${\textstyle{\mathcal{Q} \atop {A \wedge B \atop B}}} \,\Big\|\, \Gamma \;\Rightarrow\; {\textstyle{\mathcal{Q} \atop A \wedge B}} \,\Big\|\, \Gamma$

T_{eff} (3.2)

$$\left.\begin{array}{l} \|((\hat{\mathcal{Q}} \wedge (A \vee B)) \to (A \to (\hat{\mathcal{Q}} \wedge (A \vee B)))) \\ \|((\hat{\mathcal{Q}} \wedge (A \vee B)) \to (B \to (\hat{\mathcal{Q}} \wedge (A \vee B)))) \\ \quad {\textstyle{\mathcal{Q} \atop {A \vee B \atop A}}} \,\Big\|\, \Gamma \;\&\; {\textstyle{\mathcal{Q} \atop {A \vee B \atop B}}} \,\Big\|\, \Gamma \end{array}\right\} \Rightarrow {\textstyle{\mathcal{Q} \atop A \vee B}} \,\Big\|\, \Gamma$$

T_{eff} (3.3) ${\textstyle{\mathcal{Q} \atop A \to B}} \,\Big\|\, A \,\&\, {\textstyle{\mathcal{Q} \atop {A \to B \atop B}}} \,\Big\|\, \Gamma \;\Rightarrow\; {\textstyle{\mathcal{Q} \atop A \to B}} \,\Big\|\, \Gamma$

T_{eff} (3.4) ${\textstyle{\mathcal{Q} \atop \neg A}} \,\Big\|\, A \;\Rightarrow\; {\textstyle{\mathcal{Q} \atop \neg A}} \,\Big\|\, \Gamma$

T_{eff} (3.5) $\mathcal{Q} \,\|\, \Gamma \;\Rightarrow\; {\textstyle{\mathcal{Q} \atop C}} \,\Big\|\, \Gamma$

T_{eff} (3.6) $\| \,\hat{\mathcal{Q}} \to C \,\&\, {\textstyle{\mathcal{Q} \atop C}} \,\Big\|\, \Gamma \;\Rightarrow\; \mathcal{Q} \,\|\, \Gamma$

T_{eff} (4.1) $\| A \to (B \to A) \Rightarrow \| B \to (A \to B)$

T_{eff} (4.2)

$\| A_1 \to (B \to A_1) \,\&\, \| A_2 \to (B \to A_2) \Rightarrow \|(A_1 * A_2) \to (B \to (A_1 * A_2))$

T_{eff} (4.3) $\| A \to (B \to A) \Rightarrow \| \neg A \to (B \to \neg A)$

T_{eff} (5.1) $\mathcal{Q} \| A \Rightarrow \mathcal{Q} \| [A]$

T_{eff} (5.2.1) $\qquad\qquad\qquad\qquad \mathcal{C}\|A \Rightarrow \mathcal{C}\|[A, B]$

T_{eff} (5.2.2) $\qquad\qquad\qquad\qquad \mathcal{C}\|B \Rightarrow \mathcal{C}\|[A, B]$

with $* \in \{\wedge, \vee, \rightarrow\}$.

As for the connection of the tableaux-calculus T_{eff} to the dialogic semantics of the effective quantum logic, we refer to [1]. The tableaux correspond to reduced positions of success as semantic schemes of a dialog-game. Although it is suggestive to use the same notion, the dialogic tableaux, at first introduced by Lorenz [4], must be distinguished from Beth's tableaux [5] which are also semantic schemes providing a foundation of intuitionistic and classical logic.

The beginnings of T_{eff} correspond to final positions of success in the dialog-game, the constitutive rules of T_{eff} correspond (in a reverse direction) to the possible moves in the game which lead from already established positions of success to new positions of success.

Among the rules of T_{eff} one may distinguish those rules which correspond to moves where the opponent attacks a proposition of the proponent and the proponent must continue the dialog (rules T_{eff} (2) without (2.3.2, 2.3.3, 2.4.2, 2.4.3) and T_{eff} (5)) and the rules which correspond to moves of the proponent (rules T_{eff} (2.3.2, 2.3.3, 2.4.2, 2.4.3, 3, 4)).

The calculus T_{eff} represents the completely formalized dialog-game D_{eff} about quantum mechanical propositions and allows for establishing all initial positions of success, i.e., positions where the proponent has asserted a proposition as initial argument and possesses a strategy of success in the dialog about that proposition.

In the case of unrestrictedly available propositions, the calculus T_{eff} reduces to the tableaux-calculus $\emptyset_{\text{eff}}$ in Lorenz [4], p. 89, which is a calculus for the intuitionistic logic. The difference between T_{eff} and $\emptyset_{\text{eff}}$ is the following:

$\emptyset_{\text{eff}}$ inherits the beginning T_{eff} (1.1) and the rules T_{eff} (2.1, 2.2, 3.1, 3.3, 3.4, and 5). The rules T_{eff} (2.3.1, 2.4.1 and 3.2) are replaced in $\emptyset_{\text{eff}}$ by the rules:

$\emptyset_{\text{eff}}$ (2.3) $\qquad\qquad\qquad \begin{array}{c}\mathcal{C}\| \\ A \| [B]\end{array} \Rightarrow \mathcal{C}\|A \rightarrow B$

$\emptyset_{\text{eff}}$ (2.4) $\qquad\qquad\qquad \begin{array}{c}\mathcal{C}\| \\ A \| [\ ]\end{array} \Rightarrow \mathcal{C}\|\neg A$

$$\emptyset_{\text{eff}} \ (3.2) \quad \begin{array}{c} \mathfrak{A} \\ A \vee B \\ \hline A \end{array} \Bigg\| \begin{array}{c} \\ \Gamma \end{array} \quad \& \begin{array}{c} \mathfrak{A} \\ A \vee B \\ \hline B \end{array} \Bigg\| \begin{array}{c} \\ \Gamma \end{array} \quad \Rightarrow \quad \begin{array}{c} \mathfrak{A} \\ A \vee B \end{array} \Bigg\| \begin{array}{c} \\ \Gamma \end{array}$$

respectively. No further rules appear in $\emptyset_{\text{eff}}$.

Once the calculus T_{eff} is established by means of the dialogic foundation, one can investigate the equivalence of T_{eff} to other calculi which use the easier and more familiar language of propositional logic. Since the equivalence is established in the meta-logic of the formal quantum logic, we consider the meta-logic as far as necessary for the equivalence proofs.

II. SOME META-LOGICAL NOTIONS

In [1, pp. 244 and 265] the metapropositions: "The proposition A is true", "A is false", "A is satisfiable" and "A is refutable" were defined by means of the possible strategies of the proponent and the opponent in a dialog-game about A. Every proposition is associated with one of these values.

Meta-propositions (m-propositions) are also dialog-definite. Consider for instance the m-proposition "A is true". The proof procedure for this m-proposition consists, by definition, of the demonstration of a strategy of success within a dialog-game about A. If we have the m-proposition "A is false" its proof-procedure is equivalent to a demonstration of a strategy of success within a dialog-game about $\neg A$. In the following we denote these two m-propositions — the assertion of the formal truth and the assertion of the formal falsity of a proposition A — by $\vdash_{D_{\text{eff}}} A$ and $\vdash_{D_{\text{eff}}} \neg A$ respectively.

The meta-logic of m-propositions $\vdash_{D_{\text{eff}}} A$ can be formalized by means of meta-dialogs. The m-propositions of the form $\vdash_{D_{\text{eff}}} A$ are considered as elementary m-propositions which are proven by the lower order dialog-game about A. For the denotation of a meta-dialog, e.g., about $\vdash_{D_{\text{eff}}} A$, we use the following scheme:

0	[ ]		$\vdash_{D_{\text{eff}}} A$
1	$\vdash_{D_{\text{eff}}} A?$		Dialog about A

which is analogous to the dialog-schemes defined in [1]. In row 1 the opponent attacks the m-proposition $\vdash_{D_{eff}} A$ asserted by the proponent in row 0 by challenging the proponent to assert the proposition A as initial argument in a dialog (the attack is denoted by $\vdash_{D_{eff}} A?$). Upon this attack the proponent is obliged to perform a dialog about A as a defence.

In this paper we do not develop the complete dialogic approach to the meta-logic of quantum logic which will be done in some other publication, but use some meta-dialogs for defining the particular notions we need in the following. However, it can be seen already in this stage of the investigation that the meta-logic of quantum logic is some logic of unrestrictedly available propositions (all meta-propositions are formally commensurable in the meta-dialog).

Since $\vdash_{D_{eff}} A$ and $\vdash_{T_{eff}} \|A$ are equivalent in the sense of the completeness and soundness theorem, we can just as well apply the reverse directions of the rules of T_{eff} as proof-procedures for elementary m-propositions. $\vdash_{T_{eff}} \|A$ is proven if and only if each reduction of $\|A$ leads to a beginning of T_{eff}.

For the assertion that a tableau t can be deduced in the calculus T_{eff} if the beginnings $\Rightarrow t_1, \ldots, \Rightarrow t_n$ are added as hypotheses to the calculus, we write

$$t_1 \& \ldots \& t_n \vdash_{T_{eff}} t.$$

This assertion of a deduction in the calculus must be distinguished from the rule of the calculus

$$t_1 \& \ldots \& t_n \Rightarrow t.$$

If the m-proposition $t_1 \& \ldots \& t_n \vdash_{T_{eff}} t$ can be proved, $t_1 \& \ldots \& t_n \Rightarrow t$ is said to be a *deducible* rule. We assume that the class of elementary m-propositions in the meta-dialog-game consists of assertions $t_1 \& \ldots \& t_n \vdash_{T_{eff}} t$ ($t_1 \& \ldots \& t_n$ may be empty).

Meta-propositions (denoted by $\bar{t}, \bar{t}', \bar{t}_1, \bar{t}_2 \ldots$) are defined inductively by:

M1.   Elementary m-propositions are m-propositions.

M2.   If $\bar{t}$ is an m-proposition, then $\neg \bar{t}$ is an m-proposition. If $\bar{t}$

and $\bar{t}'$ are m-propositions, then $\bar{t} \land \bar{t}'$, $\bar{t} \lor \bar{t}'$, $\bar{t} \Rightarrow \bar{t}'$ are m-propositions. The symbols $\daleth$, $\land$, $\lor$, $\Rightarrow$ denote the meta-logical negation, conjunction, disjunction and material implication respectively.

Notice that the meta-logical material implication, e.g., $\vdash_{T_{eff}} (\mathfrak{A} \| [A] \land$ $\vdash_{T_{eff}} (\mathfrak{A} \| [B] \Rightarrow \vdash_{T_{eff}} (\mathfrak{A} \| A \land B$ and the rule $(\mathfrak{A} \| [A]$ & $(\mathfrak{A} \| [B] \Rightarrow (\mathfrak{A} \| A \land B$ of the calculus must be distinguished although they are denoted by means of the same sign $\Rightarrow$. The reason why we use the same sign in both cases will become clear in the following.

By means of the meta-dialog-game a proof of the m-proposition

$$\vdash_{T_{eff}} t_1 \land \cdots \land \vdash_{T_{eff}} t_n \Rightarrow \vdash_{T_{eff}} t$$

can be established as follows. (The corresponding rules do not differ from those of the material dialog-game [1] in the case of unrestricted availability of propositions):

0	[ ]	$\vdash_{T_{eff}} t_1 \land \cdots \land \vdash_{T_{eff}} t_n \Rightarrow \vdash_{T_{eff}} t$
1	$\vdash_{T_{eff}} t_1 \land \cdots \land \vdash_{T_{eff}} t_n$	[ ]
2	$\vdash_{T_{eff}} t_1$	1?
3	reduction of t_1	$\vdash_{T_{eff}} t_1$?
$\vdots$		$\vdots$
$2n$	$\vdash_{T_{eff}} t_n$	n?
$2n+1$	reduction of t_n	$\vdash_{T_{eff}} t_n$?
$2n+2$	[ ]	$\vdash_{T_{eff}} t$
$2n+3$	$\vdash_{T_{eff}} t$?	reduction of t

In row 1 the opponent (O) attacks the material m-implication by the m-proposition $\vdash_{T_{eff}} t_1 \land \cdots \land \vdash_{T_{eff}} t_n$. On this attack the proponent (P) is obliged to defend by the assertion of $\vdash_{T_{eff}} t$. P uses a strategy in which he postpones his defence and attacks the m-conjunction by challenging successively the reductions of the tableaux t_1 to t_n from O in the rows 2 to $2n + 1$. In row $2n + 2$ P cannot attack any longer but has to defend by asserting $\vdash_{T_{eff}} t$. In order to prove $\vdash_{T_{eff}} t$ P is allowed to take over appropriate tableaux from O's reductions of the t_i $(i = 1, \ldots, n)$ as hypotheses $t'_1, \ldots, t'_m$ such that he is able to deduce the tableau t (not indicated in the above general scheme). Hence P has a strategy of success for

$$\vdash_{T_{eff}} t_1 \barwedge \cdots \barwedge \vdash_{T_{eff}} t_n \Rightarrow \vdash_{T_{eff}} t$$ if and only if within each arbitrary

reduction R of the $t_2, \ldots, t_n$ there exist tableaux $t'_1, \ldots, t'_m$, dependent on R, such that $t'_1 \& \ldots \& t'_m \Rightarrow t$ is a deducible rule.

The assertion $\vdash_{T_{eff}} t_1 \barwedge \cdots \barwedge \vdash_{T_{eff}} t_n \Rightarrow \vdash_{T_{eff}} t$ must also be distinguished from the rule

$$t_1 \& \ldots \& t_n \Rightarrow t.$$

If the meta-proposition $\vdash_{T_{eff}} t_1 \barwedge \cdots \barwedge \vdash_{T_{eff}} t_n \Rightarrow \vdash_{T_{eff}} t$ is an initial position of success in a meta-dialog-game, $t_1 \& \ldots \& t_n \Rightarrow t$ is said to be an *admissible* rule.

Admissible rules may be added to the rules of the calculus without giving rise to deductions of tableaux which cannot be deduced by the constitutive rules of the calculus. In this way, admissible rules are usually characterized in the literature [6]. If a rule is deducible, then it is also admissible. The reverse conclusion does not always hold as is shown by means of the following example:

$$\mathfrak{A} \| A \Rightarrow \begin{matrix} \mathfrak{A} \\ C \end{matrix} \Big\| A.$$

Assume that the deducibility of this rule is stated by P. Upon O's request in a meta-dialog about $\mathfrak{A} \| A \vdash_{T_{eff}} \begin{matrix} \mathfrak{A} \\ C \end{matrix} \Big\| A$, P cannot give a deduction of $\begin{matrix} \mathfrak{A} \\ C \end{matrix} \Big\|_A$ in T_{eff}, starting at the tableau $\mathfrak{A} \| A$.

If, on the other hand, P asserts the admissibility of the above rule, he has a strategy of success in a meta-dialog about

$$\vdash_{T_{eff}} \mathfrak{A} \| A \Rightarrow \vdash_{T_{eff}} \begin{matrix} \mathfrak{A} \\ C \end{matrix} \Big\|_A :$$

0	[ ]	$\vdash_{\overline{T_{\text{eff}}}} \mathfrak{A} \| A \Rightarrow \vdash_{\overline{T_{\text{eff}}}} \left. \begin{matrix} \mathfrak{A} \\ C \end{matrix} \right\| A$
1	$\vdash_{\overline{T_{\text{eff}}}} \mathfrak{A} \| A$	[ ]
2		$\vdash_{\overline{T_{\text{eff}}}} \mathfrak{A} \| A \,?$

In row 2 the opponent must give a reduction of $\mathfrak{A} \| A$. If $\mathfrak{A} \| A$ is a beginning of T_{eff} then either $A \equiv a$ belongs to $\mathfrak{A}$, or $\mathfrak{A}$ is empty and

$A \equiv A_1 \rightarrow ((A_1 \rightarrow A_2) \rightarrow A_1)$. But in the first case, also $\left. \begin{matrix} \mathfrak{A} \\ C \end{matrix} \right\|_A$ is a beginning

and therefore deducible; in the second case, also $C \| A_1 \rightarrow ((A_1 \rightarrow A_2) \rightarrow A_1)$ is deducible by applying the rule T_{eff} (2.3.2) on the beginning $\| A_1 \rightarrow ((A_1 \rightarrow A_2) \rightarrow A_1)$. If $\mathfrak{A} \| A$ is not a beginning it must be a conclusion of one of the rules T_{eff} (2.). In the case $A \equiv A_1 \rightarrow A_2$ or $A \equiv \neg A_1, P$

deduces $\left. \begin{matrix} \mathfrak{A} \\ C \end{matrix} \right\|_A$ by using the rules T_{eff} (2.3.2) and T_{eff} (2.4.2) respectively.

If $A \equiv A_1 \wedge A_2, O$ must apply the rule T_{eff} (2.1) in the reduction of $\mathfrak{A} \| A$. Therefore the tableaux $\mathfrak{A} \| [A_1]$ and $\mathfrak{A} \| [A_2]$ occur in the reduction. If P takes over both tableaux and applies the rule T_{eff} (3.5) and then T_{eff} (2.1),

he can demonstrate a deduction of $\left. \begin{matrix} \mathfrak{A} \\ C \end{matrix} \right\|_A$:

$$\left. \begin{matrix} \mathfrak{A} \| [A_1] \Rightarrow \left. \begin{matrix} \mathfrak{A} \\ C \end{matrix} \right\|_{[A_1]} \\ \mathfrak{A} \| [A_2] \Rightarrow \left. \begin{matrix} \mathfrak{A} \\ C \end{matrix} \right\|_{[A_2]} \end{matrix} \right\} \Rightarrow \left. \begin{matrix} \mathfrak{A} \\ C \end{matrix} \right\|_{A_1 \wedge A_2}.$$

If $A \equiv A_1 \vee A_2, P$ can show the deducibility of $\left. \begin{matrix} \mathfrak{A} \\ C \end{matrix} \right\|_A$ by taking over the

premise of the rule T_{eff} (2.2). This proves that P has a strategy of success

for $\vdash_{\overline{T_{\text{eff}}}} \left. \begin{matrix} \mathfrak{A} \\ C \end{matrix} \right\|_A$ against all possible reductions of $\mathfrak{A} \| A$.

Deducible rules and admissible rules are frequently used in the following proofs which establish the equivalence of the tableaux-calculus T_{eff} to the calculus of sequents S_{eff} and the propositional calculus Q_{eff}.

III. THE CALCULUS OF SEQUENTS S_{eff}

The calculus S_{eff} provides a simpler scheme for establishing formally true quantum mechanical propositions. S_{eff} is obtained from T_{eff} by replacing the

tableaux figures $\begin{matrix} A_1 \\ \vdots \\ A_n \end{matrix} \bigg\| \begin{matrix} \\ \\ A \end{matrix}$ by the new figures $A_1, \ldots, A_n \| A$, and the

tableaux figures $\begin{matrix} A_1 \\ \vdots \\ A_n \end{matrix} \bigg\| \begin{matrix} \\ \\ \Gamma \end{matrix}$ (where Γ is a potential argument) by the new figures

$A_1, \ldots, A_n \| \theta$, where θ replaces the potentials arguments $[A]$, $[A,B]$, $[\ ]$ by $A, A \vee B$ and $[\ ]$ respectively. $\mathscr{A}$ denotes the sequence of propositions which corresponds to the column system $\mathfrak{C}$ in T_{eff}. The new figures are called *sequents*.

S_{eff} has the following beginnings:

S_{eff} (1.1) $\Rightarrow \mathscr{A}, a \| a$

S_{eff} (1.2) $\Rightarrow \| A \rightarrow ((A \rightarrow B) \rightarrow A)$

and the following constitutive rules:

S_{eff} (2.1) $\mathscr{A} \| A \ \& \ \mathscr{A} \| B \Rightarrow \mathscr{A} \| A \wedge B$

S_{eff} (2.2) $\| (\hat{\mathfrak{C}} \rightarrow (A \rightarrow \hat{\mathfrak{C}}) \ \& \ \mathscr{A}, A \| B \Rightarrow \mathscr{A} \| A \rightarrow B$

S_{eff} (2.3) $\| (\hat{\mathfrak{C}} \rightarrow (A \rightarrow \hat{\mathfrak{C}}) \ \& \ \mathscr{A}, A \| [\ ] \Rightarrow \mathscr{A} \| \neg A$

S_{eff} (3.1.1) $\mathscr{A}, A \wedge B, A \| \theta \Rightarrow \mathscr{A}, A \wedge B \| \theta$

S_{eff} (3.1.2) $\mathscr{A}, A \wedge B, B \| \theta \Rightarrow \mathscr{A}, A \wedge B \| \theta$

S_{eff} (3.2)

$$\left. \begin{matrix} \| ((\hat{\mathfrak{C}} \wedge (A \vee B)) \rightarrow (A \rightarrow (\hat{\mathfrak{C}} \wedge (A \vee B)))) \\ \| ((\hat{\mathfrak{C}} \wedge (A \vee B)) \rightarrow (B \rightarrow (\hat{\mathfrak{C}} \wedge (A \vee B)))) \\ \mathscr{A}, A \vee B, A \| \theta \\ \mathscr{A}, A \vee B, B \| \theta \end{matrix} \right\} \Rightarrow \mathscr{A}, A \vee B \| \theta$$

S_{eff} (3.3)

$$\mathscr{A}, A \to B \| A \ \& \ \mathscr{A}, A \to B, B \| \theta \ \Rightarrow \ \mathscr{A}, A \to B \| \theta$$

S_{eff} (3.4) $\qquad\qquad\qquad \mathscr{A}, \neg A \| A \ \Rightarrow \ \mathscr{A}, \neg A \| \theta$

S_{eff} (3.5) $\qquad\qquad\qquad\qquad \mathscr{A} \| \theta \ \Rightarrow \ \mathscr{A}, C \| \theta$

S_{eff} (3.6) $\qquad\qquad \| \hat{\mathfrak{A}} \to C \ \& \ \mathscr{A}, C \| \theta \ \Rightarrow \ \mathscr{A} \| \theta$

S_{eff} (4.1) $\qquad\qquad\qquad \| A \to (B \to A) \ \Rightarrow \ \| B \to (A \to B)$

S_{eff} (4.2)

$$\| A_1 \to (B \to A_1) \ \& \ \| A_2 \to (B \to A_2) \ \Rightarrow \ \| (A_1 * A_2) \to (B \to (A_1 * A_2))$$

S_{eff} (4.3) $\qquad\qquad \| A \to (B \to A) \ \Rightarrow \ \| \neg A \to (B \to \neg A)$

S_{eff} (5.1) $\qquad\qquad\qquad\qquad \mathscr{A} \| A \ \Rightarrow \ \mathscr{A} \| A \vee B$

S_{eff} (5.2) $\qquad\qquad\qquad\qquad \mathscr{A} \| B \ \Rightarrow \ \mathscr{A} \| A \vee B$

where θ denotes either a proposition or the empty argument [ ] and $* \in \{\wedge, \vee, \to\}$.

III.1. *Remarks on the calculus S_{eff}*

The sequents defined here remind one of Gentzen's sequents [7] although in S_{eff} sequences are allowed only on the left side with respect to $\|$ and not on both sides (as in Gentzen). In fact the calculus S_{eff} is the quantum logical equivalent of the intuitionistic Gentzen-type calculus G3 in Kleene [2, p. 481]. In our denotation the difference between S_{eff} and G3 is the following: In G3 only S_{eff} (1.1) is a beginning but not S_{eff} (1.2). The latter sequent is deducible in G3.

The rules S_{eff} (2.5, 3.6 and 4) do not occur in G3. S_{eff} (3.5, 3.6) are admissible rules and S_{eff} (4) are deducible rules of G3. In G3 the rules S_{eff} (2.2, 2.3, 3.2) are replaced by the rules

G3(2.2) $\qquad\qquad\qquad\qquad \mathscr{A}, A \| B \ \Rightarrow \ \mathscr{A} \| A \to B$

G3(2.3) $\qquad\qquad\qquad\qquad \mathscr{A}, A \| [\ ] \ \Rightarrow \ \mathscr{A} \| \neg A$

G3(3.2) $\qquad \mathscr{A}, A \vee B, A \| \theta \ \& \ \mathscr{A}, A \vee B, B \| \theta \ \Rightarrow \ \mathscr{A}, A \vee B \| \theta$

respectively, which differ from the rules of S_{eff} only by missing premises.

The missing premises are deducible sequents of G3. Thus, the main difference with respect to G3 consists of the appearance of a thinning rule S_{eff} (3.5) and a cut rule S_{eff} (3.6) in the quantum logical calculus. Because of the commensurability restrictions in S_{eff} (2.2, 2.3, 3.2), the thinning and cut rules are constitutive rules of S_{eff} and cannot be eliminated within a deduction of a sequent whereas in G3 these rules are admissible rules. The additional rules S_{eff} (4) represent special commensurability properties (cf. [1, p. 263]).

Note that in S_{eff} the beginning S_{eff} (1.1) may be equivalently replaced by the beginning:

$$S_{eff} (1.1')   \Rightarrow a \| a$$

since by means of S_{eff} (3.5) the rule $a \| a \Rightarrow \mathscr{A}, a \| a$ is deducible in S_{eff} while this is not possible in G3.

For the relation of the effective quantum logic to the intuitionistic Gentzen calculus G1, cf. Kleene [2, p. 442], we refer to note 1.

The *subformula property* (cf. [2, p. 450]) which is a characteristic feature of sequents-calculi can easily be verified also for S_{eff}: Each formula occurring in any sequent of a deduction in S_{eff} without application of the cut rule S_{eff} (3.6) is a sub-formula of some formula occuring in the end sequent.

III.2. *The equivalence of S_{eff} and T_{eff}*

The equivalence of T_{eff} and S_{eff} is established by means of the following theorem:

THEOREM I: The tableau $\| A$ can be deduced in the calculus T_{eff} if and only if the sequent $\| A$ can be deduced in the calculus S_{eff}, i.e.

$$\vdash_{T_{eff}} \| A   \text{iff}   \vdash_{S_{eff}} \| A.$$

The proof of this theorem is given in detail in [8] and will only be outlined here.

(A) *only if*: Assume that a deduction of $\| A$ in T_{eff} is given. We define the following mapping φ which maps tableaux onto sequents:

(i) $\varphi(\mathfrak{k}\|A) \equiv \mathscr{A}\|A$

(ii) $\varphi(\mathfrak{k}\|\Gamma) \equiv \begin{cases} \mathscr{A}\|[\ ] & \text{if} \quad \Gamma \equiv [\ ] \\ \mathscr{A}\|A & \text{if} \quad \Gamma \equiv [A] \\ \mathscr{A}\|A \vee B & \text{if} \quad \Gamma \equiv [A,B]. \end{cases}$

By means of this mapping we obtain:

(1) The beginnings of T_{eff} are mapped onto the beginnings of S_{eff}.

(2) The rules T_{eff} (2.2) and T_{eff} (5.1) are mapped onto the admissible rule $\mathscr{A}\|A \Rightarrow \mathscr{A}\|A$ of S_{eff}.

(3) The rules T_{eff} (2.3.2, 2.3.3, 2.4.2 and 2.4.3) are mapped onto deducible rules of S_{eff}.

(4) All remaining rules of T_{eff} are mapped onto rules of S_{eff}. Since it is shown that the images of all beginnings are deducible and all rules of T_{eff} are deducible or admissible in S_{eff}, this direction of the theorem is proven.

(B) *if*: Assume that a deduction of $\|A$ in S_{eff} is given. Then we define the following mapping ψ from sequents into tableaux:

(i) $\psi(\mathscr{A}\|A) \equiv \mathfrak{C}\|A$ if A is not an elementary proposition,

(ii) $\psi(\mathscr{A}\|a) \equiv \mathfrak{C}\|[a]$ if a is an elementary proposition,

(iii) $\psi(\mathscr{A}\|[\ ]) \equiv \mathfrak{C}\|[\ ]$

Again we have to show that all beginnings and rules of S_{eff} are mapped into deducible tableaux and admissible rules of T_{eff}. We obtain:

(1) The images of the beginnings of S_{eff} are deducible tableaux:

$$\Rightarrow \left.\begin{array}{c}\mathfrak{C}\\ a\end{array}\right\|a \quad \Rightarrow \left.\begin{array}{c}\mathfrak{C}\\ a\end{array}\right\|[a]\,,$$

$$\Rightarrow \|A \to ((A \to B) \to A).$$

(2) The images of the rules S_{eff} (2.) are deducible rules of T_{eff}.

(a) if A and B are compound propositions in the right columns of the tableaux we have:

(2.1): $\left.\begin{array}{c}\mathfrak{C}\|A \Rightarrow \mathfrak{C}\|[A] \\ \mathfrak{C}\|B \Rightarrow \mathfrak{C}\|[B]\end{array}\right\} \Rightarrow \mathfrak{C}\|A \wedge B$

(2.2): $\left.\begin{array}{c}\dfrac{\mathcal{Q}\Big\|}{A\,\Big\|\,B}\overset{\Rightarrow}{~}\dfrac{\Big\|}{A\,\Big\|\,[B]}\\[4pt] \|\,\hat{\mathcal{Q}}\to(A\to\hat{\mathcal{Q}}\,)\end{array}\right\}\Rightarrow\mathcal{Q}\,\|\,A\to B$

(2.3): $\left.\begin{array}{c}\dfrac{\mathcal{Q}\Big\|}{A\,\Big\|\,[\ ]}\\[4pt] \|\,\hat{\mathcal{Q}}\to(A\to\hat{\mathcal{Q}})\end{array}\right\}\Rightarrow\mathcal{Q}\,\|\,\neg A.$

(b) If A and B are elementary propositions the images of the rules S_{eff} (2.) satisfy the rules T_{eff} (2.).

(3) The images of the rules S_{eff} (3.) are admissible rules of T_{eff}.

(a) If the arguments θ and A are compound propositions, we prove at first the admissibility of the image of S_{eff} (3.5) and of S_{eff} (3.6):

(3.5) $\mathcal{Q}\|\theta\Rightarrow\dfrac{\mathcal{Q}\Big\|}{C\,\Big\|\,\theta}$ is already proven in Part II,

(3.6) $\|\,\hat{\mathcal{Q}}\to C\,\&\,\dfrac{\mathcal{Q}\Big\|}{C\,\Big\|\,\theta}\Rightarrow\mathcal{Q}\|\theta$ can be proven in a similar way by means of induction:

One shows that from all possible premises of the tableaux $\dfrac{\mathcal{Q}\Big\|}{C\,\Big\|\,\theta}$ the tableau $\mathcal{Q}\,\|\,\theta$ can be deduced.

To prove the admissibility of the remaining rules of S_{eff} (3) one shows first:

(i) $A\,\|\,A$ is deducible in T_{eff}, which is easily demonstrated by means of induction on the length of the proposition A.

(ii) $\mathcal{Q}\,\|\,[A]\Rightarrow\mathcal{Q}\,\|\,A$ is deducible in T_{eff}: From the premise it follows that $\|\,\hat{\mathcal{Q}}\to A$ is deducible. $\dfrac{\mathcal{Q}\Big\|}{A\,\Big\|\,A}$ is deducible, therefore $\mathcal{Q}\,\|\,A$ is deducible.

If we apply (ii) we easily obtain the admissibility of the images of the rules S_{eff} (3.1) to (3.4).

(b) If A is an elementary proposition, we must demonstrate the admissibility of the images of S_{eff} (3):

(3.3) $\dfrac{\mathcal{Q}\Big\|}{a\to B\,\Big\|\,[a]}\,\&\,a\to B\,\dfrac{\Big\|}{B}\,\dfrac{\mathcal{Q}\Big\|}{\ }\,\theta\Rightarrow\dfrac{\mathcal{Q}\Big\|}{a\to B\,\Big\|\,\theta}\,.$

Since $a \to B \left\|\begin{array}{c}\mathfrak{a}\\a\end{array}\right.$ and $\begin{array}{c}\mathfrak{a}\\a \to B\\a\\B\end{array}\left\|\vphantom{\begin{array}{c}\mathfrak{a}\\a \to B\\a\\B\end{array}}\right._{[B]}$ are deducible $a \to B \left\|\begin{array}{c}\mathfrak{a}\\a\\B\end{array}\right.$ and, therefore,

$\|(\hat{\mathfrak{a}} \wedge (a \to B) \wedge a) \to B$ are deducible.

Since from the second premise it follows that $\begin{array}{c}\mathfrak{a}\\a \to B\\a\\B\end{array}\left\|\vphantom{\begin{array}{c}\mathfrak{a}\\a \to B\\a\\B\end{array}}\right._{\theta}$ is deducible, it

follows that $a \to B \left\|\begin{array}{c}\mathfrak{a}\\a\end{array}\right._{\theta}$ is deducible. From the first premise we obtain the

deducibility of $\|(\hat{\mathfrak{a}} \wedge (a \to B)) \to a$ and, finally, of $\begin{array}{c}\mathfrak{a}\\a \to B\end{array}\left\|\vphantom{\begin{array}{c}\mathfrak{a}\\a \to B\end{array}}\right._{\theta}$

$$(3.4) \qquad \begin{array}{c}\mathfrak{a}\\\neg a\end{array}\left\|\vphantom{\begin{array}{c}\mathfrak{a}\\\neg a\end{array}}\right._{[a]} \Rightarrow \begin{array}{c}\mathfrak{a}\\\neg a\end{array}\left\|\vphantom{\begin{array}{c}\mathfrak{a}\\\neg a\end{array}}\right._{\theta}$$

Here, from the premise and T_{eff} (3.4, 3.5) it follows that $\begin{array}{c}\mathfrak{a}\\\neg a\\a\end{array}\left\|\vphantom{\begin{array}{c}\mathfrak{a}\\\neg a\\a\end{array}}\right._{[\theta]}$ and

$\|(\hat{\mathfrak{a}} \wedge \neg a) \to a$ are deducible ($[\theta] : \equiv [\ ]$ if $\theta \equiv [\ ]$). Therefore $\begin{array}{c}\mathfrak{a}\\\neg a\end{array}\left\|\vphantom{\begin{array}{c}\mathfrak{a}\\\neg a\end{array}}\right._{[\theta]}$

and by means of (ii) $\begin{array}{c}\mathfrak{a}\\\neg a\end{array}\left\|\vphantom{\begin{array}{c}\mathfrak{a}\\\neg a\end{array}}\right._{\theta}$ is also deducible.

If θ and A are elementary propositions the proofs are as above.

If θ is an elementary proposition or the empty argument, the images of the rules S_{eff} (3.) satisfy the rules T_{eff} (3.).

(4) The images of the rules S_{eff} (4.) are the rules T_{eff} (4.).

(5) The images of the rules S_{eff} (5.) are deducible rules of T_{eff}.

This proves Theorem I.

IV. THE PROPOSITIONAL CALCULUS Q_{eff}

The calculus Q_{eff} is interesting for an algebraic approach to quantum logic. The formulae of this calculus consist (as in the two preceding calculi T_{eff} and S_{eff}) of variables $a, b, \ldots$ for elementary propositions which constitute compound propositions by means of the connective signs $\wedge$, $\vee$, $\rightarrow$, $\neg$, and parentheses. In addition to these formulae we postulate the existence of a particular proposition Λ, the 'false proposition' which satisfies $\Lambda \leqslant A$ for all formulae A. By means of these formulae (denoted by $A, B, C, \dots$) the beginnings and constitutive rules for the deduction of *implications* $A \leqslant B$ are established. These new figures, when compared with the tableaux and the sequents, consist only of one proposition on the left side as well as on the right side of $\leqslant$. The empty argument is replaced by the proposition Λ.

Q_{eff} (1.1) $\Rightarrow A \leqslant A$

Q_{eff} (1.2) $A \leqslant B \ \& \ B \leqslant C \Rightarrow A \leqslant C$

Q_{eff} (2.1) $\Rightarrow A \wedge B \leqslant A$

Q_{eff} (2.2) $\Rightarrow A \wedge B \leqslant B$

Q_{eff} (2.3) $C \leqslant A \ \& \ C \leqslant B \Rightarrow C \leqslant A \wedge B$

Q_{eff} (3.1) $\Rightarrow A \leqslant A \vee B$

Q_{eff} (3.2) $\Rightarrow B \leqslant A \vee B$

Q_{eff} (3.3) $A \leqslant C \ \& \ B \leqslant C \Rightarrow A \vee B \leqslant C$

Q_{eff} (4.1) $\Rightarrow A \wedge (A \rightarrow B) \leqslant B$

Q_{eff} (4.2) $A \wedge B \leqslant C \Rightarrow A \rightarrow B \leqslant A \rightarrow C$

Q_{eff} (4.3) $A \leqslant B \rightarrow A \Rightarrow B \leqslant A \rightarrow B$

Q_{eff} (4.4) $B \leqslant A \rightarrow B \ \& \ C \leqslant A \rightarrow C \Rightarrow B * C \leqslant A \rightarrow (B * C)$

Q_{eff} (5.1) $\Rightarrow A \wedge \neg A \leqslant \Lambda$

Q_{eff} (5.2) $A \wedge B \leqslant \Lambda \Rightarrow A \rightarrow B \leqslant \neg A$

Q_{eff} (5.3) $A \leqslant B \rightarrow A \Rightarrow \neg A \leqslant B \rightarrow \neg A$

with $* \in \{\wedge, \vee, \rightarrow\}$.

IV.1. *Remarks on the calculus Q_{eff}*

(1) The effective calculus Q_{eff} is of the Brouwer-type [9]. Its affirmative part, Q_{eff} (1 to 4), can be considered as the quantum logical equivalent to the Brouwer calculus *Br*, cf. [3, p. 80].

The rules Q_{eff} (1 to 4.1) constitute also the calculus Br. Only the rule Q_{eff} (4.2) is replaced in Br by the rule

Br (4.2) $A \wedge B \leqslant C \Rightarrow B \leqslant A \rightarrow C$

(in our denotation). The rules Q_{eff} (4.3, 4.4) which are additional rules to Br represent particular commensurability properties (as in T_{eff} and in S_{eff}) and do not occur in Br.

If the Brouwer calculus Br is extended to the intuitionistic logic, we have the following additional beginning and additional rule:

Br_{eff} (5.1) $\Rightarrow A \wedge \neg A \leqslant \wedge$

Br_{eff} (5.2) $A \wedge B \leqslant \wedge \Rightarrow B \leqslant \neg A$.

The quantum logical calculus Q_{eff} is the quantum logical equivalent to the intuitionistic calculus Br_{eff} (1 to 5). While Q_{eff} (5.1) is inherited by Br_{eff} (5.1), only the rule Br_{eff} (5.2) differs from Q_{eff} (5.2) analogously to Br (4.2). The commensurability property Q_{eff} (5.3) does not occur in the intuitionistic Brouwer calculus.

The rules Q_{eff} (4.2, 4.3, 4.4, 5.2, and 5.3) are deducible in Br_{eff} whereas the rules Br_{eff} (4.2 and 5.2) obviously are not deducible in Q_{eff}.

IV.2. *Some properties of* Q_{eff}

By means of the rules Q_{eff} (5.1 and 2.1) we obtain:

(i) $\wedge = A \wedge \neg A$ (we write $A = B$ (the bi-implication or logical equivalence) if $A \leqslant B$ and $B \leqslant A$).

and by means of Q_{eff} (5.2 and 5.3):

(ii) $A \rightarrow \wedge = \neg A$.

Furthermore, it can be shown that there exists a proposition V, the 'true' proposition, defined by $A \leqslant V$ for all A, which satisfies:

(iii) $V = A \rightarrow A$ for arbitrary A

(iv) $A \leqslant B \iff V \leqslant A \rightarrow B$.

IV.3. *The equivalence of Q_{eff} and S_{eff}*

THEOREM II: The propositional calculus Q_{eff} is equivalent to the calculus of sequents S_{eff} in the following sense: The sequent $\|A$ can be deduced in the calculus S_{eff} if and only if the implication $V \leqslant A$ can be deduced in the calculus Q_{eff}, i.e.

$$\vdash_{\overline{S_{eff}}} \|A \quad \text{iff} \quad \vdash_{\overline{Q_{eff}}} V \leqslant A.$$

In the following we give an outline of the proof, the details of which are given in [8].

(A) *only if*: Assume that a deduction of $\|A$ in S_{eff} is given. Then we define the mapping Φ which maps sequents onto implications as:

(i) $\qquad \Phi(A_1, \ldots, A_n \| A) \equiv A_1 \wedge \ldots \wedge A_n \leqslant A,$

(ii) $\qquad \Phi(\ \|A) \equiv V \leqslant A$

(iii) $\qquad \Phi(A_1, \ldots, A_n \| [\ ]) \equiv A_1 \wedge \ldots \wedge A_n \leqslant \Lambda.$

By means of this mapping we can show:

(1) The images of the beginnings of S_{eff} are deducible implications in Q_{eff}.

(2) The images of the rules S_{eff} (2.2, 2.3, and 3.) are deducible rules in Q_{eff}.

(3) The image of the rule S_{eff} (2.1) is the rule Q_{eff} (2.3) and the images of the rules S_{eff} (4.1, 4.2, and 4.3) are the rules Q_{eff} (4.3, 4.4 and 5.3).

(B) *if*: Assume that a deduction of the implication $V \leqslant A$ is given. Then the following mapping Ψ is defined which maps implications into sequents:

(i) $\qquad \Psi(A \leqslant B) \equiv A \| B$ — if Λ neither occurs as a subproposition of A nor of B in $A \leqslant B$,

(ii) $\qquad \Psi(A \leqslant \Lambda) \equiv A \| [\ ]$ — if Λ does not occur as a subproposition of A,

(iii) $\qquad \Psi(\Lambda \leqslant B) \equiv C \wedge \neg C \| B$ — if Λ does not occur as a

$\qquad\quad\ \ \Psi(V \leqslant B) \equiv \| B$ — subproposition of B,

(iv) $\Psi(A(\Lambda) \leqslant B(\Lambda))$ if Λ occurs as a sub-

          $\equiv A(C \wedge \neg C) \| B(C \wedge \neg C)$ proposition of A or of B

where C is an arbitrary proposition.

By means of the mapping Ψ we obtain:

(1) The images of all beginnings of Q_{eff} are deducible sequents in S_{eff}.

(2) The images of the rules Q_{eff} (1.2, 2.3, 3.3, 4.3, 4.4, and 5.3) are deducible rules in S_{eff}.

(3) The images of the rules Q_{eff} (4.2 and 5.2) are admissible rules in S_{eff}. In order to prove the admissibility of the image of Q_{eff} (4.2) where C is not Λ:

$$A \wedge B \| C \Rightarrow A \rightarrow B \| A \rightarrow C$$

we show at first that the rule

$$\mathscr{A}, A \wedge B \| \theta \Rightarrow \mathscr{A}, A, B \| \theta$$

is admissible in S_{eff} by means of induction on the premises of $\mathscr{A}, A \wedge B \| \theta$.

(i) $\mathscr{A}, A \wedge B \| \theta$ is a beginning of S_{eff}. Then θ must be an elementary proposition a and a member of the sequence $\mathscr{A}$. But in this case $\mathscr{A}, A, B \| \theta$ is also deducible as a beginning.

(ii) $\mathscr{A}, A \wedge B \| \theta$ is the conclusion of one of the rules S_{eff} (2., 3., or 5.). By assumption, the corresponding premises are deducible. Then one can easily show that $\mathscr{A}, A, B \| \theta$ also is deducible.

By means of the above rule we have:

$$A \wedge B \| C \Rightarrow A, B \| C \Rightarrow A, A \rightarrow B, B \| C.$$

$A, A \rightarrow B \| A$ is deducible, therefore, $A, A \rightarrow B \| C$ is deducible. Since $\| (A \rightarrow B) \rightarrow (A \rightarrow (A \rightarrow B))$ is deducible, $A \rightarrow B \| A \rightarrow C$ is also deducible.

If C is the false proposition Λ, the image of the rule Q_{eff} (4.2) is the rule:

$$A \wedge B \| [\ ] \Rightarrow A \rightarrow B \| A \rightarrow (C \wedge \neg C).$$

From the premise we obtain:

$$A \wedge B \| [\ ] \Rightarrow A \wedge B, A \| [\ ].$$

$\| (A \wedge B) \rightarrow (A \rightarrow (A \wedge B))$ is deducible. Therefore $A \wedge B \| \neg A$ is deducible.

Since $A \wedge B \| A$ is deducible, $A \wedge B \| A \wedge \neg A$ is also deducible. Since $A \wedge \neg A \| C \wedge \neg C$ is deducible in S_{eff} we obtain that $A \wedge B \| C \wedge \neg C$ is deducible. As above, from this sequent the sequent $A \to B \| A \to (C \wedge \neg C)$ can be deduced.

The admissibility of the image of Q_{eff} (5.2):

$$A \wedge B \| [\ ] \Rightarrow A \to B \| \neg A$$

is proven in the following way:

$$A \wedge B \| [\ ] \Rightarrow A, B \| [\ ] \Rightarrow A, A \to B, B \| [\ ].$$

$A, A \to B \| A$ is deducible. Therefore $A, A \to B \| \theta$ is deducible. Since $\|(A \to B) \to (A \to (A \to B))$ is deducible, $A \to B \| \neg A$ is also deducible.

IV.4. *The syntactical completeness of Q_{eff} with respect to admissible rules*

THEOREM III: If a rule is admissible in Q_{eff} it is also deducible.

For didactic reasons we demonstrate the syntactical completeness at first for the intuitionistic Brouwer calculus Br_{eff} (cf. Section IV.1). By means of this demonstration we see how the above theorem can be generalized to Q_{eff}.

Assume that $i_1 \& \ldots \& i_n \Rightarrow j$ is an admissible rule of Br_{eff}, i.e. (cf. Part II) the proponent has a strategy of success for j against all possible reductions of the $i_1, \ldots, i_n$ which the opponent might give in the meta-dialog. Under this condition we have to demonstrate that $i_1 \& \ldots \& i_n \Rightarrow j$ is a deducible rule of Br_{eff}, i.e., given the hypotheses $i_1, \ldots, i_n$ the implication j is deducible in Br_{eff}.

If $i_1 \& \ldots \& i_n \Rightarrow j$ is admissible then for each reduction of the $i_1, \ldots, i_n$ to beginnings there exist implications $i_1', \ldots, i_n'$ such that $i_k' \Rightarrow i_k$ ($k = 1, \ldots, n$) are deducible and $i_1' \& \ldots \& i_n' \Rightarrow j$ is deducible. In the following, we show by means of induction on the premises i_k' of i_k that from the deducibility of $i_1' \& \ldots \& i_n' \Rightarrow j$ for all possible premises i_k' of i_k it follows that $i_1 \& \ldots \& i_n \Rightarrow j$ also is deducible.

For the proof it is sufficient to consider only one premise, i.e., $i \Rightarrow j$.

(1) If i is a beginning then $i \Rightarrow j$ is deducible since no premise i' of i exists and $i \Rightarrow j$ is assumed to be admissible.

(2) If i is not a beginning it is a conclusion of one of the constitutive rules of Br_{eff}. We must distinguish the following cases:

(2.1) $i \equiv A \leqslant C \wedge D$. If $A \leqslant C \wedge D$ is deducible then, as possible premises i', $A \leqslant C$ and $A \leqslant D$ are deducible. By assumption, $i' \Rightarrow j$ is deducible. Since $A \leqslant C \wedge D \Rightarrow i'$ is deducible, $i \Rightarrow j$ also is deducible.

(2.2) $i \equiv A \vee B \leqslant C$. The premises i' of i are $A \leqslant C$ and $B \leqslant C$, and by assumption, $i' \Rightarrow j$ is deducible. Since $A \vee B \leqslant C \Rightarrow i'$ is deducible, $i \Rightarrow j$ also is deducible.

(2.3) $i \equiv A \wedge B \leqslant C \vee D$. The premises i' of i are premises of the transitivity rule Br_{eff} (1.2). If $A \wedge B \leqslant C \vee D$ is deducible either $A \wedge B \leqslant A$ and $A \leqslant C \vee D$, or $A \wedge B \leqslant B$ and $B \leqslant C \vee D$, or $C \leqslant C \vee D$ and $A \wedge B \leqslant C$, or $D \leqslant C \vee D$ and $A \wedge B \leqslant D$ are deducible for all premises i' of i, i.e., $A \leqslant C \vee D \Rightarrow j$, $B \leqslant C \vee D \Rightarrow j$, $A \wedge B \leqslant C \Rightarrow j$ and $A \wedge B \leqslant D \Rightarrow j$ are deducible. If j is deducible from the above implications, the strongest implication which can be deduced from these implications and from which j can be deduced is $A \wedge B \leqslant C \vee D$. Hence $A \wedge B \leqslant C \vee D \Rightarrow j$ also is deducible.

(2.4) $i \equiv A \rightarrow B \leqslant C \vee D$. The premises i' of i again are premises of the transitivity rule Br_{eff} (1.2). From the deducibility of $A \rightarrow B \leqslant C \vee D$ it follows that either $C \leqslant C \vee D$ and $A \rightarrow B \leqslant C$, or $D \leqslant C \vee D$ and $A \rightarrow B \leqslant D$ are deducible premises of i. By assumption, $i' \Rightarrow j$ is deducible from all premises i' of i, i.e., $A \rightarrow B \leqslant C \Rightarrow j$ and $A \rightarrow B \leqslant D \Rightarrow j$ are deducible. The strongest implication which can be deduced from these implications and from which j can be deduced is $A \rightarrow B \leqslant C \vee D$. Hence $A \rightarrow B \leqslant C \vee D \Rightarrow j$ also is deducible.

(2.5) $i \equiv \neg A \leqslant C \vee D$. The argumentation is analogous to (2.4). By means of the assumption $\neg A \leqslant C \Rightarrow j$ and $\neg A \leqslant D \Rightarrow j$ are deducible. The strongest implication which can be deduced from these implications and from which j can be deduced is $\neg A \leqslant C \vee D$. Hence $\neg A \leqslant C \vee D \Rightarrow j$ also is deducible.

(2.6) $i \equiv A \leqslant C \to D$. The premise i' of i is the premise $A \wedge C \leqslant D$ of the rule Br_{eff} (4.2). By assumption, $i' \Rightarrow j$ is deducible. Since $A \leqslant C \to D \Rightarrow i'$ is deducible, $i \Rightarrow j$ also is deducible.

(2.7) $i \equiv A \leqslant \neg C$. The premise i' of i is the premise $A \wedge C \leqslant \Lambda$ of the rule Br_{eff} (5.2). Since by assumption $i' \Rightarrow j$ is deducible and $A \leqslant \neg C \Rightarrow i'$ is deducible it follows that $i \Rightarrow j$ also is deducible. This completes the proof for Br_{eff}.

The proof for Q_{eff} is performed exactly like the above proof for Br_{eff} as far as the cases (1) and (2.1 to 2.5) are concerned. In the cases (2.6 and 2.7) the implication i is not the conclusion of the corresponding rules Q_{eff} (4.2) and Q_{eff} (5.2). Instead of the above argumentation we have for Q_{eff}:

(2.6) $i \equiv A \leqslant C \to D$. We distinguish the following cases:
(1) $A \equiv C \to B$. If $C \to B \leqslant C \to D$ is deducible the premise $C \wedge B \leqslant D$ is deducible. By assumption $C \wedge B \leqslant D \Rightarrow j$ is deducible. Since $C \to B \leqslant C \to D \Rightarrow C \wedge B \leqslant D$ is deducible, $C \to B \leqslant C \to D \Rightarrow j$ also is deducible.
(2) $A \equiv D$. If $D \leqslant C \to D$ is deducible the following premises are deducible. In the case:

(i) $a \leqslant c \to a$ where a and c are elementary propositions, c must be a and the premise is $a \leqslant a$. By assumption, j is deducible in Q_{eff}.

(ii) $a \leqslant (C_1 * C_2) \to a$ where a is an elementary proposition and $* \in \{\wedge, \vee, \to\}$, the possible premises i' are: $V \leqslant (C_1 * C_2) \to a$, if i is the conclusion of the transitivity rule Q_{eff} (1.2), or $V \leqslant a \to (C_1 * C_2)$ or $C_1 * C_2 \leqslant \Lambda$, if i is the conclusion of the rule Q_{eff} (4.3) and the premise is the conclusion of Q_{eff} (1.2), or $C_1 \leqslant a \to C_1$ and $C_2 \leqslant a \to C_2$ if i is the conclusion of Q_{eff} (4.3) and the premise is the conclusion of Q_{eff} (4.4). By assumption j is deducible from all possible premises i' of i. The strongest implication which is deducible from all i' however is $a \leqslant (C_1 * C_2) \to a$. Therefore $a \leqslant (C_1 * C_2) \to a \Rightarrow j$ also is deducible.

(iii) $a \leqslant \neg C \to a$ where a is an elementary proposition, the possible premises i' are:
$V \leqslant \neg C \to a$, if Q_{eff} (1.2) is applied, or $V \leqslant a \to \neg C$ or

$\neg C \leqslant \wedge$, if Q_{eff} (1.2, 4.3) are applied, or $C \leqslant a \rightarrow C$, if Q_{eff} (5.3, 4.3) are applied. The strongest implication which is deducible from all i' is $a \leqslant \neg C \rightarrow a$. Therefore $a \leqslant \neg C \rightarrow a \Rightarrow j$ also is deducible.

(iv) $A_1 \times A_2 \leqslant (C_1 * C_2) \rightarrow (A_1 \times A_2)$ where $\times, * \in \{\wedge, \vee, \rightarrow\}$ the possible premises i' are: $V \leqslant (C_1 * C_2) \rightarrow (A_1 \times A_2)$ or $A_1 \times A_2 \leqslant \wedge$ if Q_{eff} (1.2) is applied, or $V \leqslant (A_1 \times A_2) \rightarrow (C_1 * C_2)$ or $C_1 * C_2 \leqslant \wedge$ if Q_{eff} (1.2, 4.3) are applied, or $A_1 \leqslant (C_1 * C_2) \rightarrow A_1$ and $A_2 \leqslant (C_1 * C_2) \rightarrow A_2$, if Q_{eff} (4.4) is applied, or $C_1 \leqslant (A_1 \times A_2) \rightarrow C_1$ and $C_2 \leqslant (A_1 \times A_2) \rightarrow C_2$ if Q_{eff} (4.4, 4.3) are applied. The strongest implication which is deducible from all i' is $A_1 \times A_2 \leqslant (C_1 * C_2) \rightarrow (A_1 \times A_2)$. Therefore $A_1 \times A_2 \leqslant (C_1 * C_2) \rightarrow (A_1 \times A_2) \Rightarrow j$ also is deducible.

(v) $\neg A \leqslant (C_1 * C_2) \rightarrow \neg A$ where $* \in \{\wedge, \vee, \rightarrow\}$ the possible premises i' are: $V \leqslant (C_1 * C_2) \rightarrow \neg A$ or $\neg A \leqslant \wedge$, if Q_{eff} (1.2) is applied, or $V \leqslant \neg A \rightarrow (C_1 * C_2)$ or $C_1 * C_2 \leqslant \wedge$ if Q_{eff} (1.2, 4.3) are applied, or $A \leqslant (C_1 * C_2) \rightarrow A$ if Q_{eff} (5.3) is applied, or $C_1 \leqslant \neg A \rightarrow C_1$ and $C_2 \leqslant \neg A \rightarrow C_2$ if Q_{eff} (4.4, 4.3) are applied. The strongest implication which is deducible from all i' is $\neg A \leqslant (C_1 * C_2) \rightarrow \neg A$. Therefore $\neg A \leqslant (C_1 * C_2) \rightarrow \neg A \Rightarrow j$ also is deducible.

(3) If A does not satisfy the cases 1 and 2 we show that the premises i' of $A \leqslant C \rightarrow D$ are $B \wedge C \leqslant D$ and $B \leqslant C \rightarrow B$ and $A \leqslant B$ with some formula B. If $A \leqslant C \rightarrow D$ is deducible the possible premises i' are premises of the transitivity rule Q_{eff} (1.2) and are implications $A \leqslant B$ and $B \leqslant C \rightarrow B$ and $C \rightarrow B \leqslant C \rightarrow D$. The latter implication is the conclusion of Q_{eff} (4.2). Therefore, if $C \rightarrow B \leqslant C \rightarrow D$ is deducible, $B \wedge C \leqslant D$ also is deducible. By assumption j is deducible from the premises i' for arbitrary formulae B which may occur in the premises. Therefore B must not occur in the conclusion. The strongest implication which is deducible from the i' and which does not contain B is the implication $A \leqslant C \rightarrow D$. Thus $A \leqslant C \rightarrow D \Rightarrow j$ also is deducible.

This completes the proof for (2.6).

(2.7) $i \equiv A \leqslant \neg C$. The argumentation is analogous to (2.6), cases
1 and 3. If A is of the form $C \rightarrow B$, the implication
$C \rightarrow B \leqslant \neg C$ is the conclusion of the rule Q_{eff} (5.2) and the
premise i' is $C \wedge B \leqslant \Lambda$. Since by assumption $i' \Rightarrow j$ is
deducible and since $C \rightarrow B \leqslant \neg C \Rightarrow i'$ is deducible
$C \rightarrow B \leqslant \neg C \Rightarrow j$ also is deducible. If A is not of the above
form, one can prove that as premises i' the implications
$B \wedge C \leqslant \Lambda$ and $B \leqslant C \rightarrow B$ and $A \leqslant B$ with some formula B
are deducible. By assumption j is deducible from the i' for all
B which might occur in the premises. Therefore B must not
occur in the conclusion. The strongest implication which is
deducible from the i' and which does not contain B is the
implication $A \leqslant \neg C$. Thus $A \leqslant \neg C \Rightarrow j$ also is deducible.

From Theorem III it follows that the material implication of the meta-
logic of Q_{eff} can be completely described by means of the language of
Q_{eff}. Since there are no admissible rules (corresponding to meta-logical
material implications) which are not deducible rules in Q_{eff}, the meta-logical
material implications can be described by means of the deducibility of the
corresponding rules in Q_{eff} without making use of the meta-language, i.e.,
the meta-dialog-game. This syntactical completeness of Q_{eff} establishes the
possibility of replacing Q_{eff} by an algebraic structure in which the axioms
are given in a language corresponding to the meta-language of Q_{eff}.

IV.5. *Effective quantum logic as an interpretation for the quasi-implicative lattice L_{qi}*

The algebraic representation of quantum logic is of interest with respect to
the conjecture, often stated in literature [10], that the lattice of subspaces
of a Hilbert space can be interpreted as a propositional calculus. The
quantum logical propositional calculus Q_{eff} does not provide an interpretation for the orthocomplemented quasi-modular lattice L_q which has the
lattice of subspaces of a Hilbert space as a model, but provides an interpretation for a weaker algebraic structure. This structure is called quasi-
implicative lattice. With the addition of the 'excluded middle' or alternatively the 'double negation' law, the lattice L_{qi} can be extended to the
lattice L_q.

The *quasi-implicative* lattice L_{qi} is defined as a pair $\langle \mathcal{L}, \leqslant \rangle_{qi}$ where $\mathcal{L}$ is a non empty set of elements $(A, B, C, \ldots)$ and $\leqslant$ is an ordering relation on $\mathcal{L}$ such that:

(1) L_{qi} (1.1) $A \leqslant A$

    L_{qi} (1.2) $A \leqslant B \curlywedge B \leqslant C \Rightarrow A \leqslant C.$

(2) For any two elements A and B of $\mathcal{L}$ the infimum $(A \wedge B)$ exists:

    L_{qi} (2.1) $A \wedge B \leqslant A$

    L_{qi} (2.2) $A \wedge B \leqslant B$

    L_{qi} (2.3) $C \leqslant A \curlywedge C \leqslant B \Rightarrow C \leqslant A \wedge B$

(3) For any two elements A and B of $\mathcal{L}$ the supremum $(A \vee B)$ exists:

    L_{qi} (3.1) $A \leqslant A \vee B$

    L_{qi} (3.2) $B \leqslant A \vee B$

    L_{qi} (3.3) $A \leqslant C \curlywedge B \leqslant C \Rightarrow A \vee B \leqslant C$

(4) For any two elements A and B of $\mathcal{L}$ the *quasi-implication* $(A \rightarrow B)$ exists:

    L_{qi} (4.1) $A \wedge (A \rightarrow B) \leqslant B$

    L_{qi} (4.2) $A \wedge C \leqslant B \Rightarrow A \rightarrow C \leqslant A \rightarrow B$

    L_{qi} (4.3) $A \leqslant B \rightarrow A \Rightarrow B \leqslant A \rightarrow B$

    L_{qi} (4.4) $B \leqslant A \rightarrow B \curlywedge C \leqslant A \rightarrow C \Rightarrow B * C \leqslant A \rightarrow (B * C)$

with $* \in \{\wedge, \vee, \rightarrow\}$.

(5) For any element A of $\mathcal{L}$ the *quasi-pseudocomplement* $(\neg A)$ exists:

    L_{qi} (5.1) $A \wedge \neg A \leqslant \wedge$

    L_{qi} (5.2) $A \wedge C \leqslant \wedge \Rightarrow A \rightarrow C \leqslant \neg A$

    L_{qi} (5.3) $A \leqslant B \rightarrow A \Rightarrow \neg A \leqslant B \rightarrow \neg A$

For simplicity we use the same denotation within the axioms of L_{qi} as for the respective interpretations given by Q_{eff}. Obviously, the elements of L_{qi} are interpreted as propositions, the ordering relation as implication, the infimum, supremum, material quasi-implication and quasi-pseudocomplement as conjunction, disjunction, material implication and negation respectively. For the denotation of the axioms of L_{qi} we use the meta-linguistic signs $\Rightarrow$ and π .

The material quasi-implication and the quasi-pseudocomplement of L_{qi} replace the material implication and the pseudocomplement of the implicative lattice L_i [11] which is well known to have the intuitionistic calculus Br_{eff} as a model. In L_i the material implication $A \rightarrow B$ and the pseudocomplement $\lnot A$ are defined by the following axioms:

L_i (4.1) $A \wedge (A \rightarrow B) \leqslant B$

L_i (4.2) $A \wedge C \leqslant B \Rightarrow C \leqslant A \rightarrow B$

L_i (5.1) $A \wedge \lnot A \leqslant \wedge$

L_i (5.2) $A \wedge C \leqslant \wedge \Rightarrow C \leqslant \lnot A.$

Properties of L_{qi}:

(1) The material quasi-implication of L_{qi} as well as the material implication of L_i satisfy the following properties:

(i) $A \wedge B \leqslant A \rightarrow B$

(ii) $A \rightarrow B \leqslant A \rightarrow ((A \wedge B) \vee C)$

(iii) $A \leqslant B \Longleftrightarrow V \leqslant A \rightarrow B.$

V denotes the unit element of $\mathcal{L}$, defined by $A \leqslant V$ for all $A \in \mathcal{L}$.
(2) The quasi-pseudocomplement of L_{qi} as well as the pseudocomplement of L_i satisfy

(i) $A \leqslant \lnot\lnot A$

(ii) $A \leqslant B \Rightarrow \lnot B \leqslant \lnot A$

(iii) $V \leqslant A \vee \lnot A \Rightarrow \lnot\lnot A \leqslant A$

(iv) $\lnot\lnot(A \vee \lnot A) \leqslant A \vee \lnot A \Rightarrow V \leqslant A \vee \lnot A$

(v) $\lnot A \wedge \lnot B = \lnot(A \vee B)$

(vi) $\lnot A \vee \lnot B \leqslant \lnot(A \wedge B).$

(3) In L_{qi} the following 'weak quasi-modularity' holds:

$$B \leqslant A \, \pi \, C \leqslant \neg A \Rightarrow A \wedge (B \vee C) = B.$$

Remark: If $\neg A$ is an orthocomplement, the above property is usually called the quasi-modularity.

(4) In L_{qi} the following 'weak distributivity' holds:

$$A \leqslant B \rightarrow A \, \pi \, A \leqslant C \rightarrow A \Rightarrow A \wedge (B \vee C) \leqslant (A \wedge B) \vee (A \wedge C).$$

Remark: 3. is a consequence of 4.

V. CALCULI OF VALUE-DEFINITE QUANTUM LOGIC

The effective quantum logic allows for establishing all formally true quantum mechanical propositions (tautologies), i.e., propositions which are true only because of their formal structure. Since the logical notions are not confined to unrestrictedly available propositions, the effective quantum logic is a more general logic than the intuitionistic logic.

Whereas in the material dialog-game, cf. [1, p. 251], the elementary propositions are assertions of proof-definite propositions which have to be proved in the dialog, they are replaced in the formal dialog-game by variables of propositions which must not be attacked and defended, cf. [1, p. 255]. However, within the formal dialogic proof procedure, particular proof conditions with respect to elementary propositions belonging to the material properties of these propositions can be taken into account in order to extend the logic.

In the following we consider an extension of effective quantum logic which assumes the material property of the value-definiteness for elementary quantum mechanical propositions.

In order to establish this extension, the postulate of value-definiteness is investigated for the dialogic approach first.

V.1. *The postulate of value-definiteness*

Consider the material dialog-game in [1]. A proposition A is defined to be *materially true* if and only if the proponent P has a strategy of success within the dialog-game about A, asserted by P, i.e., P wins the dialog about A against O's material proof procedures with respect to elementary propositions and all possible dialogic arguments of O. A proposition A is defined to be *materially false* if and only if P has a strategy of success within the dialog-game against the proposition A, asserted by O. In this case, P has a strategy of success within the dialog-game about the negation $\neg A$ of A, asserted by P.

Elementary propositions a assert affirmative material proofs. They may be attacked by the challenge a? to perform the corresponding material proof, cf. [1, p. 247]. If P succeeds in proving a, denoted by $a!$ in the dialog scheme, P wins the dialog-game about a and, furthermore, has a strategy of success for a. In this case, a is materially true. On the other hand, assume that O asserts an elementary proposition a, P attacks by a? and O fails in giving a proof for a in a finite dialog. In this case P wins the dialog and O loses the dialog. But P cannot establish a strategy of success against O's proposition a unless he knows a procedure which establishes a disproof of a. If a disproof of a exists the elementary proposition a is said to be materially false.

If a participant is able to disprove an elementary proposition a, the disproof of a can also be used as an affirmative material proof which defines a new elementary proposition $\bar{a}$. While a asserts an affirmative proof, $\bar{a}$ asserts the corresponding disproof. In a material dialog-game $\bar{a}$ may be attacked by the challenge $\bar{a}$? to prove $\bar{a}$. The corresponding defence consists in a successful proof of $\bar{a}$, i.e., a disproof of a. From the definition of $\bar{a}$ it follows that $\bar{a}$ is true whenever a is false and vice versa. A proposition a is said to be *disproof-definite* if and only if $\bar{a}$ is proof-definite.

We now extend the language of the dialog-game by assuming that for each elementary proposition a there exists a corresponding elementary proposition $\bar{a}$. We have $\bar{\bar{a}} = a$. Furthermore, $\bar{a}$ and $\neg a$ are not dialog-equivalent while $\neg\bar{a}$ and $\neg\neg a$ are dialog-equivalent and thus $\neg\bar{a} = \neg\neg a$. Two propositions are said to be *dialog-equivalent* if and only if they can be replaced in any dialog without changing the strategic possibilities of P and O. We prove the dialog-equivalence by comparing these possibilities with respect to $\neg\bar{a}$ and $\neg\neg a$:

n		$\neg\bar{a}$	n		$\neg\neg a$
$n+1\ \bar{a}\,(n)$		[ ]	$n+1$	$\neg a\,(n)$	[ ]
$n+2$		$(n+1)\,\bar{a}?$	$n+2$	[ ]	$(n+1)\,a$
			$n+3$	$a?\ (n+2)$	

P has a strategy of success against $\bar{a}$ in the left dialog if and only if P has a strategy of success for a in the right dialog. In both of the dialogs an empty argument, i.e., an obligation of defence which is impossible, is on the right side in row $n + 1$ of the dialogs. Therefore, in case P has not a strategy of success against $\bar{a}$ and for a, respectively, and there still exist obligations to defend previous to row n, he must not assume those previous obligations in both of the dialogs since he has to assume his last obligations to defend in row $n + 1$ first, cf. the frame-rule F4b) in [1, p. 243]. Thus the strategic possibilities are the same in the two cases.

However $\bar{a}$ and $\neg a$ are not dialog-equivalent as can easily be seen by means of the two cases:

$$
\begin{array}{ll|l}
n & & \bar{a} \\
n + 1 \ \ \bar{a}? \ (n) & & \\
\end{array}
\qquad
\begin{array}{ll|l}
n & & \neg a \\
n + 1 & a \ (n) & [\ ] \\
n + 2 & & (n + 1) \ a? \\
\end{array}
$$

Although P has a strategy of success for $\bar{a}$ if and only if P has a strategy of success against a, the strategic possibilities are not the same for the dialogs about $\bar{a}$ and $\neg a$. P, in case he fails to prove $\bar{a}$, may assume an obligation to defend which is previous to row n in the left dialog, cf. the argument-rule $A3_m$ in [1, p. 247], but not in the right dialog where he has to assume the last (impossible) obligation to defend in row $n + 1$ first.

If the extension of the language to elementary propositions $\bar{a}$ and the dialog-equivalence of $\neg \bar{a}$ and $\neg\neg a$ is taken into account in the formal dialog-game and if we procede to the corresponding extended calculus Q_{eff}^{+} of Q_{eff}, as shown in [1] and in the first parts of this paper, we get the additional beginning:

$$Q_{\text{eff}}^{+} \ (6) \quad \Rightarrow \ \neg \bar{a} = \neg\neg a.$$

We are now in the position to introduce the value-definiteness of elementary propositions as a second extension of the dialog-game and the quantum logical calculi.

A proposition A is defined to be *value-definite* if and only if the material dialog-game decides between the material truth and falsity of A.

If A is an elementary proposition a the value-definiteness of a means that there either exists a proof or a disproof for a. In the case of elementary propositions about physical systems which are proven by means of measurements the value-definiteness assumes the existence of yes–no experiments. Since physical knowledge is constituted by means of decidable empirical processes, we investigate the extension of effective quantum logic under this assumption.

The postulate of value-definiteness for elementary propositions leads to the deducibility of the proposition $a \vee \bar{a}$ in the material dialog-game:

0	[ ]			$a \vee \bar{a}$
1	?	(0)	$\langle 0 \rangle$	a
2	a?	(1)		[ ]
3	[ ]		$\langle 0 \rangle$	$\bar{a}$
4	$\bar{a}$?	(3)	$\langle 3 \rangle$	$\bar{a}$!

Since by assumption there either exists a proof or a disproof for a, P either has a strategy of success for a in row 2 or a strategy of success for $\bar{a}$ in row 4, as assumed in the above dialog, and thus a strategy of success for $a \vee \bar{a}$. On the other hand $a \vee \bar{a}$ is only deducible if a is value-definite.

The formal dialog-game is constituted such that P is allowed to state elementary propositions only if they have been asserted by O previously, cf. [1, p. 250]. The opponent always guarantees the material truth of the elementary propositions asserted by him in the course of the dialog (by choosing appropriate elementary propositions for the variables $a, b, \ldots$). Thus the proponent's strategy of success does not depend on his own proofs for elementary propositions. If in the framework of the formal dialog-game, the value-definiteness of elementary propositions is taken into account, the opponent must also guarantee this material property of elementary propositions. Since the value-definiteness of a is equivalent to the material truth of the propositions $a \vee \bar{a}$, P is always allowed to refer to $a \vee \bar{a}$ as a proposition of O. Furthermore O is not allowed to state $\bar{a}$ if he asserted a previously and a is still available, and he is not allowed to state a if he asserted $\bar{a}$ previously and $\bar{a}$ is still available.

If these possibilities are introduced into the formal dialog-game one can show the dialog-equivalence of $\bar{a}$ and $\neg a$:

m	$\bar{a}$			m	$\bar{a}$	
:				:		
:				:		
n		$\bar{a}$		n		$\neg a$

If P's proposition $\bar{a}$, which only occurs if P can take $\bar{a}$ over from O, is replaced by $\neg a$, O is not allowed to attack the negation by a since he asserted $\bar{a}$ previously. Thus, the strategic possibilities do not change in the two dialogs. On the other hand, $\neg a$ can be eliminated in a dialog:

$$
\left.
\begin{array}{c|c|c}
n & \neg a \\
n+1\ a\ (n) & [\ ]
\end{array}
\right|
\left(
\begin{array}{cc|cc|c}
n & a \vee \bar{a} & [\ ] & a \vee \bar{a} & [\ ] \\
n+1\ a\ \langle n \rangle & (n)\,?, & \bar{a} & \langle n \rangle & (n)\,?
\end{array}
\right)
$$

Instead of asserting $\neg a$, P challenges $a \vee \bar{a}$ from O. O must defend the attack against $a \vee \bar{a}$ in row $n + 1$ either by the assertion of a or by the assertion of $\bar{a}$. In both of the cases we have the same situation concerning the further dialog as in the left dialog.

Because of the dialog-equivalence of $\bar{a}$ and $\neg a$, the elementary propositions $\bar{a}$ introduced in the first extension of the logic can be eliminated *a posteriori* in the second extension with respect to the value-definiteness of elementary propositions. The value-definiteness of the elementary proposition a then is assumed by the proposition $a \vee \neg a$ which is also called the *principle of excluded middle* in logic. In another paper [12] it is shown that the principle of excluded middle for elementary propositions a which is an empirical concept in physics is inherited by all finite compound propositions A. This fact is well known for the usual logic. By replacing $\bar{a}$ by $\neg a$, we obtain the following additional argument-rule in the formal dialog-game:

A_f (value-def.):

> P is allowed in any position of the dialog to challenge the proposition $A \vee \neg A$ from O, where P chooses the proposition A.

In order to guarantee the finiteness of the dialog-game we have to restrict the number of possible applications of the rule A_f(value-def.) to a maximal number. As the extension of the effective logic we consider the union of all dialog-games with respect to these finite numbers.

In order to illustrate the extended game we consider the dialog about the proposition $(a \rightarrow b) \rightarrow (\neg a \vee (a \wedge b))$ as an example:

−1	$a \vee \neg a$			
0	[ ]			$(a \rightarrow b) \rightarrow (\neg a \vee (a \wedge b))$
1	$a \rightarrow b$	(0)	$\langle 0 \rangle$	$\neg a \vee (a \wedge b)$
2	?	(1)		[ ]
3	a	$\langle -1 \rangle$	(-1)	?
4	[ ]		$\langle 1 \rangle$	$a \wedge b$
5	2?	(4)		[ ]
6	b	$\langle 1 \rangle$	(1)	a
7	[ ]		$\langle 4 \rangle$	b

In row 2 P does not defend on the attack ? by O but makes use of the rule A_f(value-def.) by challenging the hypothesis $a \vee \neg a$ from O. The hypothesis is placed into row − 1 of the dialog scheme. Without A_f(value-def.), P must defend on O's attack in row 2:

0	[ ]			$(a \rightarrow b) \rightarrow (\neg a \vee (a \wedge b))$
1	$a \rightarrow b$	(0)	$\langle 0 \rangle$	$\neg a \vee (a \wedge b)$
2	?	(1)	$\langle 1 \rangle$	$\neg a$
3	a	(2)		[ ]

If P defends by the assertion of $\neg a$, O attacks the negation by asserting a in row 3. P is not allowed to defend on O's attack in row 2 a second time, now by asserting $a \wedge b$ and taking back the first attempt to defend by $\neg a$, since he must assume his last obligation to defend in row 3 first, cf. the frame-rule F4b) in [1, p. 243]. If P defends by the assertion of $a \wedge b$ in row 2, P also loses the dialog since he cannot defend the proposition $a \wedge b$. In the extended game O must defend the hypothesis $a \vee \neg a$ by the assertion of either a or $\neg a$. In the latter case P, then, defends his proposition in row 1 by $\neg a$ and, as easily can be seen, wins the dialog. In the first case, as assumed above, P defends by the assertion of $a \wedge b$. O has to continue by challenging the conjunction by either 1? or 2? in row 5. In the first case P defends by a since he can take a over from O after having attacked the material implication of O in row 1. Thus, P has a strategy of success for the above proposition in the extended game.

Another example is the dialog about the proposition $((a \rightarrow b) \rightarrow a) \rightarrow a$ which corresponds to Peirce's law:

-1	$a \vee \neg a$			
0				$((a \to b) \to a) \to a$
1	$(a \to b) \to a$	(0)		$[\ ]$
2	$\neg a$	$\langle -1 \rangle$	(-1)	?
3	$[\ ]$		(1)	$a \to b$
4	a	(3)		$[\ ]$
5	$[\ ]$		(2)	a

In row 2 P challenges the hypothesis $a \vee \neg a$ from O. In the effective dialog-game P must attack in row 2 by the assertion of $a \to b$. Thereupon, O attacks by a. But now P cannot take a over in order to defend his initial proposition since by means of F4b) he must assume his last obligation to defend in row 4 first. Since this is impossible he loses the dialog. In the extended game O has to defend the hypothesis by either a or $\neg a$. In case O defends by a P can take a over and defend on O's attack against the material implication. In case O defends by $\neg a$, as assumed above, P attacks the material implication of O in row 1. In case O defends in row 4, P can take a over and either defend his initial proposition or attack O's negation in row 2 whereupon no defence is possible for O. In case O attacks in row 4, as assumed above, P takes a over and attacks O's negation in row 2. P wins since O cannot defend on this attack. In case O defends the hypothesis by a, P can take a over and defend his initial proposition in row 0. Since P wins in all cases he has a strategy of success for the above proposition.

A proposition A is said to be *formally true* if and only if P has a formal strategy of success for A, i.e., P wins the formal dialog-game irrespective of all arguments of O.

A proposition is *formally false* if and only if P has a formal strategy of success against A, i.e., when O asserts A, P wins the further dialog irrespective of all arguments of O.

Hence the proposition corresponding to Peirce's law is formally true in the extended dialog-game.

It is obvious that the proposition $A \vee \neg A$ is formally true in the extended dialog-game. However the formal truth of $A \vee \neg A$ does not imply the formal truth of either A or $\neg A$, i.e., does not imply that any proposition A is either formally true or formally false. This conclusion only holds for the material truth-values.

It is well known from the usual logic that all classically true propositions can be deduced from appropriate excluded middle hypotheses in the intuitionistic logic. In fact, in the case of unrestricted availability of propositions the dialog-game extended in the above way leads to the classical logic. In case of restricted availabilities of propositions investigated here, the above extension which takes into account the value-definiteness of propositions leads to a logic which is the quantum logical equivalent to classical logic and is called full quantum logic.

V.2. *The calculi T, S and Q of value-definite quantum logic*

The formal dialog-game extended by A_f(value-def.) leads to the following extensions T, S and Q of the effective quantum logical calculi T_{eff}, S_{eff} and Q_{eff}.

(1) The *tableaux-calculus* T which is complete and consistent with respect to the extended dialog-game D is the calculus T_{eff} with the additional rule:

$$T\,(3.7) \qquad \frac{\mathfrak{C}}{A \vee \neg A}\bigg\|_\theta \;\Rightarrow\; \mathfrak{C}\|\theta.$$

θ denotes a proposition or a potential argument.

It can immediately be seen that the rule $T\,(3.7)$ replaces the argument-rule A_f(value-def.) in D. The completeness and consistency proofs are similar to the proofs in [1] and are straightforward.

(2) The *calculus of sequents* S which is equivalent to T is the calculus S_{eff} with the additional rule:

$$S\,(3.7) \qquad \mathscr{A}, A \vee \neg A\|\theta \;\Rightarrow\; \mathscr{A}\|\theta.$$

θ denotes a proposition or the empty argument [ ].

For the equivalence proof we use the same mappings φ and ψ as defined in part III and see immediately that the φ-image of the rule $T\,(3.7)$ is the rule $S\,(3.7)$ and the ψ-image of $S\,(3.7)$ is the rule $T\,(3.7)$.

Remark: A sequents calculus which corresponds to the classical Gentzen-type calculus G3 in [2, p. 481], and which is equivalent to S can also be obtained as is pointed out in the note 2.

(3) The *propositional calculus* Q which is equivalent to S (and to T) is the calculus Q_{eff} with the additional beginning:

Q (5.4) $\Rightarrow V \leqslant A \vee \neg A.$

V denotes the 'true' proposition which is defined by $B \leqslant V$ for all propositions B.

For the equivalence proof we extend the mapping Ψ of part IV to

(v) $\Psi(A(V) \leqslant B(V)) \equiv A(C \vee \neg C) \| B(C \vee \neg C)$

where C is an arbitrary proposition.

In one direction we have to show that the Ψ-image of the implication $V \leqslant A \vee \neg A$ is deducible in S. The image is the sequent $C \vee \neg C \| A \vee \neg A$ and is deducible in S. In the other direction we show that the Φ-image of the rule S (3.7), namely $B \wedge (A \vee \neg A) \leqslant C \Rightarrow B \leqslant C$, is deducible in Q:

From $V \leqslant A \vee \neg A$ it follows that $B \leqslant A \vee \neg A$ and, therefore, $B \leqslant B \wedge (A \vee \neg A)$ are deducible. Since by hypothesis $B \wedge (A \vee \neg A) \leqslant C$ is deducible, we obtain by means of the transitivity rule Q (1.2) that $B \leqslant C$ also is deducible.

Remark: If the elementary propositions $\bar{a}$ introduced by the first extension of the logic are not replaced by the dialog-equivalent propositions $\neg a$ in the second extension, the latter extension leads to the additional beginning

$Q_{\text{eff}}^{++}(7)$ $\Rightarrow V \leqslant a \vee \bar{a}.$

In Q_{eff} with $Q_{\text{eff}}^{+}(6)$ and $Q_{\text{eff}}^{++}(7)$ the bi-implication $\bar{a} = \neg a$, however, is deducible and, thus, the proposition $\bar{a}$ can be eliminated *a posteriori*:

From $V \leqslant a \vee \bar{a}$ and $a \leqslant \neg\neg a$ and $\neg\bar{a} = \neg\neg a$ it follows that $V \leqslant \bar{a} \vee \neg\bar{a}$. Therefore $\neg\neg\bar{a} = \bar{a}$ and, since $\neg\neg\bar{a} = \neg a$, we obtain $\bar{a} = \neg a$.

It can be shown that the Theorem III also holds with respect to the propositional calculus Q, i.e., all admissible rules in Q are also deducible. This establishes the possibility to replace the logical calculus Q by the following lattice structure.

V.3. *Value-definite quantum logic as a model for the orthocomplemented quasi-modular lattice L_q*

If the principle of excluded middle Q (5.4) is added to the axioms (5.) of the quasi-implicative lattice L_{qi} we obtain a lattice L_q. L_q is defined as a pair $\langle \mathcal{L}, \leqslant \rangle_q$ where (the interpretation of) $\mathcal{L}$ is the set of all value-definite quantum mechanical propositions and $\leqslant$ is the ordering relation. The pseudocomplement of L_{qi} becomes an orthocomplement in L_q:

(5) For any element A of $\mathcal{L}$ the orthocomplement ($\neg A$) exists:

L_q (5.1) $A \wedge \neg A \leqslant \wedge$

L_q (5.2) $A \wedge C \leqslant \wedge \Rightarrow A \to C \leqslant \neg A$.

L_q (5.3) $A \leqslant B \to A \Rightarrow \neg A \leqslant B \to \neg A$

L_q (5.4) $V \leqslant A \vee \neg A$.

Remark: The element $\neg A$ is unique by means of L_q (5.3), i.e., by means of the restriction of the $\neg$-operation to the commensurable orthocomplement $\neg A$ of A.

The transition from the effective quantum logic to the value-definite quantum logic is analogous to the transition from the usual effective logic to the classical logic. Under the addition of the principle of excluded middle or, alternatively, the *double negation law* $\neg\neg a = a$, the implicative lattice L_i of effective logic becomes the Boolean lattice of classical logic.

L_q has the following properties:

(1) $\neg\neg A = A$

(2) $A \leqslant B \iff \neg B \leqslant \neg A$

(3) $A \to B = \neg A \vee (A \wedge B)$.

(4) The De Morgan-rules:

(i) $\neg A \wedge \neg B = \neg(A \vee B)$

(ii) $\neg A \vee \neg B = \neg(A \wedge B)$.

Remark: By means of the properties (3) and (4) the material implication ($\to$-operation) can be replaced by the $\neg$-operation and one of the operations $\wedge$ and $\vee$, i.e.:

$$A \to B = \neg(A \wedge \neg(A \wedge B))$$

$$A \to B = \neg A \vee \neg(\neg A \vee \neg B).$$

(5) $A \wedge B = \neg((A \to B) \to \neg A)$

$$A \vee B = (\neg A \to \neg B) \to A.$$

Remark: From (3), (4), and (5) it follows that each of the two-place relations $\wedge$, $\vee$, $\rightarrow$ can be replaced by one of them and by the $\neg$-operation. This possibility is well-known with respect to the classical logic.

(6) The quasi-modularity:

$$B \leqslant A \, \bar{\wedge} \, C \leqslant \neg A \, \Rightarrow \, A \wedge (B \vee C) = B.$$

(7) For the commensurability relation $A \leqslant B \rightarrow A$, the following equivalence holds in L_q:

$$A \leqslant B \rightarrow A \Longleftrightarrow A \leqslant (A \wedge B) \vee (A \wedge \neg B).$$

Remark: In an orthocomplemented lattice the commensurability of propositions A and B is usually defined by means of the relation $A \leqslant (A \wedge B) \vee (A \wedge \neg B)$.

The lattice L_q is isomorphic to the orthocomplemented quasi-modular lattice. By means of the axiom L_q (5) and the property (6) of L_q we obtain that L_q is an orthocomplemented quasi-modular lattice. On the other hand, it can be shown that every orthocomplemented quasi-modular lattice satisfies the axioms of L_q. Therefore, we are justified to consider the calculus Q of value-definite quantum logic as an interpretation of the orthocomplemented quasi-modular lattice which has the lattice of the subspaces of a Hilbert space as a model.

The further properties of the lattice of the subspaces of a Hilbert space, namely the atomicity and the covering law, are not formal properties of quantum mechanical propositions (as the value-definiteness of the elementary propositions is not a formal property), but are material properties which result from further material conditions with respect to propositions about quantum mechanical systems. They are subject to a more extended quantum logical interpretation of quantum mechanics.

ACKNOWLEDGEMENT

I would like to thank Professor P. Mittelstaedt for interesting discussions and correspondence concerning this paper.

University of Western Ontario

NOTES

* Permanent address: Institut für Theoretische Physik der Universität zu Köln, Zülpicherstr. 77, 5000 Köln 41, Germany.

[1] A relation of the effective quantum logic to the intuitionistic Gentzen calculus G1 in Kleene [2, p. 442], can also be established.

Reconsider the formal dialog-game D_n as constituted in [1]. The index n determines the maximal number of possible attacks of P against the same argument of O. In the special case of $n = 1$ the dialogic procedure leads via the tableaux-calculus T_1 in [1, p. 267], to a sequents-calculus S_1. The difference between S_1 and S_{eff} consists in the non-occurrence of the logical connectives $A \wedge B, A \vee B, A \to B$ and $\neg A$ in the premises of the rules S_{eff} (3.). The calculus S_1 essentially is the quantum logical equivalent to the intuitionistic calculus G1 without the contraction rule. The thinning rule and the cut rule of G1 are admissible rules of S_1.

By adding the contraction rule:

$$S_1^+(6) \qquad \mathscr{A}, A, A \parallel \theta \Rightarrow \mathscr{A}, A \parallel \theta$$

to the rules of S_1, the calculus S_1 can be extended to a calculus S_1^+ which is equivalent to S_{eff}. The subformula property also holds in S_1^+ unless the cut rule $S_1^+(3.6)$ is applied within a deduction.

[2] The extension of the effective dialog-game D_{eff} to a value-definite dialog-game D can also be performed by replacing the frame-rule F4b), cf. [1, p. 243], of the dialog-game by the weaker rule:

F4b)(value-def.):

> On the other hand, having been attacked, the participants are obliged to defend, at the latest when there is no opportunity of attack left.

It can be proved that the dialog-game extended by means of F4b)(value-def.) is equivalent to the dialog-game extended by means of A_f(value-def.) in Section V of this paper. This means that both extensions establish the same initial positions of success.

The tableaux- and sequents-calculi T' and S' which correspond to the extension of D_{eff} by means of F4b)(value-def.) are equivalent to T and S. The characteristic difference of T' with respect to T is the occurrence of finite systems of zero or more potential arguments on the right side of the tableaux. Since by means of the weaker frame-rule the proponent may assume obligations to defend which are previous to the last obligation to defend, those previous obligations have to be taken into account as potential arguments of P in the positions of the dialog-game. The rule $T(3.7)$ does not occur in T'.

If one proceeds from T' to the sequents-calculus S' in a way analogous to Section III of this paper, the systems of formulae on the right side of the figures are inherited by S'. The rule $S(3.7)$ does not occur in S'. In this way we obtain a sequents-calculus S' which can be considered as the quantum logical equivalent to the classical Gentzen-type calculus G3 in [2, p. 480].

A relation to the classical Gentzen calculus G1 in [2, p. 442], can also be established in a way analogous to the way pointed out in note 1 for the intuitionistic calculus. Besides the restriction of P's possibilities to only one attack against each argument as in T_1, the number of the possible defenses of the same proposition is restricted to only one defense for P. In this way, together with F4b)(value-def.), a tableaux-calculus $T_{1.1}$ and a sequents-calculus $S_{1.1}$ respectively are defined. The difference between $S_{1.1}$ and S' consists in the non-occurrence of the logical connectives $A \wedge B, A \vee B, A \to B$ and $\neg A$ (on the right side of the sequents) in the premises of the rules $S'(2.$ and $5.)$. The calculus $S_{1.1}$ is the quantum logical equivalent to the classical calculus G1 without the contraction rules for sequences on the left side and the right side of a sequent and the thinning rule for sequences on the right side of a sequent. The thinning rule for sequences on the left side of a sequent and the cut rule are admissible in $S_{1.1}$. By adding the contraction rules and the thinning rule to $S_{1.1}$ a sequents-calculus $S_{1.1}^{+}$ is obtained which is equivalent to S' and to S and which can be considered as a quantum logical equivalent to the classical Gentzen calculus G1.

If one procedes from the sequents-calculus S' or $S_{1.1}^{+}$ to a classical propositional calculus in a way analogous to Section IV of this paper the sequents $A_1, \ldots, A_n \| B_1, \ldots, B_m$ are mapped on the implications $A_1 \wedge \ldots \wedge A_n \leqslant B_1 \vee \ldots \vee B_m$ In this way the equivalence to S' or $S_{1.1}^{+}$ respectively can be established.

BIBLIOGRAPHY

[1] Stachow, E. -W., *Journal of Philosophical Logic* 5 (1976).
[2] Kleene, S. D., *Introduction to Metamathematics*, North-Holland Publ. Co., Amsterdam, 1964.
[3] Lorenzen, P., *Formale Logik*, de Gruyter & Co., Berlin, 1970, p. 80, (Engl. ed.: *Formal Logic*, Reidel Publ. Co., Dordrecht, 1965, p. 56).
[4] Lorenz, K., *Arch. f. Math. Logik und Grundlagenforschung* 11 (1968).
[5] Beth, E. W., *The Foundations of Mathematics*, Amsterdam, 1959.
[6] Lorenzen, P., *Einführung in die operative Logik und Mathematik*, Springer-Verlag, Berlin, 1969.
[7] Gentzen, G., 'Untersuchungen über das logische Schließen', *Math. Z.* 39 (1934).
[8] Stachow, E. -W., Dissertation, Universität zu Köln, 1975.
[9] Brouwer, L. E. J., Over de grondslagen der wiskunde (Dissertation), Amsterdam, 1907.
[10] Mittelstaedt, P., *Z. Naturforsch.* 27a (1972) 1358; Mittelstaedt, P. and E. -W. Stachow, *Found. of Physics* 4 (1974) 335.
[11] Curry, H. B., *Foundations of Mathematical Logic*, McGraw-Hill Book Co., New York, 1963; Rasiova, H., *An Algebraic Approach to Non-Classical Logics*, North-Holland Publ. Co., Amsterdam, 1974.
[12] Mittelstaedt, P. and E. -W. Stachow, 'The Principle of Excluded Middle in Quantum Logic', *Journal of Philosophical Logic* 7 (1978).

E.-W. STACHOW

AN OPERATIONAL APPROACH TO QUANTUM PROBABILITY

INTRODUCTION

In the two preceding papers 'Completeness of Quantum Logic' (CQL) and 'Quantum Logical Calculi and Lattice Structures' (QLC) an operational approach to formal quantum logic was developed. Beginning with a pragmatic definition of quantum mechanical propositions by means of material dialogs a formal dialog-game was introduced for establishing formally true propositions. It was shown in CQL that the formal dialog-game can be replaced by a calculus T_{eff} of effective (intuitionistic) quantum logic which is complete and consistent with respect to the dialogic procedure. In QLC we showed that T_{eff} is equivalent to a propositional calculus Q_{eff}. Since the calculus Q_{eff} is a model for a certain lattice structure, called quasi-implicative lattice (L_{qi}), the connection between quantum logic and the quantum theoretical formalism is provided. L_{qi} is a weaker algebraic structure than the orthomodular lattice of the subspaces of a Hilbert space L_q which can be interpreted as the propositional calculus of value-definite quantum logic. This establishes a quantum logical interpretation of L_q,

In CQL we considered the question whether a certain position in a dialog is a position of success, i.e. whether all continuations of the dialog, starting at the position, are successful for the proponent. The answer to this question leads to the calculus of formal quantum logic.

In this paper we are concerned with the more general question whether a position in a dialog can be associated with a certain probability for the proponent to win the continuation of the dialog. Since we consider arbitrary positions in a dialog the formal dialog-game of CQL cannot be applied. We have to use material dialogs which are now precisely defined by means of a *material dialog-game* (part I. of the paper).

This material dialog-game is a result of the investigations in CQL where material dialogs were introduced for a foundation of the formal dialog-game. For this purpose a game theoretical formulation of a material game was not necessary. For simplicity we assume in this paper that quantum mechanical propositions are *value-definite*, i.e. that their proof pro-

285

Hooker (ed.), *Physical Theory as Logico-Operational Structure*, 285–321.
All Rights Reserved.
Copyright © 1978 by *D. Reidel Publishing Company, Dordrecht, Holland.*

cedures decide between truth and falsity. (This assumption is usually made in physics and is justified in principle by ideal measurements). This hypothesis leads to a weaker version of the dialog-game which is already discussed in QLC.

The dialog-rules of the material game are summarized in a systematic representation. Their operational concept is given in CQL and needs no repetition here. Some remarks are intended to illustrate the rules.

In part II. of the paper *frequencies* of successful dialogs for a participant are considered. A dialog about a physical proposition is always performed with respect to an individual system, i.e. the material proofs of elementary propositions (which are measurements) always refer to the same system under consideration. Propositions about frequencies are defined by dialogs with respect to collections of systems. As we shall see, not all possible dialogs are reasonable dialogs for the determination of frequencies. By means of an additional argument-rule to the material game the dialogic possibilities are restricted to those in which both participants apply their most advantageous strategies.

The next step (part III.) consists in the dialogic introduction of *probability*. The probability is determined as the limiting frequency of successful dialogs with respect to *ensembles* of physical systems. An ensemble is a countable collection of systems equally prepared by means of an unrestricted series of dialogs, each with respect to an individual system. Two preparation procedures are equal if and only if they involve the same information about the proofs and disproofs of material propositions (contained in the dialog), but no other information is involved. The probability should not depend on a particular series of dialogs but, rather, is established by means of a strategy of success for the proponent with respect to each ensemble, characterized by the same preparation procedure. We assume here that material propositions satisfy the property of *probability-definiteness*. From this hypothesis it follows the all dialog-definite propositions are probability-definite. The proof is established by means of the *calculus of quantum probability* which is founded by the material dialog-game.

For a survey of the different approaches to the foundations of probability see for instance [1].

In the final part IV. of the paper the connection to the quantum theoretical probability formula is investigated. For their interpretation we at first introduce the notions of a physical observable and of a state of a physical system. An *observable* is interpreted as a set of elementary propositions which are mutually exclusive and the total disjunction of which is the true proposition. A *state* of a physical system is interpreted as a proposition which represents a maximal *material knowledge* which is possible about a system. The material knowledge about a system is obtained in material dialogs and consists only of material propositions and negations of those propositions (corresponding to proofs and disproofs of material propositions). If we assume that the material knowledge can be maximized the atomicity of the lattice L_q of subspaces of a Hilbert space follows.

The states considered here are associated with individual systems. A *probabilistic description* can be given by means of the quantum probabilities with respect to a system which is prepared in a certain state. We show that the usual description of the state as a real-valued function, defined on the lattice elements (see for instance Jauch [2]) is satisfied by the quantum probabilities with respect to a certain state.

The state propositions are represented by the "pure" states of quantum theory. With respect to the pure states the quantum theory allows the calculation of the expectation values $\langle \alpha | P_A | \alpha \rangle$ where P_A is the projection operator with respect to the subspace M_A. The lattice properties of the subspaces can be expressed formally in terms of algebraic properties of the corresponding projections. This correspondence is verified by means of the quantum theoretical probability formulas. We show that the quantum theoretical expectation values of projections with respect to a pure state can be interpreted as quantum probabilities with respect to an ensemble of systems which are prepared in the state $|\alpha\rangle$. In the case of the sequential expectation value $\langle \alpha | P_{A \lor B} \cdot P_C \cdot P_{A \lor B} | \alpha \rangle$ the corresponding quantum probability $P_{\langle \alpha \rangle}(A \lor B, C)$ does not give the right interpretation because $P_{\langle \alpha \rangle}(A \lor B, C)$ has no interference terms but refers to a mixture in contradiction to $\langle \alpha | P_{A \lor B} \cdot P_C \cdot P_{A \lor B} | \alpha \rangle$. However, if one postulates that the algebra of projections only applies in the case that every projection $P_{A \lor B}$ corresponds to a material proposition $e_{A \lor B}$ the contradiction resolves. This is because the propositions $A \lor B$ and $e_{A \lor B}$ are generally *not dialog-equivalent* although $A \lor B = e_{A \lor B}$ holds in L_q. Therefore the dialogic approach yields $p_{\langle \alpha \rangle}(A \lor B, C) \neq p_{\langle \alpha \rangle}(e_{A \lor B}, C)$. The probability $p_{\langle \alpha \rangle}(e_{A \lor B}, C)$ is purely empirical in $e_{A \lor B}$ and can be interpreted as the above expectation value. But the decomposition of this expectation value by means of the algebraic properties of the projections (in which the interference terms occur) is a formal mathematical representation and has no operational interpretation. We shall see that the difference between $p_{\langle \alpha \rangle}(A \lor B, C)$ and $p_{\langle \alpha \rangle}(e_{A \lor B}, C)$ which is mathematically expressed by the interference term can be founded by means of the different material content of the two propositions $A \vee B$ and $e_{A \vee B}$ (which establishes the dialogic inequivalence of $A \vee B$ and $e_{A \vee B}$) but not by formal properties of quantum logic (since $A \vee B = e_{A \vee B}$ holds in L_q).

I. A MATERIAL DIALOG-GAME ABOUT VALUE-DEFINITE QUANTUM MECHANICAL PROPOSITIONS

a) *The Rules of the Material Game*

1. *The frame-rules:*

$F1$:   At the beginning of a dialog, one of the participants, the proponent (P), asserts the initial argument. In this way the initial position of a dialog is established.

$F2$:   The second participant, the opponent (O), attempts to refute the initial assertion. The dialog then consists of a series of arguments which are stated in turn by the two participants and which obey certain rules.

$F3$:   Arguments are either attacks on or defences of previous arguments, but not both.

$F4$: a. The participants have the right to attack any attackable argument in the dialog.

   b. On the other hand, having been attacked, the participants are obliged to defend. Defences may be postponed but have to be performed at the latest when there remains no opportunity of attack.

$F5$:   If one of the participants has no argument to continue he loses the dialog, and the second participant wins the dialog. In this way the final position of the dialog is established.

2. The *argument-rules:*

A_m1: If an elementary proposition a is asserted it may be attacked by a challenge (a?) to prove it. The defence consists in a demonstration of a. (For a successful proof we write a!.) I.e.

elementary proposition	attack	defence
a	a?	a!

A_m2: If a commensurability proposition $k(A, B)$ is asserted it may be attacked by a challenge ($k(A, B)$?) to to prove it. The defence consists in a demonstration of $k(A, B)$. (For a successful proof we write $k(A, B)$!.) I.e.:

commensurability proposition	attack	defence
$k(A, B)$	$k(A, B)$?	$k(A, B)$!

A_m3: a. If a conjunction $A \wedge B$ is asserted the commensurability of A and B may be attacked (by $k(A, B)$?). The defence on this attack consists in a demonstration of the commensurability of A and B. (For a successful proof we write $k(A, B)$!). After the commensurability attack each of the two subpropositions may be challenged (by 1? and 2?). The corresponding defences are the assertions of A and B.

b. If a disjunction $A \vee B$ is asserted it may be attacked by a challenge (?) to defend. Upon this attack the incommensurability of $\neg A$ and $\neg B$ (i.e. $\neg k(\neg A, \neg B)$) has to be stated first. After this defence, there are two further possibilities to defend by asserting A and B.

c. If a material implication $A \to B$ is asserted it may be attacked by the assertion of A. Thereupon either a defence by B or an attack on A are possible. After B is stated the commensurability of A and B may be attacked (by $k(A, B)$?). The defence on this attack is a demonstration of the commensurability of A and B. (For a successful proof we write $k(A, B)$!). The next argument may be an attack against B.

d. If a negation $\neg A$ is asserted it may be attacked by the assertion of A. Thereupon only an attack against A is possible. The attack and defence schemes which correspond to this argument-rule are the following:

connective	attack	defence
$A \wedge B$	1. $k(A, B)$? 2. 1? 3. 2?	$k(A, B)$! A B
$A \vee B$	?	1. $\neg k(\neg A, \neg B)$ 2. A 3. B
$A \to B$	1. A 2. $k(A, B)$?	B $k(A, B)$!
$\neg A$	A	

A_m4: The participants are allowed to attack a previous proposition A only if A is still available, i.e. if all subsequent propositions are commensurable with A.

Remarks to the Dialog-Rules:

To F4: b.: The frame-rule F4: b. corresponds to the frame-rule F4: b. (value-def.) in note 2 of QLC. There it is shown that the additional postulate of the value-definiteness of quantum mechanical propositions leads to a weaker formulation of the dialog-rules which is established by the frame-rule F4: b. (value-def.).

To the argument-rules: In CQL we showed how the dialog-rules can be introduced in a genetic way, i.e. a way in which additional rules of a dialog are established whenever they are motivated by the praxis of the dialogic procedure. Following this principle we first introduced the argument-rule A_f1 in CQL which defines the logical connectives by means of the different possibilities to attack and to defend compound propositions. A_f1 is still a formal rule in a dialog because it does not depend on the content of the subpropositions of a connective. To obtain practically useful dialogs we required the dialogic procedure to be finite. All other argument-rules are consequences of this important postulate. It entails the existence of elementary propositions and, for restrictedly available propositions (as in quantum mechanics), the existence of commensurability propositions. By means of commensurability proofs and disproofs which replace infinite strategies in a dialog a finite dialogic procedure for all propositions can be obtained. Technically this can be established by additional possibilities to attack and to defend logical connectives, such as those given in the above argument-rule A_m3.

Since we are not concerned again with the genetic introduction of the dialog-rules in this paper we give a slighly different representation of the argument-rules for systematic reasons. We start with the argument-rules concerning the material propositions (elementary propositions and commensurability propositions). In A_m3 the previous rule A_f1 and the additional possibilities to attack and to defend by commensurability arguments are melted together. The previous rule A_m3 in CQL is unnecessary because of the hypothesis of value-definiteness and is eliminated.

To A_m2: In CQL (Section B. 2.a.) the commensurability $k(A, B)$ is dialogically defined by means of a series of alternating dialogs about A and B. $k(A, B)$ is proven if and only if in an arbitrarily long sequence of dialogs about A and B the final positions of the respective dialogs belong to the same success-class. Since infinite strategies should be excluded with respect to the dialogic procedure we require a *confirmation strategy* for the commensurability proof. Propositions the proofs of which are performed separately in a dialog are called material propositions. In dialogs about quantum mechanical propositions not only elementary propositions which are proven by measurements but also commensurability propositions belong to the class of material propositions.

It follows from the dialogic definition of the commensurability of A and B that $k(A, B)$ is dialog-equivalent with the proposition $(A \wedge B) \vee (B \wedge \neg A) \vee (\neg A \wedge B) \vee (\neg A \wedge \neg B)$. As defined in CQL two propositions are called dialog-equivalent if and only if in an arbitrary dialog the possibilities of success do not change for the two participants in case one of the two propositions is replaced by the other. The dialog-definiteness of the above propositions is demonstrated by comparing the possibilities in the respective dialogs:

	$k(A, B)$			$(A \wedge B) \vee (A \wedge \neg B) \vee$ $(\neg A \wedge B) \wedge (\neg A \wedge \neg B)$
$k(A, B)?$			$?$	$A \wedge B$
$[\ ]$	A		$1?$	A
$A?$	$A!$	$\leftrightarrow$	$A?$	$A!$
$[\ ]$	B		$2?$	B
$B?$	$B!$		$B?$	$B!$
$[\ ]$	A		$1?$	A
$\vdots$	$\vdots$		$\vdots$	$\vdots$

	$k(A, B)$			$(A \wedge B) \vee (A \wedge \neg B) \vee$ $(\neg A \wedge B) \vee (\neg A \wedge \neg B)$
$k(A, B)?$			$?$	$(A \wedge \neg B)$
$[\ ]$	A		$1?$	A
$A?$	$A!$	$\leftrightarrow$	$A?$	$A!$
$[\ ]$	B		$2?$	$\neg B$
$B?$	$\neg B!$		B	$[\ ]$
$[\ ]$	A		$\neg B!$	$B?$
$\vdots$	$\vdots$		$1?$	A
			$\vdots$	$\vdots$

We see that in the above pairs of dialogs the same propositions have to be proved. A? denotes a strategy to refute the proposition A and A! or $\neg A$!

denote a successful or unsuccessful dialogical proof respectively. The two remaining possibilities in the dialogs about the above propositions also correspond to each other, as can be seen by analogy with the first two pairs. Therefore, the propositions can be replaced by each other without changing the possibilities of success in an arbitrary dialog. This proves their dialog-equivalence.

In CQL we showed that two propositions A and B are formally commensurable if and only if the proposition $A \rightarrow (B \rightarrow A)$ is formally true, i.e.

$$V \leq k(A, B) \Leftrightarrow V \leq A \rightarrow (B \rightarrow A).$$

It can easily be seen that $(A \wedge B) \vee (A \wedge \neg B) \vee (\neg A \wedge B) \vee (\neg A \wedge \neg B)$ satisfies this property for $k(A, B)$.

Note: We have three different kinds of equivalences between two propositions in a dialog:

1. dialog-equivalence $A \leftrightarrow B$
2. formal (logical) equivalence $A = B$ which holds if and only if the implications $A \leq B$ and $B \leq A$ hold in the q-implicative lattice L_q (see QLC)
3. equivalence with respect to the truth-values:

   A true $\Leftrightarrow B$ true
   A false $\Leftrightarrow B$ false.

The strongest equivalence is the dialog-equivalence as defined above. Whereas dialog-equivalent propositions can be interchanged in any position of a dialog without altering the subsequent dialogic possibilities the formal equivalence between two propositions A and B holds if and only if the following positions are positions of success in the formal dialog-game:

$$A \,|\, [B] \quad \text{and} \quad B \,|\, [A].$$

For instance the propositions $a \vee \neg a$ and $b \vee \neg b$ are formally equivalent because both propositions are formally true. But generally they are not dialog-equivalent since in the material proof of $a \vee \neg a$ a measurement of a is involved which might have a different influence on a subsequent dialog than a measurement of b.

In a value-definite logic there is no difference between the equivalences 2. and 3. whereas in an effective (intuitionistic) logic there exist propositions which are always simultaneously true or false but not formally equivalent. For instance A and $\neg\neg A$ are both simulateneously true or false but the position:

$$\neg\neg A \mid [A]$$

is not a position of success in a formal dialog.

To A_m3: The following examples for dialogs about the connectives $a \wedge b$, $a \vee b$ and $a \rightarrow b$ are intended to illustrate the above argument-rules. The denotations in the dialogs are taken over from the representation of the formal game in CQL.

1. 0	[ ]			$a \wedge b$
1	$k(a, b)$?	(0)	$\langle 0 \rangle$	$k(a, b)$!
2	1?	(0)	$\langle 0 \rangle$	a
3	a?	(2)	$\langle 2 \rangle$	a!
4	2?	(0)	$\langle 0 \rangle$	b
5	b?	(4)	$\langle 4 \rangle$	b!

In this dialog P succeeds in defending against all attacks of O. Since a and b are commensurable, subsequent proofs of a and b will reproduce the same results. Therefore, the dialog can be restricted to the above length. In case a and b are incommensurable the proponent loses the dialog in row 1 where he cannot defend. In the original dialogic procedure, only specified by means of A_f1 and A_m2 (see CQL), the incommensurability of a and b means that the proponent would not be able to defend a and b successfully in an arbitrarily long series of challenges by O. Therefore, P is sure to lose the unrestricted dialog about $a \wedge b$. In this way it can be seen that the commensurability attack in the above dialog indeed replaces a dialog with unrestricted possibilities to attack by 1? and 2?. It is obvious that the possibilities of success in a dialog about a conjunction do not depend on the succession of the attacks 1? and 2?.

2. 0	[ ]			$a \vee b$
1	?	(0)	$\langle 0 \rangle$	$\neg k(\neg a, \neg b)$
2	$k(\neg a, \neg b)$	(1)		[ ]
3	$k(\neg a, \neg b)$!	$\langle 2 \rangle$	(2)	$k(\neg a, \neg b)$?
4	[ ]		$\langle 0 \rangle$	a
5	a?	(4)	$\langle 0 \rangle$	b
6	b?	(5)	$\langle 5 \rangle$	b!

In this dialog P does not succeed in his first two attempts to defend by $\neg k(\neg a, \neg b)$ and a. Since he succeeds in proving b he wins the dialog. Because of the commensurability of $\neg a$ and $\neg b$ the results of subsequent proofs of a and b would be reproduced. Therefore it is sufficient to restrict the dialog to the above length. If P does not succeed in proving b in row 6 he loses the dialog as in the unrestricted procedure where the commensurability of $\neg a$ and $\neg b$ implies that in each series of proof attempts of a and b the outcomes are negative. On the other hand, if $\neg a$ and $\neg b$ are incommensurable P wins the dialog in row 3 because O cannot defend. In an unrestricted dialog P also wins in this case because he is certain to prove one of the two propositions in a sufficiently long series of proofs of a and b. In this way we see that the commensurability proposition $k(\neg a, \neg b)$ in the above dialog indeed replaces an unrestricted series of assertions of a and b by P.

Again it is obvious that the possibilities of success in a dialog about a disjunction do not depend on the succession of the defences by the two subpropositions.

					$a \rightarrow b$
3.	0	[ ]			
	1	a	(0)		[ ]
	2	$a!$	$\langle 1 \rangle$	(1)	$a?$
	3	[ ]		$\langle 0 \rangle$	b
	4	$k(a, b)?$	(0)	$\langle 0 \rangle$	$k(a, b)!$
	5	$b?$	(3)	$\langle 3 \rangle$	$b!$

Here the commensurability attack replaces the unrestricted strategy where O may repeat his attack by a an arbitrary number of times. Since P succeeds in demonstrating $k(a, b)$ and also b in the above dialog he wins as in the unrestricted dialog because there he is certain to prove b after every attack by a. Therefore, by means of the above commensurability attack the dialog can be restricted to one attack by a without changing the possibilities of success for the two participants. If P does not succeed in demonstrating $k(a, b)$ he loses the dialog. This corresponds again to the situation in the unrestricted procedure where in the case of the incommensurability of a and b P is certain to fail in a proof of b after a sufficiently long series of attacks by a. If O cannot prove a in row 2 he loses the dialog because, since P did not defend by b, he has no right to continue the dialog.

The four logical connectives are not dialogically independent of each other if the value-definiteness of propositions is assumed. It is shown in QLC that in this case the dialogically independent connectives reduce to the negation and one two-place connective which is either the conjunction, the disjunction or the material implication. With respect to the possible generalisation to not value-definite propositions we maintain the four logical connectives throughout this paper.

Since the value-definiteness is assumed each attack and defence is restricted to only one attempt without reducing the possibilities of success in comparison with more than one attempt. In the general case of intuitionistic dialogs a repetition of the same attack or the same defence might improve the possibilities of success if a participant takes over some proof procedures (previously unknown to him) from the second participant.

To A_m4: A participant who asserts a proposition A should assume the obligation to defend A against an attack only as long as the proposition is still available in the dialog, i.e. as long as, after the assertion of A, no dialogic manipulations have been performed which affect the result of a proof of A. As demonstrated in CQL a proposition A is still available in a certain position of a dialog if and only if A is commensurable with all propositions which have been proven after the assertion of A.

b) A Game Theoretical Representation of Material Dialogs

A game theoretical representation for a decidable two person game requires the following specification (see for instance Berge [3]):

1. A class of game positions $z_0, z_1, \ldots, \in Z$,
2. a two place relation R, defined on Z, which is established by means of the game rules. (If the relation R holds between two positions z_i and z_{i+1} we write $z_i R z_{i+1}$ where z_i is called the R-predecessor of z_{i+1} and z_{i+1} is called the R-successor of z_i. Positions without R-predecessor are initial positions and positions without R-successor are final positions. A game then consists in a series of positions z_i ($0 \leq i$) where z_0 is the initial position and $z_i R z_{i+1}$ is always satisfied.)
3. a division of Z into two disjoint move-classes M_o and M_p. (If z_i belongs to the move-class M_o this means that O has the right to continue the game by an R-successor z_{i+1}. In an anagous way the move-class M_p is defined. The initial position z_0 always belongs to M_o.)
4. a division of the class of final positions E into two disjoint successclasses E_o and E_p. (If a final position belongs to E_o then O wins the game and P loses the game. In an analogous way E_p is defined.)

Material dialogs can be represented as a two person game which is decidable. The above requirements are satisfied in the following way:

G1: The positions z of the material dialog-game are defined as a class of
   column pairs $z_o \mid z_p$.
   a. The left column consists of all arguments of O and the right column
      consists of all arguments of P. The succession of arguments is enu-
      merated by rows which start at 0.
   b. The argument of P in row 0 is always a proposition. The argument of
      O in row 0 is always the empty argument [ ].
   c. The arguments are assigned the following numbers: A number in
      parentheses (i) on the right side of an argument of O or on the left
      side of an argument of P indicates that the respective argument
      attacks an argument in row i.
      A number in carets $\langle i \rangle$ at the above position indicates that the re-
      spective argument is a defence against an attack on an argument in
      row i.
      If a participant does not defend against an attack the empty argu-
      ment is placed in the respective row. The dialog then has to be con-
      tinued in the next row.
G2: The relation R is established by means of the above dialog-rules.
G3: The move-classes are defined as follows:
   If the last empty argument of a position z belongs to the right column
   of the column pair then $z \in M_p$. If it belongs to the left column then
   $z \in M_o$.
G4: The success-classes are defined as follows:
   Final positions $z \in M_o$ belong to E_p, final positions $z \in M_p$ belong
   to E_o.

Propositions and arguments of a dialog are defined in the formal language by:

G5: Assume that a class of elementary propositions S_e and the connec-
   tive signs $\wedge$, $\vee$, $\rightarrow$, $\neg$ and the symbol k (,) are given. Then propositions (denoted by $A, B, ...$) are defined recursively by:
   a. Elementary propositions are propositions,
   b. if A is a proposition then $\neg A$ is a propositions. If A and B are pro-
      positions then $A \wedge B$, $A \vee B$, $A \rightarrow B$ and $k(A, B)$ are propositions
G6: Arguments are defined as:
   a. propositions
   b. the doubts $1?$, $2?$, $?$, $a?$ and $k\ (A, B)?$
   c. the proof symbols $a!$, $k(A, B)!$
   d. the empty argument [ ].

III. FREQUENCIES OF SUCCESSFUL DIALOGS FOR P

a) *Dialogs about an Individual System*

In our approach the proof-definiteness of propositions is the basic concept. This concept means briefly:
Propositions are defined by means of certain processes which are considered as proof procedures for the respective propositions. We assume that there exists a criterion C which decides on the correctness of a given proof attempt π for a proposition A. C can be represented by a mapping C_A: $\{\pi\} \rightarrow \{\pi_A\}$ from the class of all proofs $\{\pi\}$ onto the class of proofs $\{\pi_A\}$ which are admissible proofs of A. For dialog-definite propositions the criterion C consists in the dialog-rules which establish admissible moves in the dialog-game. For an elementary proposition the criterion is given (in the physical case) by a certain kind of measurements which are considered as admissible proof attempts for the elementary proposition under consideration.

For instance, consider the standard example of an elementary proposition: "The electron (under consideration) has spin-up in x^+-direction". The class of admissible measurements can be characterized in the following way: An admissible measurement of a physical property (here the spin of an electron) is a measurement which is affected by the property in such a way that the outcome of the measurement is conclusive. We see that the question, if a given measurement is admissible, cannot be answered without any knowledge about the physical laws which describe the interaction between the system under consideration and the measuring apparatus. Since we know that the path of an electron through an inhomogeneous magnetic field is dependent on the spin direction of the electron we are justified in considering a Stern-Gerlach experiment as an admissible measurement for the above spin proposition.

For the following we assume that the class S_e of proof-definite elementary propositions is always associated with a certain kind of a physical system in the sense that the elementary propositions refer to the properties represented by projections on subspaces of the Hilbert space of the system. Furthermore, we assume that a dialog can always be directed to an individual system of a certain kind of a system, i.e. the proofs of material propositions are measurements with respect to the same system under consideration.

Because of the dialog-definiteness of propositions with respect to an individual system all propositions, defined by G5, are meaningful propositions about an individual system. Since the proof conditions are taken into account in the definition of propositions we say that all propositions are *objective* with respect to a physical system. If a dialog is carried out the truth or falsity of a proposition can be established.

b) *Dialogs about Collections of Systems*

Until now we considered dialogs about individual systems. The material dialog-game, as well as the formal dialog-game (which is established in CQL), are assumed to be performed with respect to a proposition about an individual system (in the physical case). Quantum logic, as far as it concerns the logic of individual systems, is only one part of the language about quantum mechanical systems. A next step in the extension of quantum logic is the inclusion of propositions about probabilities which are propositions about collective properties of systems. In the following we show how probability propositions can be introduced dialogically as meta-propositions about an appropriate collection of systems.

To prepare the dialogic foundation of probability propositions we make the following assumption:

It is assumed that dialogs can be directed an unlimited number of times to individual systems of a certain kind.

This means that in practice an unlimited number of systems of a certain kind can be produced. By a *production* of a system we mean an operation by means of which *only* the knowledge is obtained that the system exists. Note that here the production of a system does not mean that the system is prepared in a certain state (the notion of a state will be defined later). The knowledge that a system of a certain kind is produced is based on the preceding empirical experience that a certain operation always produces a system of the same kind. This preceding experience can only be established by means of measurements with respect to elementary propositions which characterize a physical system. If a proposition is asserted and previous dialogs have already been carried out with respect to the system under consideration, we call these previous dialogs a *preparation* of the system. Every preparation of a system consists of a sequence of proofs or disproofs of material propositions. Two preparations are defined to be identical if and only if they can be performed by the same sequence of proofs and disproofs of material propositions. The operation of the production of a system is represented by the projection on the full Hilbert space of the system whereas a preparation is represented by a projection on a particular subspace.

For the introduction of probability propositions we need a sufficiently large number of systems of the same kind, i.e. systems which are identically produced and prepared. We define a countable number of such systems γ_i as an *ensemble*, denoted by $\{\gamma_1, \gamma_2, ...\}$. We do not assume here that an ensemble of quantum mechanical systems of the same kind can be established in praxis. The possibility of an ensemble is considered as the *operational* possibility to produce and to prepare systems an unrestricted number of times.

Assume that an ensemble $\{\gamma_1, \gamma_2, ...\}$ is produced. A preparation of this ensemble with respect to the material proposition a, then, is defined by means of the mapping:

$$P_a: \{\gamma_1, \gamma_2, ...\} \to \{\gamma_{a_1}, \gamma_{a_2}, ...\}$$

in the following way. The material proposition a is subsequently asserted with respect to the systems of the sequence $\gamma_1, \gamma_2, ...$. The first successful proof within the sequence determines the system γ_{a_1}, the second successful proof determines γ_{a_2} and so on. If by means of this procedure an ensemble of systems $\gamma_{a_1}, \gamma_{a_2}, ...$ can be established the above mapping is defined. If the preparation of an ensemble involves a proof of a further material proposition b the mapping:

$$P_b: \{\gamma_{a_1}, \gamma_{a_2}, ...\} \to \{\gamma_{b_1}, \gamma_{b_2}, ...\}$$

is defined in the same way as the above mapping P_a. Analogously also a preparation of an ensemble with respect to a disproof of a material proposition can be defined.

For the following we assume that for any finite set of material propositions $a, b, ..., c$ a preparation of an ensemble of systems with respect to $a, b, ..., c$ exists, i.e. that for all ensembles the mapping $P_c \circ ... \circ P_b \circ P_a$ is defined.

c) *The Dialogic Determination of Frequencies of Successful Dialogs*

Assume that n dialogs with respect to n systems of an ensemble arrive at the same position z. If a preparation of the systems has been performed we assume furthermore that the preparation is identical with respect to the n systems. For specifying the preparation we denote a proof of a material proposition a by $a!$ (as usual) and a disproof of a by $\neg a!$. If the preparation of a system consists in the sequence $\mathscr{A}$ of proofs $a!$ and disproofs $\neg b!$ of material propositions and a subsequent dialog leads to the position z we indicate the whole dialogic position by $_{\langle\mathscr{A}\rangle}z$.

Now assume that the above n dialogs are continued until a final position is established. If $n(_{\langle\mathscr{A}\rangle}z)$ is the number of dialogs which lead to a final position of success for P the frequency of those dialogs is given by:

$$f(_{\langle\mathscr{A}\rangle}z): = \frac{n\,(_{\langle\mathscr{A}\rangle}z)}{n}.$$

The proposition: "The frequency of successful dialogs for P with respect to n dialogs starting at the position $_{\langle\mathscr{A}\rangle}z$ is the real number ν" is a proof-definite proposition because by means of the above definition a proof procedure for the frequency proposition is established.

However, for the determination of the frequency not all dialogic possibilities within a continuation of a dialog, starting at the position $_{\langle\mathscr{A}\rangle}z$, should be allowed. For instance, it would not be reasonable to count dialogs in which the participants do not use all possibilities, given by the argument-rules, to be successful or even give up the dialog. On the other hand, the participants should not state more arguments than necessary to attack or to defend a proposition successfully because these redundant arguments, if they involve material proofs, might influence the availability of previous propositions. Therefore we postulate that in dialogs which are used to determine the frequency of successful dialogs, the participants must apply their most advantageous strategies. Thereby the manipulation of the system by means of material proofs must be restricted to a minimal number of proofs.

This means in particular:

A_p a. A proposition which is attackable must be attacked as long as it is available in a dialog.

   b. If a participant attacks or defends he must use all possibilities given by the argument-rules.

   c. If an attack or a defence was successful for a participant he must not use further possibilities to attack or to defend the same proposition.

   d. The participants are not allowed to attack a proposition which is formally true in the position of the dialog. On the other hand, the participants are not allowed to defend against an attack on a proposition which is formally false in the position of the dialog.

Since in formal dialogs (which are defined in CQL) the formal truth and the formal falsity of propositions can be established without performing material proofs with repect to the system, d. requires all propositions in a material dialog to be checked with respect to formal truth and falsity. For this, either the formal dialog-game or one of the calculi given in QLC can be applied.

In this way the dialog-equivalence of all formally true and false propositions is established in a material dialog.

III. PROBABILITIES OF SUCCESSFUL DIALOGS

a) *Specification of a Certain Kind of Positions in a Dialog*

As we shall see later the following kind of positions in a dialog is of particular interest:

In O's column no proposition which is attackable remains. Furthermore no obligation to defend which consists in the assertion of a proposition remains.

In P's column at most one attackable proposition remains but no obligation to defend which consists in the assertion of a proposition. For simplicity we reduce positions of the above kind to the content which is important for establishing these positions in a dialog and for continuing the dialog from these positions. The reduced positions are obtained in the following way:

All propositions which are no longer attackable are eliminated in the dialog scheme. Empty arguments are eliminated if they are placed in row 0 or after an attack against a negation (whereupon no defence is possible). Empty arguments are replaced by $\neg a!$ if they designate a disproof of the material proposition a after an attack $a?$ in a dialog. Furthermore all doubts and all attack and defence numbers are eliminated in the dialog scheme. The remaining arguments then are systems $\mathfrak{A}!$ and $\mathfrak{A}'!$ of arguments $a!\ b!, \ldots$ (where a and b are material propositions) in both columns and a proposition A in P's column. The general form of such a position is:

$$\mathfrak{A}! \mid \mathfrak{A}'! \\ \mid A.$$

If a preparation $\langle \mathscr{A}' \rangle$ of the system has been performed in previous dialogs $\langle \mathscr{A}' \rangle$ has to be taken into account for the position under consideration and we have:

$$\mathfrak{A}! \mid \mathfrak{A}'! \\ \mid A. \\ \langle \mathscr{A}' \rangle$$

Since the systems $\mathfrak{A}$ and $\mathfrak{A}'$ can also be considered as belonging to the preparation of the position $|A$ we write for simplicity instead of the above position the position:

$$\langle \mathscr{A} \rangle \qquad | \, A$$

where $\mathscr{A}$ now indicates the propositions of $\mathscr{A}'$ as well as of $\mathfrak{A}$ and $\mathfrak{A}'$. Notice that because of the general incommensurability of the propositions the succession of the propositions in $\mathscr{A}$ is important.

b) *Probability-Definite Propositions*

Consider frequencies of successful dialogs starting at a position $\langle \mathscr{A} \rangle \, | \, A$ of the above kind. If the subsequent dialog leads to a final position of success for P we write $\langle \mathscr{A} \rangle \, | \, A!$ where $A!$ denotes a successful dialog about A. A frequency of successful dialogs then is given by:

$$f(\langle \mathscr{A} \rangle | A) = \frac{n\,(\langle \mathscr{A} \rangle | A!)}{n}.$$

Probability propositions are established by means of the following definition:

DEFINITION: A proposition A is said to be *probability-definite* if and only if for each finite preparation $\mathscr{A}$ there exists a real number $P_{\langle \mathscr{A} \rangle}(A)$ such that for each ensemble $\{\gamma_{\mathscr{A}_1}, \gamma_{\mathscr{A}_2}, \ldots\}$ the limiting frequency $n(\langle \mathscr{A} \rangle \, | \, A!)/n$ exists and equals $p_{\langle \mathscr{A} \rangle}(A)$. $p_{\langle \mathscr{A} \rangle}(A)$ is called the probability that a dialog about the proposition A with respect to the preparation $\langle \mathscr{A} \rangle$ is successful for P. Therefore $p_{\langle \mathscr{A} \rangle}(A)$ is interpreted as the conditional probability of A with respect to the sequence $\mathscr{A}$.

By means of the above definition the probability proposition: "The probability that a dialog, starting at the position $\langle \mathscr{A} \rangle | A$, leads to a final position of success for P is the real number ν" is a proof-definite proposition if and only if A is probability-definite.

In the following we postulate that material propositions are probability-definite. This means that each ensemble can be characterized by the probabilities of the material propositions.

HYPOTHESIS: *Material propositions are probability-definite.*

THEOREM: *Dialog-definite propositions are probability-definite.*

The proof of this theorem is established by means of the following probability calculus.

c) *The Calculus of Quantum Probabilities*

For establishing formally true or formally false propositions A we use the order relation $V \leq A$ or $V \leq \neg A$ respectively of the q-implicative lattice (see QLC section V. 3)). The calculus then reads:

$$(1.1) \quad V \leq A_1 \to (A_2 \to (\cdots(A_m \to A)\cdots)) \Leftrightarrow p_{\langle\mathscr{A}\rangle}(A) = 1$$
$$(1.2) \quad V \leq A_1 \to (A_2 \to (\cdots(A_m \to \neg A)\cdots)) \Leftrightarrow p_{\langle\mathscr{A}\rangle}(A) = 0$$

where $\mathscr{A}$ stands for the sequence $A_1, \ldots, A_m$ and it is assumed that $\forall_{i \in \{1, \cdots, m\}} \Lambda < A_i$ where Λ is the least element of L_q.

$$(2) \quad \left.\begin{array}{l} A_1 \to (A_2 \to (\cdots(A_m \to A)\cdots)) < V \\ A_1 \to (A_2 \to (\cdots(A_m \to \neg A)\cdots)) < V \end{array}\right\} \text{ in } p_{\langle\mathscr{A}\rangle}(A) \Rightarrow$$

$$(2.1) \quad p_{\langle\mathscr{A}\rangle}(A \wedge B) = p_{\langle\mathscr{A}\rangle}(k(A, B)) \cdot p_{\langle\mathscr{A}, k(A,B)\rangle}(A) \cdot$$
$$\cdot p_{\langle\mathscr{A}, k(A, B), A\rangle}(B)$$

$$(2.2) \quad p_{\langle\mathscr{A}\rangle}(A \vee B) = p_{\langle\mathscr{A}\rangle}(\neg k(\neg A, \neg B)) + p_{\langle\mathscr{A}\rangle}(k(\neg A, \neg B)) \cdot$$
$$\cdot p_{\langle\mathscr{A}, k(\neg A, \neg B)\rangle}(A) + p_{\langle\mathscr{A}\rangle}(k(\neg A, \neg B)) \cdot$$
$$\cdot p_{\langle\mathscr{A}, k(\neg A, \neg B)\rangle}(\neg A) \cdot p_{\langle\mathscr{A}, k(\neg A, \neg B), \neg A\rangle}(B)$$

$$(2.3) \quad p_{\langle\mathscr{A}\rangle}(A \to B) = p_{\langle\mathscr{A}\rangle}(\neg A) + p_{\langle\mathscr{A}\rangle}(A) \cdot$$
$$\cdot p_{\langle\mathscr{A}, A\rangle}(k(A, B)) \cdot p_{\langle\mathscr{A}, A, k(A,B)\rangle}(B)$$

$$(2.4) \quad p_{\langle\mathscr{A}\rangle}(\neg A) = 1 - p_{\langle\mathscr{A}\rangle}(A)$$

$$(3.1) \quad V < A_1 \to (A_2 \to \cdots(A_m \to A)\cdots)) \Rightarrow p_{\langle\mathscr{A}, A, \mathscr{B}\rangle}(C) = p_{\langle\mathscr{A}, \mathscr{B}\rangle}(C)$$

$$(3.2) \quad V \leq A_1 \to (A_2 \to (\cdots(A_m \to \neg A)\cdots)) \Rightarrow p_{\langle\mathscr{A}, A, \mathscr{B}\rangle}(C) = 1$$

where it is assumed that $\forall_{i \in \{1, \cdots, m\}} \Lambda < A_i$.

$$(4) \quad \left.\begin{array}{l} A_1 \to (A_2 \to (\cdots(A_m \to A)\cdots)) < V \\ A_1 \to (A_2 \to (\cdots A_m \to \neg A)\cdots)) < V \end{array}\right\} \text{ in } p_{\langle\mathscr{A}\rangle}(A) \Rightarrow$$

$$(4.1) \quad p_{\langle\mathscr{A}\rangle}(A \wedge B) \cdot p_{\langle\mathscr{A}, A\wedge B\rangle}(A_1') \cdots p_{\langle\mathscr{A}, A\wedge B, A_1', \cdots, A_{n-1}'\rangle}(A_n') \cdot$$
$$\cdot p_{\langle\mathscr{A}, A\wedge B, \mathscr{A}''\rangle}(C) = p_{\langle\mathscr{A}\rangle}(k(A, B)) \cdot p_{\langle\mathscr{A}, k(A,B)\rangle}(A) \cdot$$
$$\cdot p_{\langle\mathscr{A}, k(A, B), A\rangle}(B) \cdot p_{\langle\mathscr{A}, k(A,B), A, B\rangle}(A_1') \cdots$$
$$\cdot p_{\langle\mathscr{A}, k(A, B), A, B, A_1', \cdots A_{n-1}'\rangle}(A_n') \cdot p_{\langle\mathscr{A}, k(A,B), A, B, \mathscr{A}''\rangle}(C)$$

$$(4.2) \quad p_{\langle\mathscr{A}\rangle}(A \vee B) \cdot P_{\langle\mathscr{A}, A\vee B\rangle}(A_1') \cdots p_{\langle\mathscr{A}, A\vee B, A_1', \cdots, A_{n-1}'\rangle}(A_n') \cdot$$
$$\cdot p_{\langle\mathscr{A}, A, \vee B, \mathscr{A}''\rangle}(C) = p_{\langle\mathscr{A}\rangle}(\neg k(\neg A, \neg B)) \cdot p_{\langle\mathscr{A}, \neg k(\neg\mathscr{A}, \neg B)\rangle}(A_1') \cdot$$
$$\cdots p_{\langle\mathscr{A}, \neg k(\neg A, \neg B), \mathscr{A}''\rangle}(C) + p_{\langle\mathscr{A}\rangle}(k(\neg A, \neg B)) \cdot$$
$$\cdot p_{\langle\mathscr{A}, k(\neg A, \neg B)\rangle}(A) \cdot p_{\langle\mathscr{A}, k(\neg A, \neg B), A\rangle}(A_1') \cdots$$
$$\cdot p_{\langle\mathscr{A}, k(\neg A, \neg B), A, \mathscr{A}''\rangle}(C) + p_{\langle\mathscr{A}\rangle}(k(\neg A, \neg B)) \cdot$$
$$\cdot p_{\langle\mathscr{A}, k(\neg A, \neg B)\rangle}(\neg A) \cdots p_{\langle\mathscr{A}, k(\neg A, \neg B), \neg A, B, \mathscr{A}''\rangle}(C)$$

$$(4.3) \quad p_{\langle \mathscr{A} \rangle}(A \to B) \cdot p_{\langle \mathscr{A}, A \to B \rangle}(A_1') \cdots p_{\langle \mathscr{A}, A \to B, \mathscr{A}' \rangle}(C) =$$
$$p_{\langle \mathscr{A} \rangle}(\neg A) \cdot p_{\langle \mathscr{A}, \neg A \rangle}(A_1') \cdots p_{\langle \mathscr{A}, \neg A, \mathscr{A}' \rangle}(C) +$$
$$p_{\langle \mathscr{A} \rangle}(A) \cdot p_{\langle \mathscr{A}, A \rangle}(k(A, B)) \cdot p_{\langle \mathscr{A}, A, k(A, B) \rangle}(B) \cdot$$
$$\cdot p_{\langle \mathscr{A}, A, k(A, B), B \rangle}(A_1') \cdots p_{\langle \mathscr{A}, A, k(A, B), B, \mathscr{A}' \rangle}(C)$$

$$(4.4) \quad p_{\langle \mathscr{A}, \neg(A \wedge B), \mathscr{A}' \rangle}(C) = p_{\langle \mathscr{A}, \neg A \vee \neg B, \mathscr{A}' \rangle}(C)$$

$$(4.5) \quad p_{\langle \mathscr{A}, \neg(A \vee B), \mathscr{A}' \rangle}(C) = p_{\langle \mathscr{A}, \neg A \wedge \neg B, \mathscr{A}' \rangle}(C)$$

$$(4.6) \quad p_{\langle \mathscr{A}, \neg(A \to B), \mathscr{A}' \rangle}(C) = p_{\langle \mathscr{A}, A \wedge \neg(A \wedge B), \mathscr{A}' \rangle}(C)$$

$$(4.7) \quad p_{\langle \mathscr{A}, \neg(\neg A), \mathscr{A}' \rangle}(C) = p_{\langle \mathscr{A}, A, \mathscr{A}' \rangle}(C)$$

where $\mathscr{A}$ and $\mathscr{A}'$ in (4) are sequences which are not necessarily material propositions and negations of material propositions. $\mathscr{A}'$ stands for the sequence $A_1', \ldots, A_n'$.

Remarks about the Calculus:

The above calculus represents a twofold recursive scheme, one with respect to the propositions A in $p_{\langle \mathscr{A} \rangle}(A)$ (given by means of the rules (1) and (2)), the other with respect to the propositions in the sequence $\mathscr{A}$ (given by (3) and (4)). As easily can be seen by induction on the length of compound propositions every probability $p_{\langle \mathscr{A} \rangle}(A)$ can be established by a finite arithmetical formula consisting of probabilities $p_{\langle \mathscr{A} \rangle}(a)$ the existence of which is assumed by means of the above hypothesis. The sequence $\mathscr{A}$ in $p_{\langle \mathscr{A} \rangle}(a)$ contains only material propositions and negations of material propositions. Therefore every dialog-definite proposition A is also probability-definite which proves the above theorem.

The validity of the probability calculus can be shown by means of the material dialog-game with addition of the argument-rule A_p.

To (1): If the proposition A is formally true in the position $_{\langle \mathscr{A} \rangle} | A$ i.e. $V \leq A_1 \to (A_2 \to (\cdots(A_m \to A)\cdots))$, O is not allowed to attack A. Since, in this case, he cannot continue the dialog he loses and P wins every dialog starting at this position. Therefore the probability is equal to one.

If the proposition A is formally false P is not allowed to defend against O's attack. Therefore he loses every dialog starting at the position $_{\langle \mathscr{A} \rangle} | A$ and the probability is equal to zero.

On the other hand, if the proponent wins every dialog, starting at this position, then the position is a position of success in the formal dialog-game. By means of the completeness of the q-implicative lattice with respect to the formal dialog-game (see CQL and QLC) the relation $V \leq A_1 \to (A_2 \to (\cdots(A_m \to A)\cdots))$ follows.

If P loses every dialog, starting at this position, he has a strategy of success in a formal dialog, starting at the position $_{\langle\mathscr{A}\rangle}|A$. Therefore the relation $V \leq A_1 \rightarrow (A_2 \rightarrow (\cdots(A_m \rightarrow \neg A)\cdots))$ holds.

To (2.1): In a dialog, starting at the position $_{\langle\mathscr{A}\rangle} | A \wedge B$, O has to attack the proposition $A \wedge B$. The first attack consists in a challenge to demonstrate the commensurability of A and B. If P is able to prove the commensurability $k(A, B)$ the opponent must continue by asking 1? whereupon P has to state A. For the continuation of the dialog it is irrelevant if O attacks A or if he attacks the conjunction again by 2?. Since A and B are proved to be commensurable the respective moves can be interchanged in the dialog. Therefore, we assume that at first a dialog about A is performed. The probability that P succeeds in defending $k(A, B)$ and, thereafter, A is the product

$$p_{\langle\mathscr{A}\rangle}(k(A, B)) \cdot p_{\langle\mathscr{A}, k(A, B)\rangle}(A).$$

If $k(A, B)$ is formally false, i.e. $V \leq A_1 \rightarrow (A_2 \rightarrow (\cdots(A_m \rightarrow \neg k(A, B))\ldots))$ the probability $p_{\langle\mathscr{A}\rangle}(k(A, B))$ is zero and the second factor of the above product does not exist. Since the probability $p_{\langle\mathscr{A}, k(A, B)\rangle}(A)$ always stands in a product with $p_{\langle\mathscr{A}\rangle}(k(A, B))$ we define arbitrarily $p_{\langle\mathscr{A}, k(A, B)\rangle}(A) := 1$ if $V \leq A_1 \rightarrow (A_2 \rightarrow (\cdots(A_m \rightarrow \neg k(A, B))\cdots))$ Generally we have the rule (3.2).

If P succeeds in demonstrating A the opponent has to use his third and last opportunity to attack the conjunction, namely by 2?, whereupon P has to demonstrate B. Thus we see that the probability for P to succeed in a dialog, starting at the position $_{\langle\mathscr{A}\rangle} | A \wedge B$ is given by the product:

$$p_{\langle\mathscr{A}\rangle}(k(A, B)) \cdot p_{\langle\mathscr{A}, k(A, B)\rangle}(A) \cdot p_{\langle\mathscr{A}, k(A, B), A\rangle}(B).$$

To (2.2): In a dialog, starting at the position $_{\langle\mathscr{A}\rangle}|A \vee B$, the opponent must attack the disjunction and, thereafter, P must defend by stating $\neg k(\neg A, \neg B)$ at first. If P succeeds he wins the dialog. On the other hand, if P loses the dialog about $\neg k(\neg A, \neg B)$, i.e. O proves $k(\neg A, \neg B)$, P must defend the disjunction again, this time by stating A. If P wins the dialog about A the opponent has no further argument and P wins the dialog about $A \vee B$. If P loses the dialog he has to use his last opportunity to defend the disjunction by stating B. If P succeeds in defending B against O's attack he wins the dialog about $A \vee B$. Thus we see that the total probability for P to win a dialog, starting at the position $_{\langle\mathscr{A}\rangle} | A \vee B$, is given by the following sum:

$$p_{\langle \mathscr{A} \rangle}(\neg k(\neg A, \neg B)) + p_{\langle \mathscr{A} \rangle}(k(\neg A, \neg B)) \cdot p_{\langle \mathscr{A}, k(\neg A, \neg B) \rangle}(A) +$$
$$p_{\langle \mathscr{A} \rangle}(k(\neg A, \neg B)) \cdot p_{\langle \mathscr{A}, k(\neg A, \neg B) \rangle}(\neg A) \cdot p_{\langle \mathscr{A}, k(\neg A, \neg B), \neg A \rangle}(B).$$

To (2.3): In a dialog, starting at the position $_{\langle \mathscr{A} \rangle} | A \twoheadrightarrow B$, O has to attack the material implication by stating A. If P succeeds in refuting O's attack A he wins the dialog about $A \twoheadrightarrow B$. But in this case he also wins the dialog about the negation $\neg A$. If P does not succeed in refuting O's attack A he must defend by asserting B. Thereupon O must attack by challenging the commensurability proof of A and B. If P succeeds in demonstrating $k(A, B)$ the opponent must continue the dialog by an attack against P's proposition B. Therefore the probability for P to win a dialog, starting at the position $_{\langle \mathscr{A} \rangle} | A \twoheadrightarrow B$, is given by the sum:

$$p_{\langle \mathscr{A} \rangle}(\neg A) + p_{\langle \mathscr{A} \rangle}(A) \cdot p_{\langle \mathscr{A}, A \rangle}(k(A, B)) \cdot p_{\langle \mathscr{A}, A, k(A, B) \rangle}(B).$$

To (2.4): In a position $_{\langle \mathscr{A} \rangle} | \neg A$ the opponent must attack by asserting A. Therefore, the probability for P to win a dialog, starting at this position, is equal to the probability to win a dialog, starting at the position $_{\langle \mathscr{A} \rangle} A |$. But, since the value-definiteness of A is assumed, P wins this dialog if and only if O wins the dialog, starting at the position $_{\langle \mathscr{A} \rangle} | A$. Therefore, we have:

$$p_{\langle \mathscr{A} \rangle}(\neg A) = 1 - p_{\langle \mathscr{A} \rangle}(A).$$

To (3) and (4): It is because of the simplicity of the particular kind of positions $_{\langle \mathscr{A} \rangle} | C$ for which the probability is established that an additional recursive scheme is required with respect to the propositions A occurring in the sequence $\mathscr{A}$.

If we have a position $_{\langle \mathscr{A} \rangle} | C$ in a material dialog a probability can be defined for a successful proof of A and, thereafter, of C. We call this probability a *sequential* probability and write for the sequential probability $p_{\langle \mathscr{A} \rangle}(A, C)$. For this sequential probability the following equation holds:

$$p_{\langle \mathscr{A} \rangle}(A, C) = p_{\langle \mathscr{A} \rangle}(A) \cdot p_{\langle \mathscr{A}, A \rangle}(C).$$

In the general case of more than one attackable probositions in P's column:

$$\begin{vmatrix} A \\ A_1' \\ \vdots \\ A_n' \\ C \end{vmatrix}$$

the sequential probability is

$p_{\langle \mathscr{A} \rangle}(A, \mathscr{B}, C)$ where $\mathscr{B}$ denotes the sequence of propositions $A'_1,..., A'_1$. Then the equation

$$(*) \quad p_{\langle \mathscr{A} \rangle}(A, \mathscr{B}, C) = p_{\langle \mathscr{A} \rangle}(A) \cdot p_{\langle \mathscr{A}, A \rangle}(A'_1) \cdots$$
$$\cdot p_{\langle \mathscr{A}, A, A'_1, \cdots, A'_{n-1} \rangle}(A'_n) \cdot p_{\langle \mathscr{A}, A, \mathscr{B} \rangle}(C)$$

holds.

It can easily be shown that the probability rules (2.1) to (2.3) can be applied with respect to A in the sequential probability $p_{\langle \mathscr{A} \rangle}(A, \mathscr{B}, C)$. This leads to the rules (4.1) to (4.3).

If we generalized the recursive scheme, given by means of the rules (1) and (2) of the above calculus, to sequential probabilities the rules (3) and (4) of the calculus would become redundant because of the equation (*). Instead of this we can maintain the probabilities of the simple form $P_{\langle \mathscr{A} \rangle}(C)$. But then we have to introduce a recursive scheme with respect to the propositions A in the sequence $\mathscr{A}$.

To (4.4)–(4.7): These rules follow from the dialog-equivalence of the respective propositions. Two propositions are defined to be dialog-equivalent if and only if the possibilities to attack and to defend are the same in the subsequent dialog when one proposition is replaced by the other.

To prove the dialog-equivalence of $\neg(A \wedge B)$ and $\neg A \vee \neg B$ which leads to the rule (4.4) we have to compare the possibilities in the respective dialogs:

[ ]	$\neg(A \wedge B)$		[ ]	$\neg A \vee \neg B$
$A \wedge B$	[ ]	$\longleftrightarrow$	?	$\neg k(\neg A. \ \neg B)$
$k(A, B)!$	$k(A, B)?$		$k(\neg A, \neg B)$	[ ]
			$k(\neg A, \neg B)!$	$k(\neg A, \wedge B)?$

(Note that, in the case of the value-definiteness of A and B, $k(A, B)$ is proven if and only if $k(\neg A, \neg B)$ is proven.)

[ ]	$\neg(A \wedge B)$		[ ]	$\neg A \vee \neg B$
$A \wedge B$	[ ]	$\longleftrightarrow$	?	$\neg k(\neg A, \neg B)$
$k(A, B)!$	$k(A, B)?$		$k(\neg A, \neg B)$	[ ]
A	1?		$k(\neg A, \neg B)!$	$k(\neg A, \neg B)?$
			[ ]	$\neg A$
			A	

[ ]	$\neg(A \wedge B)$		[ ]	$\neg A \vee \neg B$
$A \wedge B$	[ ]	$\longleftrightarrow$	?	$\neg k(\neg A, \neg B)$
$k(A, B)!$	$k(A, B)?$		$k(\neg A, \neg B)$	[ ]
$A!$	1?		$k(\neg A, \neg B)!$	$k(\neg A, \neg B)?$
B	2?		[ ]	$\neg A$
			$A!$	[ ]
			[ ]	$\neg B$
			B	

In the last two dialog schemes $A!$ is an abbreviation for a successful dialog about A.

By means of the above correspondences we see that the possibilities to attack and to defend are the same in the dialogs about $\neg(A \wedge B)$ and $\neg A \vee \neg B$. Therefore, both propositions are dialog-equivalent.

The dialog-equivalences of $\neg(A \vee B)$ and $\neg A \wedge \neg B$, $\neg(A \rightarrow B)$ and $A \wedge \neg(A \wedge B)$, $\neg(\neg A)$ and A can be shown in an analogous way.

The rules of the probability calculus are not independent from each other since the four logical connectives are not independent if the value-definiteness of quantum mechanical propositions is assumed.

The following properties hold:

$$(1) \quad A \leq B \Rightarrow p_{\langle \mathscr{A} \rangle}(A) \leq p_{\langle \mathscr{A} \rangle}(B)$$

where '$\leq$' on the left side denotes the lattice order relation and '$\leq$' on the right side denotes the order relation between the real numbers.
The *proof* is the following:

$$A \leq B \Rightarrow V \leq k(A, B) \Rightarrow p_{\langle \mathscr{A} \rangle}(A \wedge B) = p_{\langle \mathscr{A} \rangle}(A) \cdot p_{\langle \mathscr{A}, A \rangle}(B)$$
$$= p_{\langle \mathscr{A} \rangle}(B) \cdot p_{\langle \mathscr{A}, B \rangle}(A)$$
$$p_{\langle \mathscr{A} \rangle}(\neg(\neg A \wedge B)) = 1 - p_{\langle \mathscr{A} \rangle}(\neg A \wedge B) = 1 - p_{\langle \mathscr{A} \rangle}(B) \cdot p_{\langle \mathscr{A}, B \rangle}(A)$$
$$= 1 - p_{\langle \mathscr{A} \rangle}(B) + p_{\langle \mathscr{A} \rangle}(B) \cdot p_{\langle \mathscr{A}, B \rangle}(A)$$
$$= 1 - p_{\langle \mathscr{A} \rangle}(B) + p_{\langle \mathscr{A} \rangle}(A) \cdot p_{\langle \mathscr{A}, A \rangle}(B)$$
$$\text{since } A \leq B \Rightarrow p_{\langle \mathscr{A}, A \rangle}(B)$$
$$= 1 + p_{\langle \mathscr{A} \rangle}(A) - p_{\langle \mathscr{A} \rangle}(B)$$
$$\Rightarrow p_{\langle \mathscr{A} \rangle}(A) \leq p_{\langle \mathscr{A} \rangle}(B).$$

$$(2) \quad (\forall_{\mathscr{A}} \, p_{\langle \mathscr{A} \rangle}(A) = p_{\langle \mathscr{A} \rangle}(B)) \Rightarrow A = B$$

where $A = B$ denotes $A \leq B$ and $B \leq A$.
Proof: For $\mathscr{A}$ we first take A. Then, it follows:

$$1 = p_{\langle A \rangle}(A) = p_{\langle A \rangle}(B) \Rightarrow V \leq A \rightarrow B \Rightarrow A \leq B.$$

Secondly we take B:

$$1 = p_{\langle B \rangle}(B) = p_{\langle B \rangle}(A) \Rightarrow V \leq B \rightarrow A \Rightarrow B \leq A.$$

Therefore $A = B$.

IV. APPLICATION TO QUANTUM THEORY

In the following we show that the probability formulas of quantum theory can be interpreted as quantum probabilities established in the above calculus. For this application we first introduce two notions: the notion of a physical observable and the notion of a state of a physical system. If the system is prepared in a certain state $|\alpha\rangle$ a probability function exists which satisfies the probabilities $p_{\langle\alpha\rangle}(A)$ in the calculus. This function is given in quantum theory by $\langle\alpha|P_A|\alpha\rangle$ which is the expectation value of the projection operator P_A with respect to the state $|\alpha\rangle$. In this paper we restrict ourselves to the interpretation of the expectation values with respect to the pure states $|\alpha\rangle$ although the calculus of quantum probabilities allows an interpretation with respect to arbitrary states (statistical states).

a) *The Notion of a Physical Observable*

Physical elementary propositions $e \in S_e$ are defined by measuring processes. In praxis not all measuring processes with respect to elementary propositions are independent of each other. There are measurements which allow a proof not only for one elementary proposition but for a certain set $O \subseteq S_e$ of elementary propositions. The propositions of O then can be interpreted as different outcomes of a single measurement. As an example consider different regions of a photographic plate as outcomes of a position measurement with respect to a photon. This experimental relationship between elementary propositions leads to an interpretation of an observable of a physical system:

An observable O is a set of elementary propositions which is given by

$$ O = \{e_i|\ i \in I,\ e_i \in S_e,\ \underset{i \neq j \in I}{\forall}\ e_i \leq \neg\ e_j,\ \underset{i \in I}{\bigvee} e_i = V\} $$

(I may be an infinite index set since the propositional logic can be extended to a logic of quantifiers. In this case $\bigvee_{i \in I} e_i$ is an existential quantifier.) The following two postulates are justified by means of the praxis of quantum mechanical measurements:

POSTULATE 1.: S_e *is a disjoint union of observables*, i.e. $S_e = \bigcup_i O_i$.

POSTULATE 2. :

$$ \left\{ \underset{\substack{i \in I_1,\ j \in I_2 \\ e_i^1 e_j^2 \neq \Lambda, V}}{\exists}\ k(e_i^1, e_j^2) \right\} \Rightarrow \left\{ \underset{k \in I_1,\ m \in I_2}{\forall}\ k(e_k^1, e_m^2) \right\} $$

(where the indices 1 and 2 refer to the observables O_1 and O_2).
By means of the postulate 2 the notion of commensurability can be transferred from propositions to observables. Two observables O_1 and O_2

are said to be commensurable if an arbitrary series of measurements of O_1 and O_2 always yields the same results. This is the case if and only if the premise of postulate 2. holds.

Since elementary propositions e_i of an observable are interpreted as different outcomes of a single measurement every disjunction $\bigvee_i e_i$ is formally equivalent to an elementary propositions $\check{e}$.

Furthermore, by means of the definition of an observable, we have the following formal equivalence:

$$\neg e_i = \bigvee_{i \neq j,\, j \in I} e_j.$$

As pointed out in I.a) this equivalence in general does not imply the dialog-equivalence of $\neg e_i$ and $\bigvee_{i \neq j,\, j \in I} e_j$. For instance assume that an observable (take the momentum of a particle) can be presented by a three-slit experiment:

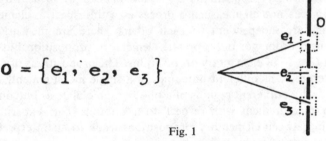

Fig. 1

Besides the propositions e_i which read: "The particle is registered at the slit i" we have the elementary propositions $\check{e}_{1,2,3} = e_1 \vee e_2 \vee e_3 = V$. $\check{e}_{1,2} = \neg e_3$, $\check{e}_{1,3} = \neg e_2$ and $\check{e}_{2,3} = \neg e_1$. For instance $\check{e}_{1,2}$ corresponds to a dectector with respect to both slits 1 and 2:

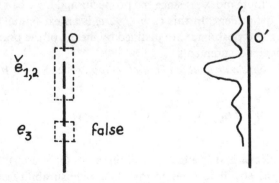

Fig. 2

and it follows that $\neg e_3$ and $\check{e}_{1,2}$ are dialog-equivalent. However, $\check{e}_{1,2}$ and $e_1 \vee e_2$ are not dialog-equivalent. $e_1 \vee e_2$ has to be proved in a dialog by means of a measurement of either e_1 or e_2 which corresponds to two detectors at each of the two slits 1 and 2. That indeed a measurement of $\check{e}_{1,2}$ and a proof of $e_1 \vee e_2$ may have different influences on a further dialog is a well-known fact. Namely, if a further observable O' is measured which is incommensurable with O (take the position of a particle) we observe a different probability distribution for the outcomes of the O'-measurements in both cases:

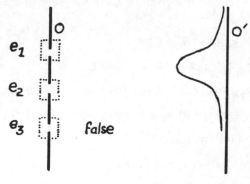

Fig. 3

Since an observable is a set of propositions and describes possible outcomes of a measurement we can represent an observable formally by a homomorphic mapping from the propositions of O onto the scale of possible outcomes of the measuring apparatus. The reverse mapping is sometimes called a random variable. For this representation see for instance [4].

Since the $e_i \in O$ are pairwise commensurable the logical structure of an observable is a Boolean lattice with atoms e_i.

b) *Material Knowledge and Dialogical Knowledge about a System*

If a proposition A is not formally true or false the material dialog about A always leads to a final (reduced) position $\mathfrak{A}!|\mathfrak{A}'!$ where $\mathfrak{A}$ and $\mathfrak{A}'$ are systems of material (elementary and commensurability propositions and negations of those propositions). We write these systems also as the sequence

$\mathscr{A}$ which represents the succession of the propositions in the dialog. If the propositions are proven in the sequence $\mathscr{A}$ we call this knowledge a *material knowledge* about the system under consideration. In this sense a material knowledge is established by a *sequence* of proofs of material propositions. Since in general the propositions of $\mathscr{A}$ are not pairwise commensurable the sequence cannot be replaced by the conjunction of the respective propositions as this is possible in the case of unrestrictedly available propositions. However, it can be shown that the sequence $\mathscr{A}$ can be replaced by a disjunction of conjuctions of material propositions and negations of those propositions. To prove this we first consider the simple case of a sequence A, B of two propositions. The maximal knowledge after the proof of A and B is given by the knowledge of the proposition C which satisfies:

1. $V \leq A \rightarrow (B \rightarrow C)$

2. $\forall_{C'} \{V \leq A \rightarrow (B \rightarrow C') \Rightarrow C \leq C'\}$.

LEMMA: *The proposition C which satisfies* 1. *and* 2. *is unique and is*

$$C = (A \vee \neg B) \wedge B = (B \wedge A) \vee (B \wedge \neg k(A, B)).$$

Proof: From 1. it follows that

$$(A \vee \neg B) \wedge B \leq ((B \rightarrow C) \vee \neg B) \wedge B \leq C.$$

Since $A \leq A \vee \neg B \leq B \rightarrow (A \vee \neg B) \leq B \rightarrow (B \wedge (\neg B \vee A))$ the proposition $B \wedge (\neg B \vee A)$ satisfies 1. which proves the lemma.

$$B \wedge (\neg B \vee A) = \{(B \wedge (\neg B \vee A)) \wedge (\neg B \vee \neg A)\} \vee (B \wedge A)$$

$$= \{B \wedge ((B \vee A) \wedge (B \vee \neg A) \wedge (\neg B \vee A) \wedge$$

$$(\neg B \vee \neg A))\} \vee (B \wedge A)$$

$$= (B \wedge \neg k(B, A)) \vee (B \wedge A).$$

Furthermore, it can be easily shown that the knowledge after A, B is also dialog-equivalent to the proposition $(B \wedge A) \vee (B \wedge \neg k(A, B))$:

			$(B \wedge A) \vee$ $(B \wedge \neg k(A, B))$
	A	?	$B \wedge A$
A?	A!	$k(B, A)$?	$k(B, A)$!
	B	2?	A
B?	B!	A?	A!
$k(B, A)$?	$k(B, A)$!	1?	B
		B?	B!

			$(B \wedge A) \vee (B \wedge \neg k(A, B))$
	A	?	$B \wedge \neg k(A, B)$
$A?$	$A!$	$1?$	B
	B	$B?$	$B!$
$B?$	$B!$	$2?$	$\neg k(A, B)$
$k(A, B)?$	$\neg k(A, B)!$	$k(A, B)$	$[\ ]$
		$k(A, B)!$	$\neg k(A, B)?$

Note that because of the incommensurability proof in the left dialog knowledge of the A is lost.

Thus the material knowledge after the sequential proof of A and B can be represented by a disjunction of two conjuctions. This result can be generalized to any finite series of material propositions in a dialog.

THEOREM: *Every proposition $A \in \mathscr{L}_q$ is formally equivalent to a finite disjunction $\bigvee \hat{A}$ of propositions $\hat{A}$ which are conjuctions of material propositions and negations of those propositions*, i.e.:

$$\underset{A \in \mathscr{L}_q}{\forall} \underset{\bigvee \hat{A}}{\exists} A = \bigvee \hat{A}.$$

Proof: The proof is performed by means of an induction on the length of the proposition A:

a) If A is a material proposition or a negation of a material proposition the theorem is trivially satisfied.

b) Assume that the theorem is proven with respect to the propositions A and B. We show that under this assumption the theorem is satisfied for $A \vee B, A \wedge B$ and $\neg A$.

1. $A \vee B$ follows immediately.
2. The proof for $A \wedge B$ is obtained in the following way:

By assumption we have $B = \bigvee \hat{B} = \hat{B}_1 \vee \bigvee \hat{B}'$.

Since $V \leq k(A \wedge k(A, B), B)$ we have by means of the above lemma:

$$
\begin{aligned}
A \wedge (\hat{B}_1 \vee \bigvee \hat{B}') &= A \wedge k(A, \hat{B}_1 \vee \bigvee \hat{B}') \wedge \{\neg A \vee \neg k(A, \hat{B}_1 \wedge \\
&\quad \bigvee \hat{B}') \vee \hat{B}_1 \vee \bigvee \hat{B}'\} \\
&= \{A \wedge k(A, \hat{B}_1 \vee \bigvee \hat{B}') \wedge (\neg A \vee \neg k(A, \hat{B}_1 \vee \\
&\quad \bigvee \hat{B}') \vee \hat{B}_1)\} \\
&\quad \vee \{A \wedge k(A, \hat{B}_1 \vee \bigvee \hat{B}') \wedge (\neg A \vee \neg k(A, \\
&\quad \hat{B}_1 \vee \bigvee \hat{B}') \vee \bigvee \hat{B}')\} \\
&= \{A \wedge k(A, \hat{B}_1 \vee \bigvee \hat{B}') \wedge \hat{B}_1\} \\
&\quad \vee \{A \wedge k(A, \hat{B}_1 \vee \bigvee \hat{B}') \wedge \bigvee \hat{B}'\} \\
&\quad \vee \{A \wedge k(A, \hat{B}_1 \vee \bigvee \hat{B}') \wedge \neg k(A \wedge k(A, \\
&\quad \hat{B}_1 \vee \bigvee \hat{B}'), \hat{B}_1)\} \\
&\quad \vee \{A \wedge k(A, \hat{B}_1 \vee \bigvee \hat{B}') \wedge \neg k(A \wedge k(A, \\
&\quad \hat{B}_1 \vee \bigvee \hat{B}'), \bigvee \hat{B}')\}
\end{aligned}
$$

$$= \{A \wedge \hat{B}_1\} \vee \{A \wedge \bigvee \hat{B}'\}$$
$$\vee \{A \wedge k(A, \hat{B}_1 \vee \bigvee \hat{B}') \wedge \neg k(A \wedge k(A,$$
$$\hat{B}_1 \vee \bigvee \hat{B}'), \hat{B}_1)\}$$
$$\vee \{A \wedge k(A, \hat{B}_1 \vee \bigvee \hat{B}') \wedge \neg k(A \wedge k(A,$$
$$\hat{B}_1 \vee \bigvee \hat{B}'), \bigvee \hat{B}')\}$$

By means of the iterative application of the above equivalence with respect to $\bigvee \hat{B}'$ and A the theorem is proved for $A \wedge B$.

3. The proof for $\neg A$ reduces to the preceding cases by means of the formal equivalences:

$$\neg(A \wedge B) = \neg A \vee \neg B$$
$$\neg(A \vee B) = \neg A \wedge \neg B$$
$$\neg \neg A = A$$

c) *The Notion of a State of a Physical System*

The finiteness of the dialog-game implies that the material knowledge about a system is always finite. In addition we assume that a material knowledge about a physical system can always be maximized, i.e. POSTU-LATE: *For each material knowledge $C \neq \Lambda$ there exists a maximal material knowledge $\alpha \leq C, \alpha \neq \Lambda$.*

By means of the above theorem this postulate directs itself to the propositions $\hat{A}$ which are conjunctions of material propositions and negations of those propositions. Hence we have the following lemma:

LEMMA:

$$\underset{\substack{\hat{A} \in \mathscr{L}_q \\ A > \Lambda}}{\forall} \underset{\alpha > \Lambda}{\exists} \{\alpha \leq \hat{A} \barwedge \underset{B \leq \alpha}{\forall} (\hat{B} = \Lambda \veebar \hat{B} = \alpha\}$$

where '$\barwedge$' denotes the meta-linguistic "and" and '$\veebar$' the meta-linguistic "or".

This leads to an interpretation of a state of a physical system:

The proposition α which satisfies the above lemma corresponds to a state of the physical system in the sense that the system is said to be in the state $|\alpha\rangle$ if and only if the proposition α is proved.

The consequence for the formal quantum logic is the atomicity of the q-implicative lattice L_q:

Atomicity of L_q:

$$\underset{\substack{A \in \mathscr{L}_q \\ A > \Lambda}}{\forall} \underset{\alpha > \Lambda}{\exists} \{\alpha \leq A \barwedge \underset{B \leq \alpha}{\forall} (B = \Lambda \veebar B = \alpha)\}$$

where the atoms α correspond to the possible states of the physical system.

Proof: By means of the above theorem we have:

$$\bigvee_{\substack{A \in \mathscr{L}_q \\ A > \Lambda}} \exists_{\alpha > \Lambda} \; \alpha \le A.$$

The second part of the atomicity follows from:

$$B = \Lambda \Rightarrow B = \alpha \underline{\vee} B = \Lambda$$

$$B \ne \Lambda \Rightarrow \exists_{\hat{B} > \Lambda} \hat{B} \le B \le \alpha.$$

Since α is assumed to be maximal we have $\hat{B} = \alpha$ and thus $B = \alpha$.

COROLLARY TO THE ATOMICITY: *Every proposition $A \in \mathscr{L}_q$ is logically equivalent to the disjunction $\bigvee_{\alpha \le A} \alpha$ of all atoms of A, i.e.*

$$\bigvee_{\substack{A \in \mathscr{L}_q \\ A > \Lambda}} A = \bigvee_{\alpha \le A} \alpha.$$

Proof: We have $\bigvee_{\alpha \le A} \alpha \le A$. Therefore A is commensurable with $\bigvee_{\alpha \le A} \alpha$ and with $\neg \bigvee_{\alpha \le A} \alpha$ also. Hence the distributivity law can be applied:

$$A = A \wedge (\bigvee_{\alpha \le A} \alpha \vee \neg \bigvee_{\alpha \le A} \alpha) = (A \wedge \bigvee_{\alpha \le A} \alpha) \vee (A \wedge \neg \bigvee_{\alpha \wedge A} \alpha)$$

$$= \bigvee_{\alpha \le A} \alpha \vee (A \wedge \neg \bigvee_{\alpha \le A} \alpha).$$

If $A \wedge \neg \bigvee_{\alpha \le A} \alpha = \Lambda$ the assertion follows.

If $A \wedge \neg \bigvee_{\alpha \le A} \alpha \ne \Lambda$ it follows from the atomicity of L_q that there exists an atom β with $\beta \le A \wedge \neg \bigvee_{\alpha \le A} \alpha$. From this relation it follows that $\beta \le A$, i.e. that β is an atom of A, and $\beta \le \neg \bigvee_{\alpha \le A} \alpha$. Hence β is not contained in $\bigvee_{\alpha \le A} \alpha$ which leads to a contradiction since it is assumed that $\bigvee$ ranges over all atoms of A. Therefore $A \wedge \neg \bigvee_{\alpha \le A} \alpha = \Lambda$.

The propositions α correspond to the "pure" states of quantum theory. Since α is associated with an individual quantum mechanical system the pure states are interpreted as properties of individual systems. In this sense every individual system can be proven to be in a certain state $|\alpha\rangle$. The connection with the probabilistic interpretation of a state is established in the following.

d) *Connection with the Probabilistic Description of States*

Remember that we defined a preparation of a system as a dialog (or, generally, as a series of dialogs) which is previous to a proposition under consideration. By a preparation of a system in the state $|\alpha\rangle$ we then mean a previous dialog in which the state proposition α has been proven.

In the probabilistic description (see for instance [5]) the state of a quantum mechanical system is defined as a real-valued function $p(A)$ on all elements A of the σ-complete, orthomodular lattice of subspaces of a Hilbert space. $p(A)$ is determined by the following properties:

1. $0 \leq p(A) \leq 1$

2. $p(\Lambda) = 0$; $p(V) = 1$

3. If $\{A_i\}$ is a sequence of elements which satisfy $A_i \leq \neg A_j$ $(i \neq j)$ then $p(\bigvee_i A_i) = \sum_i p(A_i)$.

4. For any sequence $\{A_i\}$
   $\{\forall_i\, p(A_i) = 1\} \Rightarrow p(\bigwedge_i A_i) = 1$ holds.

5. a. $A \neq \Lambda \Rightarrow \exists_p\, p(A) \neq 0$
   b. $A \neq B \Rightarrow \exists_p\, p(A) \neq p(B)$

It can easily be seen that the quantum probabilities $p_{\langle\alpha\rangle}(A)$ with respect to the state proposition α satisfy the above properties in case we take only finite sequences $\{A_i\}$ into account.
1. follows from the definition of $p_{\langle\alpha\rangle}(A)$,
2. from the rules (1.1) and (1.2) of the probability calculus.
3. The assumption $A_i \leq \neg A_j$ implies $V = k\,(A_i, A_j)$. From rule (1.1) it follows that

$$P_{\langle\alpha\rangle}(k(A_i, A_j)) \cdot p_{\langle\alpha, k(A_i, A_j)\rangle}(\neg A_j) \cdot p_{\langle\alpha, k(A_i, A_j)\neg A_j\rangle}(A_i)$$
$$= p_{\langle\alpha\rangle}(k(A_i, A_j)) \cdot p_{\langle\alpha, k(A_i, A_j)\rangle}(A_i) \cdot p_{\langle\alpha, k(A_i, A_j)A_i\rangle}(\neg A_j).$$

Therefore we have with $V = k(A_i, A_j)$ and $A_i \leq \neg A_j$ by means of the rule (1.2)

$$p_{\langle\alpha\rangle}(\bigvee_{i-} A_i) = \sum_i p_{\langle\alpha\rangle}(A_i).$$

4. follows immediately from (1.1).
5. a. follows from the atomicity of L_q:
   $A \neq \Lambda \Rightarrow \exists_\alpha\, \alpha \leq A \Rightarrow p_{\langle\alpha\rangle}(A) = 1$.
5. b. follows from the above theorem:
   Assume that $\forall_\alpha\, p_{\langle\alpha\rangle}(A) = p_{\langle\alpha\rangle}(B)$. If we consider all states α which

satisfy $A = \bigvee \alpha$ we have $p_{\langle \alpha \rangle}(A) = p_{\langle \alpha \rangle}(B) = 1$ and therefore $A = \bigvee \alpha \leq B$.

If, on the other hand, we consider all states β which satisfy $B = \bigvee \beta$ we have $p_{\langle \beta \rangle}(B) = p_{\langle \beta \rangle}(A) = 1$ and therefore $B = \bigvee \beta \leq A$ This proves 5.b.

e) *Quantum Theoretical Expectation Values*

In QCL we showed that quantum logic can be regarded as an interpretation of the orthocomplemented quasi-modular lattice of subspaces of a Hilbert space. Quantum mechanical propositions can be represented by closed linear manifolds M_A, M_B, ... of the Hilbert space or, equivalently, by the corresponding projections P_A, P_B, ...

The lattice operations and the order relation with respect to the subspaces can be expressed in terms of algebraic properties of the corresponding projections.

For $P_{A \wedge B}$ which is the projection on the intersection $M_A \cap M_B$ of the subspaces M_A and M_B we have:

$$1. \quad P_{A \wedge B} = \lim_{n \to \infty} (P_A \cdot P_B)^n.$$

For $P_{\neg A}$ which is the projection on the orthogonal space $M_A^{\perp}$ of M_A we have:

$$2. \quad P_{\neg A} = \hat{1} - P_A \text{ where } \hat{1} \text{ is the unit operator.}$$

Since $P_{A \vee B} = P_{\neg(\neg A \wedge \neg B)} = \hat{1} - P_{\neg A \wedge \neg B}$ it follows

$$3. \quad P_{A \vee B} = \hat{1} - \lim_{n \to \infty} ((\hat{1} - P_A) \cdot (\hat{1} - P_B))^n.$$

The projection which corresponds to the commensurability proposition $k(A, B)$ is given by:

$$4. \quad P_{k(A, B)} = \lim_{n \to \infty} (P_A \cdot P_B)^n + \lim_{n \to \infty} (P_A \cdot (\hat{1} - P_B))^n$$
$$+ \lim_{n \to \infty} ((\hat{1} - P_A) \cdot P_B)^n + \lim_{n \to \infty} (\hat{1} - P_A) \cdot (\hat{1} - P_B))^n$$

which follows from the formal equivalence

$$k(A, B) = (A \wedge B) \vee (A \wedge \neg B) \vee (\neg A \wedge B) \vee (\neg A \wedge \neg B).$$

If for two elements A and B the order relation $A \leq B$ holds, i.e. the inclusion $M_A \subseteq M_B$ is valid for the corresponding subspaces, we have:

5. $P_A \cdot P_B = P_A$ and vice versa.

The quantum theoretical expectation values for projections P_A with respect to a system in a state $|\alpha\rangle$ are real-valued functions $\langle\alpha|P_A|\alpha\rangle$ which satisfy the above probabilistic description of quantum mechanical states. The functions of the form $\langle\alpha|P_A P_B P_A|\alpha\rangle$ describe sequential expectation values, i.e. that the projection $P_A|\alpha\rangle$ is in M_A and the subsequent projection $P_B P_A|\alpha\rangle$ is in M_B. In the following we show that the functions $\langle\alpha|$ $P_A P_B P_A|\alpha\rangle$ can be interpreted as sequential probabilities defined by means of the calculus of quantum probabilities. We assume here that the ensemble under consideration is prepared in a pure state $|\alpha\rangle$. Since the calculus is no longer used as a recursive scheme for establishing the probability-definiteness of propositions but only its mathematical determination is important we obtain the following simplified representation:

(1) $0 \le p_{\langle\alpha\rangle}(\mathscr{A}, A, \mathscr{A}') \le 1$

(2) $V \le \alpha \to A \Rightarrow p_{\langle\alpha\rangle}(A) = 1$

(3) $V \le \alpha \to (A_1 \to (\cdots(A_m \to A)\cdots)) \Rightarrow$
$p_{\langle\alpha\rangle}(\mathscr{A}, A, \mathscr{A}') = p_{\langle\alpha\rangle}(\mathscr{A}, \mathscr{A}')$
$V \le \alpha \to (A_1 \to (\cdots(A_m \to \neg A)\cdots)) \Rightarrow$
$p_{\langle\alpha\rangle}(\mathscr{A}, A, \mathscr{A}') = 0$

where by assumption $\alpha > \Lambda$.

(4) $\left.\begin{array}{l} V > \alpha \to (A_1 \to (\cdots(A_m \to A)\cdots)) \\ V > \alpha \to (A_1 \to (\cdots(A_m \to \neg A)\cdots)) \end{array}\right\}$ in $p_{\langle\alpha\rangle}(A) \Rightarrow$

(4.1) $p_{\langle\alpha\rangle}(\mathscr{A}, A \wedge B, \mathscr{A}') = p_{\langle\alpha\rangle}(\mathscr{A}, k(A, B) A, B, \mathscr{A}')$

(4.2) $p_{\langle\alpha\rangle}(\mathscr{A}, A \vee B, \mathscr{A}') = p_{\langle\alpha\rangle}(\mathscr{A}, \neg k(A, B), \mathscr{A}')$
$+ p_{\langle\alpha\rangle}(\mathscr{A}, k(A,B), A, \mathscr{A}') + p_{\langle\alpha\rangle}(\mathscr{A}, k(A,B), \neg A, B, \mathscr{A}')$

(4.3) $p_{\langle\alpha\rangle}(\mathscr{A}, \neg A) = p_{\langle\alpha\rangle}(\mathscr{A}) - p_{\langle\alpha\rangle}(\mathscr{A}, A)$

where $\mathscr{A}$ stands for the sequence $A_1, A_2, \ldots, A_m$.
The functions $\langle\alpha|P_{A_1} \cdots P_{A_m} \cdot P_A \cdot P_{A_1'} \cdots P_{A_n'} \cdots P_{A_1'} \cdot P_A \cdot P_{A_m} \cdots P_{A_1}|\alpha\rangle$ which are abbreviated in the following by $\langle\alpha|P_{\mathscr{A}} \cdot P_A \cdot P_{\mathscr{A}'} \cdot P_{\mathscr{A}'*} \cdot P_A \cdot P_{\mathscr{A}*}|\alpha\rangle$ satisfy (1).
By means of the property 5. for the projections the rule (2) holds:

$$P_\alpha \cdot P_A = P_\alpha \Rightarrow \langle\alpha|P_A|\alpha\rangle = \langle\alpha|P_\alpha \cdot P_A|\alpha\rangle = \langle\alpha|P_\alpha|\alpha\rangle = 1.$$

Rule (3) is satisfied. We give a demonstration here for the special case:

$$V \leq \alpha \to (B \to A) \quad \Rightarrow \langle \alpha | P_B P_A P_B | \alpha \rangle = \langle \alpha | P_B | \alpha \rangle$$
$$V \leq \alpha \to (B \to \neg A) \Rightarrow \langle \alpha | P_B P_A P_B | \alpha \rangle = 0$$

Proof:

$$\alpha \leq B \to A \Rightarrow \alpha \leq \neg B \vee (A \wedge B) \Rightarrow P_\alpha = P_\alpha \cdot (P_{\neg B} + P_{A \wedge B})$$
$$\Rightarrow \langle \alpha | P_B P_A P_B | \alpha \rangle = \langle \alpha | P_\alpha \cdot P_B \cdot P_A \cdot P_B | \alpha \rangle = \langle \alpha | P_\alpha \cdot P_{A \wedge B} | \alpha \rangle$$
$$P_\alpha = P_\alpha \cdot (P_{\neg B} + P_{A \wedge B}) = P_\alpha - P_\alpha \cdot P_B + P_\alpha \cdot P_{A \wedge B}$$
$$\Rightarrow P_\alpha \cdot P_B = P_\alpha \cdot P_{A \wedge B} \Rightarrow \langle \alpha | P_\alpha \cdot P_{A \wedge B} | \alpha \rangle = \langle \alpha | P_B | \alpha \rangle.$$
$$\alpha \leq B \to \neg A \Rightarrow \alpha \leq \neg B \vee (\neg A \wedge \neg B)$$
$$\Rightarrow P_\alpha = P_\alpha \cdot P_{\neg B} + P_\alpha \cdot P_{\neg A \wedge B} \Rightarrow \langle \alpha | P_B \cdot P_A \cdot P_B | \alpha \rangle = 0.$$

Rule (4.1) is satisfied because the following equation holds between projections:

$$P_{A \wedge B} = (P_{A \wedge B} + P_{A \wedge \neg B} + P_{\neg A \wedge B} + P_{\neg A \wedge \neg B}) \cdot P_A \cdot P_B$$
$$= P_{k(A, B)} \cdot P_A \cdot P_B.$$

Rule (4.3) is satisfied because of the equation:

$$P_{\neg A} = \hat{1} - P_A.$$

The rule (4.2) is not generally satisfied by the quantum theoretical expectation values, as we see in the following simple example:

$$(\text{I}) \quad p_{\langle \alpha \rangle}(A \vee B, C) = p_{\langle \alpha \rangle}(A, C) + p_{\langle \alpha \rangle}(B, C)$$

where it is assumed that $A \leq \neg B$ and $A \vee B \neq V$.
For the respective quantum theoretical expectation value the following equation holds:

$$(\text{II}) \quad \langle \alpha | P_{A \vee B} \cdot P_C \cdot P_{A \vee B} | \alpha \rangle$$
$$= \langle \alpha | (P_A + P_B) \cdot P_C \cdot (P_A + P_B) | \alpha \rangle$$
$$= \langle \alpha | P_A \cdot P_C \cdot P_A | \alpha \rangle + \langle \alpha | P_B \cdot P_C \cdot P_B | \alpha \rangle$$
$$+ \langle \alpha | P_A \cdot P_C \cdot P_B + P_B \cdot P_C \cdot P_A | \alpha \rangle$$

which differs from the equation (I) by the "interference"-term $\langle \alpha | P_A P_C P_B + P_B P_C P_A | \alpha \rangle$.

Indeed, the equations (I) and (II) do not lead to a contradiction since both equations describe two different experimental situations. In (I) the disjunction $A \lor B$ is proved by means of measurements with respect to A and B. In (II) the projection $P_{A \lor B}$ corresponds to an elementary proposition $e_{A \lor B}$ which is formally equivalent to the proposition $A \lor B$ but which is not dialog-equivalent to $A \lor B$. As we have already seen in IV.a) both the propositions are distinguished by the respective experimental arrangements.

In an idealized two-slit experiment the case (I) corresponds to the experimental situation:

(I)

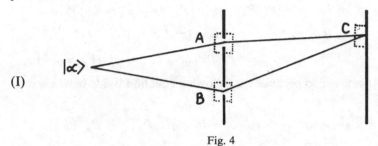

Fig. 4

where the proposition $A \lor B$ is proved "locally" by measurements of A and B.
In case (II) a "non local" measurement with respect to the propositions A and B is used which corresponds to the material proposition $e_{A \lor B}$:

(II)

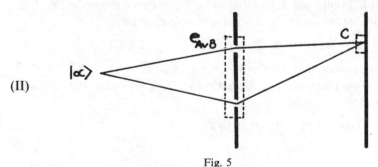

Fig. 5

Thus we see that the expectation values can only be interpreted as quantum probabilities if it is assumed that projections of the form $P_{A \lor B}$ always correspond to material propositions $e_{A \lor B}$. Since equation (I) does not hold for $p_{\langle \alpha \rangle}(e_{A \lor B}, C)$ there is no contradiction to equation (II). If $P_{A \lor B}$ does not correspond to a material proposition $e_{A \lor B}$ but to an arrangement where the propositions A and B are measured separately the right side of equation (II) reduces to the expectation value associated with a "mixture":

$$\langle \alpha | P_A P_C P_A | \alpha \rangle + \langle \alpha | P_B P_C P_B | \alpha \rangle$$

and satisfies the equation (I).

In this way we see that quantum probabilities of the form $p_{\langle \alpha \rangle}(\mathscr{A}, A \vee B, \mathscr{A}')$ always represent expectation values associated with mixtures. They must be distinguished from the probabilities $p_{\langle \alpha \rangle}(\mathscr{A}, e_{A \vee B}, \mathscr{A}')$ where $e_{A \vee B}$ is a material proposition and, although $A \vee B = e_{A \vee B}$ holds, is not dialog-equivalent to $A \vee B$ in the material dialog-game. The representation of the quantum theoretical expectation value, corresponding to $p_{\langle \alpha \rangle}(\mathscr{A}, e_{A \vee B}, \mathscr{A}')$:

$$\langle \alpha | P_{\mathscr{A}} \cdot P_{A \vee B} \cdot P_{\mathscr{A}'} P_{\mathscr{A}'*} \cdot P_{A \vee B} \cdot P_{\mathscr{A}*} | \alpha \rangle$$

$$= \langle \alpha | P_{\mathscr{A}} \cdot P_A \cdot P_{\mathscr{A}'} \cdot P_{\mathscr{A}'*} \cdot P_A \cdot P_{\mathscr{A}*} | \alpha \rangle$$

$$+ \langle \alpha | P_{\mathscr{A}} \cdot P_B \cdot P_{\mathscr{A}'} \cdot P_{\mathscr{A}'*} \cdot P_B \cdot P_{\mathscr{A}*} | \alpha \rangle$$

$$+ \langle \alpha | P_{\mathscr{A}} \cdot P_{A} \cdot P_{\mathscr{A}'} \cdot P_{\mathscr{A}'*} \cdot P_B \cdot P_{\mathscr{A}*}$$

$$+ P_{\mathscr{A}} \cdot P_B \cdot P_{\mathscr{A}'} \cdot P_{\mathscr{A}'*} \cdot P_A \cdot P_{\mathscr{A}*} | \alpha \rangle$$

is a formal (mathematical) representation and has no operational interpretation in our approach. The existence of the interference term in the above equation and, hence, the difference between this equation and the equation (4.2) of the calculus is not a consequence of formal quantum logical relations between propositions but a consequence of the different (dialogically not equivalent) material conditions of the proofs of $A \vee B$ and $e_{A \vee B}$.

University of Western Ontario

REFERENCES:

[1] Fine, T. L., *Theories of Probability: An Examination of Foundations,* Academic Press, New York (1973)
[2] Jauch, J. M., *Foundations of Quantum Mechanics,* Addison-Wesley Publ. Co., Reading, Mass. (1968)
[3] Berge, C., *Théorie Générale des Jeux à n Personnes,* Paris (1957)
[4] Ref. [2], Ch. 6.4
[5] Ref. [2], Ch. 6.3

ACKNOWLEDGEMENT

I wish to thank the Department of Philosophy of the University of Western Ontario for the kind hospitality extended to me during the winter and summer terms 1975-76.

INDEX OF NAMES

323

INDEX OF SUBJECTS

*The term quasi-modularity is used by Mittelstaedt for orthomodularity.

THE UNIVERSITY OF WESTERN ONTARIO SERIES IN PHILOSOPHY OF SCIENCE

A Series of Books on Philosophy of Science, Methodology, and Epistemology published in connection with the University of Western Ontario Philosophy of Science Programme

8. J. M. Nicholas (ed.), *Images, Perception, and Knowledge.* Papers deriving from and related to the Philosophy of Science Workshop at Ontario, Canada, May 1974. 1977, ix + 309 pp.
9. R. E. Butts and J. Hintikka (eds.), *Logic, Foundations of Mathematics, and Computability Theory.* Part One of the Proceedings of the Fifth International Congress of Logic, Methodology and Philosophy of Science, London, Ontario, Canada, 1975. 1977, x + 406 pp.
10. R. E. Butts and J. Hintikka (eds.), *Foundational Problems in the Special Sciences.* Part Two of the Proceedings of the Fifth International Congress of Logic, Methodology and Philosophy of Science, London, Ontario, Canada, 1975. 1977, x + 427 pp.
11. R. E. Butts and J. Hintikka (eds.), *Basic Problems in Methodology and Linguistics.* Part Three of the Proceedings of the Fifth International Congress of Logic, Methodology and Philosophy of Science, London, Ontario, Canada, 1975. 1977, x + 321 pp.
12. R. E. Butts and J. Hintikka (eds.), *Historical and Philosophical Dimensions of Logic, Methodology and Philosophy of Science.* Part Four of the Proceedings of the Fifth International Congress of Logic, Methodology and Philosophy of Science, London, Ontario, Canada, 1975. 1977, x + 336 pp.
13. C. A. Hooker (ed.), *Foundations and Applications of Decision Theory,* 2 volumes. Vol. I: *Theoretical Foundations.* 1978, xxiii+446 pp. Vol. II: *Epistemic and Social Applications.* 1978, xxiii+208 pp.
14. R. E. Butts and J. C. Pitt (eds.), *New Perspectives on Galileo.* Papers deriving from and related to a workshop on Galileo held at Virginia Polytechnic Institute and State University, 1975. 1978, xvi+262 pp.